55

Topics in Organometallic Chemistry

Aims and Scope

The series *Topics in Organometallic Chemistry* presents critical overviews of research results in organometallic chemistry. As our understanding of organometallic structure, properties and mechanisms increases, new ways are opened for the design of organometallic compounds and reactions tailored to the needs of such diverse areas as organic synthesis, medical research, biology and materials science. Thus the scope of coverage includes a broad range of topics of pure and applied organometallic chemistry, where new breakthroughs are being achieved that are of significance to a larger scientific audience.

The individual volumes of *Topics in Organometallic Chemistry* are thematic. Review articles are generally invited by the volume editors. All chapters from Topics in Organometallic Chemistry are published OnlineFirst with an individual DOI. In references, Topics in Organometallic Chemistry is abbreviated as Top Organomet Chem and cited as a journal.

More information about this series at http://www.springer.com/series/3418

Pierre H. Dixneuf • Henri Doucet
Editors

C-H Bond Activation and Catalytic Functionalization I

With contributions by

L. Ackermann · F. Bellina · C. Bruneau · K.J.T. Carr · S. Chang · S. Dana · S. De Sarkar · P.H. Dixneuf · H. Doucet · F. Glorius · J.G. Kim · J. Li · S.A. Macgregor · C.L. McMullin · F.W. Patureau · A.K. Sahoo · K. Shin · J.-F. Soulé · G.-W. Wang · J. Wencel-Delord · M.R. Yadav

Springer

Editors
Pierre H. Dixneuf
Université de Rennes
Rennes
France

Henri Doucet
Université de Rennes
Rennes
France

ISSN 1436-6002 ISSN 1616-8534 (electronic)
Topics in Organometallic Chemistry
ISBN 978-3-319-24628-4 ISBN 978-3-319-24630-7 (eBook)
DOI 10.1007/978-3-319-24630-7

Library of Congress Control Number: 2015959030

Springer Cham Heidelberg New York Dordrecht London

Printed on acid-free paper

Springer International Publishing AG Switzerland is part of Springer Science+Business Media (www.springer.com)

Preface

The metal-catalysed C–H bond dual activation and functionalisation have brought in the last two decades a revolution for the direct synthesis of complex molecules and molecular materials. Now the functionalisation of sp^2C–H bond for cross-coupled C–C or C-heteroatom bond formation presents advantages to replace, with better atom economy, the classical catalytic cross-coupling reactions involving a stoichiometric amount of an organometallic. In parallel the sp^3C–H bond activation, besides a faster access to natural products, is offering the possibility to functionalise alkanes in connection with renewable energy.

Whereas functional groups have shown efficiency to direct activation of neighbouring C–H bonds, as molecules containing multiple C–H bonds, the successive activations of several of these C–H bonds remain a challenge. Initially expensive metal catalysts have shown their efficiency to activate C–H bonds, but now many examples of cheap and environment-tolerant first-row metal catalysts are promoting useful activations. Examples of C–H bond functionalisation can now be performed in green solvents and even in water.

This volume gathers innovative contributions for a wide range of catalytic C–H bond functionalisations. They involve a variety of metal catalysts from Pd, Rh, Ir and Ru complexes to Fe, Ni, Cu and Ag derivatives, including surface organometallics, and they point out the importance of ancillary or transient ligands forcing the metal site to activate C–H bonds by several complementary processes. In addition this volume presents many new applications for cross C–C and C-heteroatom bond couplings and new synthetic methods, supported by mechanistic and computational studies, and examples of functionalisation of cyclopropanes or fullerenes and addresses problems of regioselectivity. The sp^3C–H bond activation reveals crucial aspects for the synthesis of natural products and for the dehydrogenation and functionalisation of alkanes.

The wide range of innovations presented here, on the concepts of C–H bond activations and their multiple profits, should be a source of inspiration for researchers and industry engineers to discover more efficient catalysts or to transfer the processes to industrial applications. They should attract teachers and students motivated by innovations, catalysis and sustainable development. They are expected to initiate new ideas to discover new catalytic and cascade transformations.

We are grateful to all the chapter authors, experts in various complementary fields, who have contributed to create this multiple-facet volume.

We dedicate this volume to all chemists and students who are contributing, via C–H bond activation and functionalisation, to discover safe, catalytic transformations that will be profitable for our society.

Rennes, France

Pierre H. Dixneuf
Henri Doucet

Contents

Top Organomet Chem (2016) 55: 1–28
DOI: 10.1007/3418_2015_140

Published online: 18 August 2015

Rh(III)- and Ir(III)-Catalyzed C–C Bond Cross Couplings from C–H Bonds

Joanna Wencel-Delord, Frederic W. Patureau, and Frank Glorius

Abstract Over the recent years, dicationic [Cp*Ir(III)] and more particularly [Cp*Rh(III)] complexes have established themselves as extremely powerful catalysts enabling direct C–H activation of various aromatic and vinylic compounds. During such transformations, a common metallacyclic intermediate [Cp*M(C^X)] (M=Ir, Rh) is formed and undergoes further transformations, according to the nature of the coupling partner, to finally afford a myriad of valuable, often complex, and otherwise difficult to access scaffolds like heterocycles and polyunsaturated skeletons. The major advances achieved in this field clearly showcase the potential of these catalysts to functionalize latent C–H bonds under surprisingly mild reaction conditions. The aim of this chapter is to present the latest and most representative contributions in the field of Rh(III)- and Ir(III)-catalyzed C–H activation by focusing on the reactivity of the corresponding metallacyclic intermediates.

Keywords Cp*Ir · Cp*Rh · Dehydrogenative cross coupling · Direct functionalization · Mild C–H activation · Oxidizing directing group · Redox neutral · Rhodacycle

J. Wencel-Delord (✉)
Laboratoire de Chimie Moléculaire (UMR CNRS 7509), Université de Strasbourg, ECPM, 25 Rue Becquerel, 67087 Strasbourg, France
e-mail: wenceldelord@unistra.fr

F.W. Patureau
Technische Universität Kaiserslautern, Erwin-Schrödinger-Straße Geb. 52, 67663 Kaiserslautern, Germany

F. Glorius
Westälische Wilhelms-Universität Münster, International Graduate, School of Chemistry and Organisch-Chemisches Institut, Corrensstrasse 40, 48149 Münster, Germany

Contents

1 Introduction and Background of Organometallic Chemistry

The extraordinary advances achieved over the past decade in the field of catalytic C–H activation have progressively established this modern catalytic approach as a valuable tool for organic synthesis. In particular Rh(III)-based catalysts bearing a pentamethylcyclopentadienyl ligand (Cp* ligand) unambiguously stand out as a privileged catalyst for direct functionalization of inert C–H bonds [1–8]. The story of half-sandwich Rh(III) and Ir(III) complexes commenced at the end of the 1960s, when Maitlis isolated and fully characterized $[Cp^*RhCl_2]_2$ and $[Cp^*IrCl_2]_2$ complexes [9, 10]. Rapidly, these organometallic structures have become an interesting subject of scientific study for inorganic chemists. The reactivity of these pseudo-octahedral complexes toward various ligands has been investigated enabling isolation and characterization of many interesting organometallic structures. Moreover, following the Maitlis work [11] and since the potential of $[Cp^*IrPR_3X_2]$ and $[Cp^*RhPR_3X_2]$ complexes to cleave C–H bonds [12, 13] had been known already in the 1980s, an extensive study of the stoichiometric reactions between the dicationic [Cp*M(III)] precatalysts (M=Rh, Ir) and aromatic substrates bearing coordinating substituents (such as imines, ketones, amines, etc.) was undertaken. Rapidly, it was proved unequivocally that such late transition metals, in the presence of sodium acetate as additive, are particularly potent catalysts for C–H activation affording cyclometalated complexes termed [Cp*M(C^X)] (M=Rh, Ir) [14]. Importantly, these Ir- and Rh-cyclometalated species could be frequently prepared with high efficiency and under relatively mild reaction conditions [15]. A further breakthrough in the field of dicationic [Cp*M] complexes was achieved at the beginning of this century, with the seminal studies targeting catalytic C–H activation. The pioneering work of Matsumoto and Periana [16],

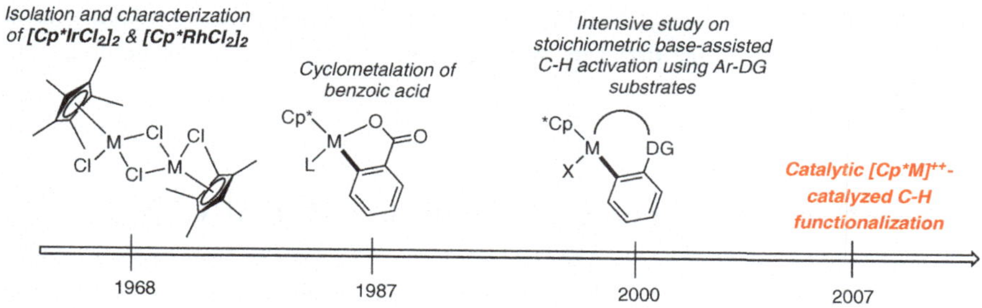

Scheme 1 Key steps in the development of $[Cp^*MCl_2]_2$-catalyzed C–H activation

followed by the key contributions by Satoh and Miura [17–20], was the first to clearly show the potential of $[RhCp^*Cl_2]_2$ precursors as underestimated catalysts for dehydrogenative cross couplings (Scheme 1).

Since those days, extraordinary advances have been achieved, and the [RhCp*(III)] complex established itself as a highly potent catalyst for a large panel of direct functionalization reactions enabling not only C–C but also C–X (X=N, S, Br, I, etc.) bond formations. In contrast, although the $[Cp^*IrCl_2]_2$ precatalyst is recognized to be a more efficient metalating agent than the Rh congener [14, 21], the use of this high-oxidation-state third-row transition metal catalyst to construct C–C bonds is clearly less documented [22]. The limited potential of the $[IrCp^*Cl_2]_2$ precatalyst in C–C bond forming reactions might be due to the high stability of the metallacyclic intermediates and the unfavorable final product formation (e.g., by means of an oxidative coupling of the C–N bond) which might compromise an efficient catalytic turnover [23]. In contrast, due to the less thermodynamically favored metalation in the case of the [RhCp*(III)] species (increased sensitivity to the steric constraints), isolation and characterization of the Rh-based metallacyclic intermediates is far from trivial. However, failure of stoichiometric C–H activation does not prevent catalytic transformations if further steps of a target catalytic cycle are thermodynamically favored and hence overcome the difficult cyclometalation event.

A vast majority of the $[Cp^*MCl_2]_2$-catalyzed transformations follow catalytic cycles with very similar initial steps. Firstly, an active dicationic catalyst is formed by silver salt-induced abstraction of the Cl ligands. Coordination between the catalyst and a directing group of a substrate brings these two partners in proximity to facilitate a base-assisted C–H bond cleavage delivering key metallacyclic intermediates. The fate of such metallacyclic intermediates depends on the nature of the coupling partner, which determines the structure of the final product. Consequently, this review is divided into several sections following the reactivity of these key metallacyclic intermediates. Taking into consideration an impressive number of publications concerning Rh(III)-catalyzed C–H activation published over the last 5 years and several excellent recent reviews dealing with this topic, we propose to selectively present the different types of the C–C bond forming reactions which can be achieved. We wish to provide the reader with an easy-to-understand overview of how the cyclometalated Rh(III) and Ir(III) intermediates may react to afford a large panel of structurally very different and valuable products, such as heterocyclic scaffolds.

2 Insertion of C≡C and C=C Bonds into [Cp*M(C^X)]

The Rh(III)- and Ir(III)-catalyzed couplings with olefins and alkynes are, by far, the most documented transformations. Insertion of such unsaturated coupling partners into the C–M bond of metallacyclic species was firstly extensively studied [23–25]. Subsequently, the judicious choice of a terminal oxidant enables to complete a catalytic cycle, thus unlocking the door toward catalytic transformations.

*2.1 Insertion of C≡C Bonds into [Cp*M(C^X)]*

Rh(III) and Ir(III) oxidative couplings of aromatic and vinylic substrates with alkynes are particularly appealing from the synthetic point of view as they might be considered as straightforward and highly atom- and step-economic synthetic routes toward a myriad of heterocyclic scaffolds. Indeed, due to the protic character of some frequently applied directing groups (DGs) (such as amide, carboxylate, benzhydroxamic acids, etc.) bearing X–H moieties, a commonly proposed catalytic cycle consists of four fundamental steps: (1) DG-orientated metalation via base-assisted C–H activation, (2) coordination of an alkyne and its consecutive insertion into the C–M bond, (3) oxidative C–X bond forming reductive elimination and release of the reduced catalyst, and (4) oxidative regeneration of the catalyst (Scheme 2). Following such a catalytic scenario, highly valuable scaffolds such as indoles [26], pyrroles [27], isoquinolines [28, 29], isoquinolones [30–32], pyridinium salts [33], isocoumarins [34], carbazoles [35], indolo[2,1-*a*] isoquinolines [36], isochromenes [37], pyridones [38, 39], pyrones [40], phospha-isocoumarins [41], and indenones [42] could be constructed.

Alternatively, if aromatic substrates bearing nonpolar DGs are employed, C–H activation followed by double insertion of an alkyne is observed, hence delivering naphthalene scaffolds (Scheme 2) [19, 43, 44].

[Cp*Ir(C^X)] species are also prompt to undergo the insertion of an alkyne, but a subsequent reductive elimination yielding annulated products is more troublesome [23, 45]. Besides, a complementary reactivity of $[Cp^*RhCl_2]_2$ and $[Cp^*IrCl_2]_2$ precursors may also be observed sometimes, as reported by Satoh and Miura in the context of oxidative coupling of benzoic acids with alkynes (1) [18, 34]. In the presence of a second-row metal catalysts in combination with $Cu(OAc)_2{\bullet}H_2O$ as oxidant, a C–H activation/alkyne insertion and oxidative C–O bond formation sequence yield isocoumarins, whereas a third-row transition metal catalysts, when used in combination with Ag_2CO_3 as oxidant and at elevated temperature, catalyze

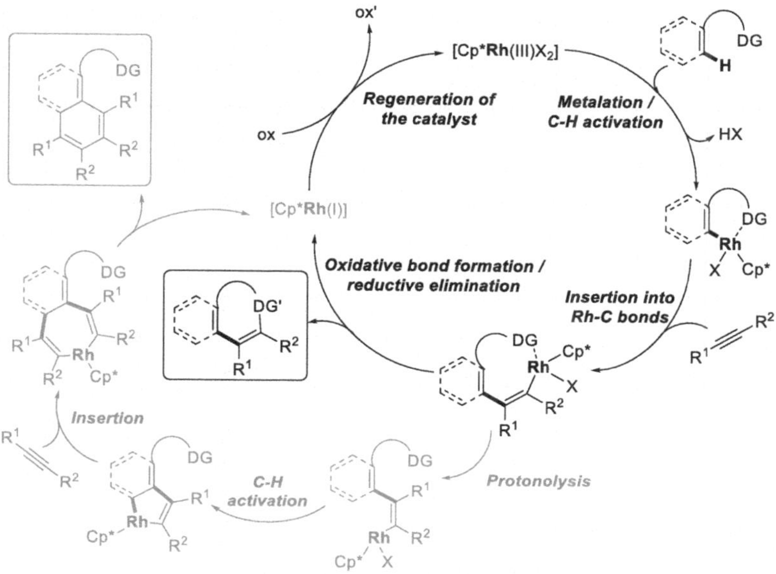

Scheme 2 General catalytic cycle of [Cp*Rh(III)]-catalyzed C–H activation/annulative coupling with alkynes

decarboxylative twofold C–H activation/alkyne insertion affording naphthalenes. Notably, Ir-catalyzed oxidant-induced oxyfunctionalization enabling isocoumarin synthesis was reported recently, under milder reaction conditions and in the presence of AgOAc as oxidant in methanol medium [46].

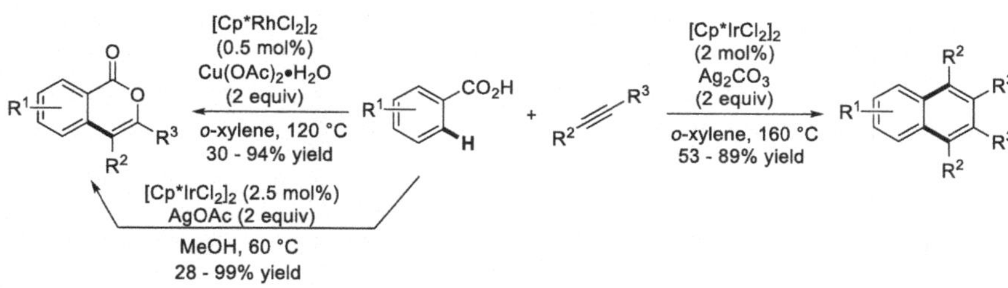

(1)

These pioneering works in the field of Rh(III)-catalyzed C–H activation showcased the exceptional potential of this catalyst, even if harsh reaction conditions (temperature above 100°C) are usually required. An astute use of oxidizing DGs, independently reported by Fagnou [32, 47] and Glorius in the context of oxidative Heck reactions [48], allows to circumvent the temperature limitation. Indeed, when benzhydroxamic acid esters are used as coordinating moieties, N–OR bond cleavage occurring at a final stage of the catalytic cycle enables to restore the active species of Rh(III) catalyst in an absence of an external oxidant and under astonishingly mild reaction conditions (room temperature) (2). Noteworthy, this catalytic strategy improves not only reactivity but also selectivity, and challenging coupling partners like terminal and unsymmetrical heterocycle-substituted alkynes could be efficiently used. Following these pioneering reports, several other N–O- and N–N-based oxidizing directing groups (such as *N*-phenoxyacetamides, *O*-acyloximes, hydrazones, etc.) were designed and applied in redox-neutral couplings with alkynes [49–54]. In addition, an intramolecular version of this transformation was used as a key step allowing construction of indolizidine-based natural products like (±)-antofine and (±)-septicine [55] and (±)-goniomitine, bearing a pyrido[1,2-a] indole core [56, 57] (3).

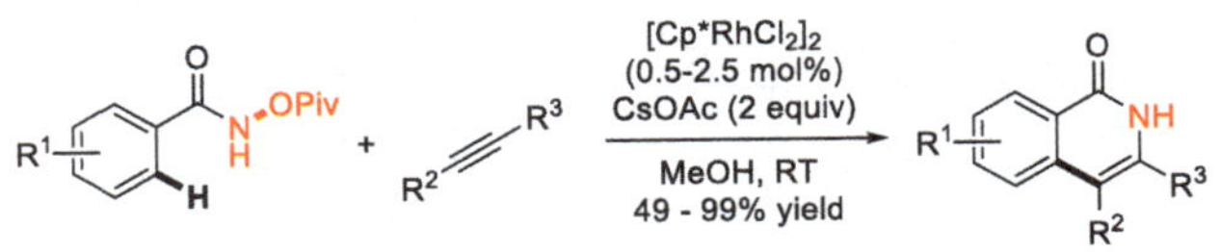

(2)

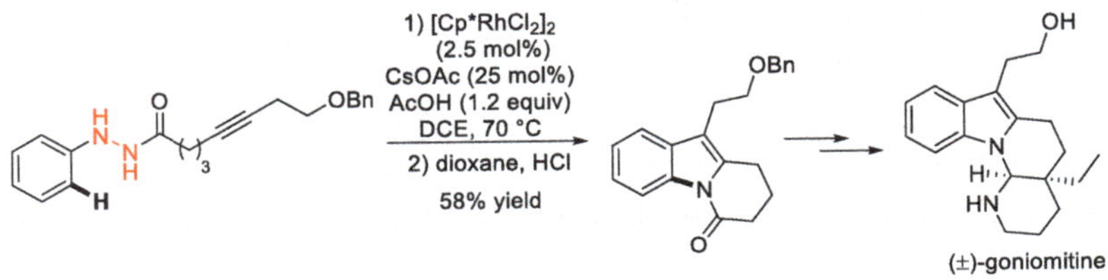

(3)

An alternative outcome of a redox-neutral C–H activation/alkyne insertion reaction was observed in the case of substrates bearing N–O polar bonds like quinoline *N*-oxides [58, 59] or arylnitrones [60]. The Rh(III) complex turned out to be a potent catalyst to realize a cascade process, C–H activation and intramolecular O-atom transfer, hence delivering 2-(quinolin-8-yl)-1,2-diarylethanone and indoline derivatives, respectively (4). Noteworthy, and contrary to the redox-neutral transformations presented above, these C–H activation/O-atom transfer couplings occur with 100% atom economy. However, the products were generally obtained as a mixture of two diastereomers.

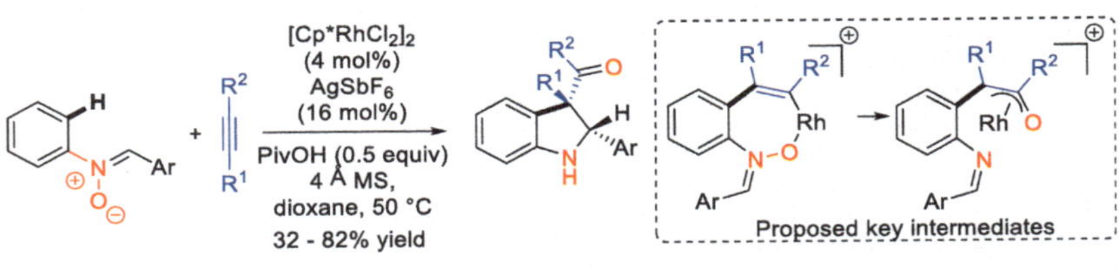

(4)

Finally, an electrophilic character of several DGs such as ketones [61], urea [62], azomethine ylides [63], or tertiary benzamides [64] is another appealing feature. As a Rh–C(alkenyl) bond, generated via migratory insertion of an alkyne into a Rh–C bond, may be considered as a nucleophile, an intramolecular addition to such an electrophilic moiety as the directing group can be envisioned (5). Consequently, original cyclized products are generated in this redox-neutral reaction.

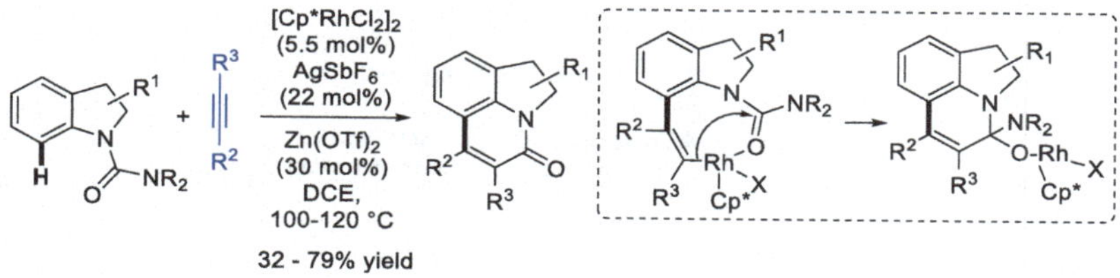

(5)

2.2 *Insertion of C=C Bonds into [Cp*M(C^X)]*

Initially underestimated compared to Pd-based complexes, the potential of the [Cp*Rh(III)] catalyst for catalytic C–H activation was fully recognized at the beginning of this decade when several research groups reported its great capacity to enable oxidative Heck (Fujiwara–Moritani) reactions [17, 20, 65, 66]. Thereafter, an impressive number of exciting articles dealing with Rh(III)-catalyzed dehydrogenative olefinations appeared in the literature [1, 3, 5]. These transformations initially follow a catalytic cycle similar to the one proposed for couplings with alkynes, i.e., base-assisted C–H activation and subsequent formation of key metallacyclic intermediate followed by coordination and regioselective insertion of alkene into the Rh–C bond. The final *E*-product results from β-H elimination (Scheme 3). Alternatively, if a polar directing group is applied, olefination–Michael cyclization takes place [40, 67, 68].

An original example of a direct olefination reaction delivering a functionalized product not via the β-H elimination but a cyclative capture of the Rh–C(sp^3) intermediate was disclosed by Glorius [69]. The authors discovered that a direct *ortho*-C–H activation of aromatic diazenecarboxylates with a range of alkenes affords 1-aminoindoline products (6). Formation of these compounds was rationalized by an addition of the nucleophilic Rh–C(sp^3) bond to a polarized double N=N bond. Importantly, chelation of Rh with the Boc substituent of the DG in a five-membered metallacyclic intermediate prevented a generally observed β-H elimination, hence triggering the cyclization event.

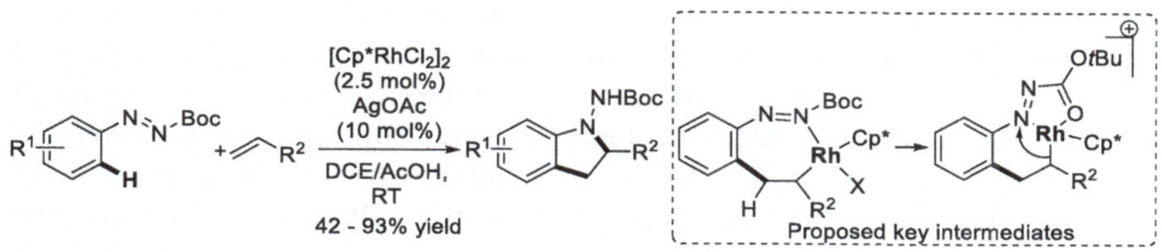

(6)

Exceptional DG tolerance, compatibility with a variety of functionalities, and relatively low catalyst loadings showcase the superiority of the Rh-based catalytic system over other catalysts. However, originally such oxidative Heck reactions were mostly limited to activated alkenes, such as acrylates and styrenes. Progressively, improved catalytic systems compatible now with unactivated alkane-substituted terminal olefins have been designed [70–74]. The key to success

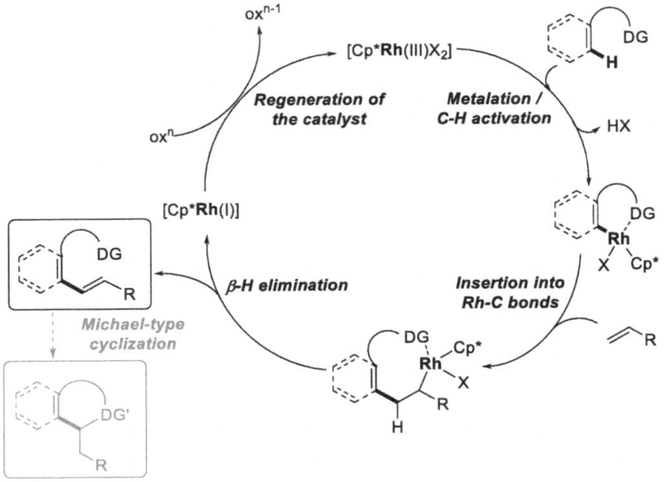

Scheme 3 General catalytic cycle for Rh(III)-catalyzed C–H activation and oxidative coupling with olefins

lies on a finely adjusted combination between a potent substrate (like naphthalene bearing *O*-methyl oxime or sulfonamide DGs) and reaction conditions, a design of a more potent, electron-deficient Cp*-derived Rh(III) catalyst, or performing intramolecular couplings.

Subsequently, significant efforts have been engaged to enlarge the scope of these dehydrogenative couplings. Marsden and coworkers endeavored to design a catalytic cycle compatible with electron-rich vinyl esters as novel coupling partners [75]. Gratifyingly, a targeted C–H activation/annulation sequence could be achieved when reacting benzoyl hydroxamate (an "internal-oxidant-containing" C–H substrate) with either vinyl acetate or enol ether and enamide (Scheme 4a). Importantly, the vinyl acetate behaves as a formal acetylene equivalent, and the reaction occurs under mild and external-oxidant-free conditions. Noteworthy, the regioselectivity of the olefin's insertion into the Rh–C bond is probably controlled via coordination between the carbonyl moiety of the vinylic acetate and the metal atom. A few months later, Ellman employed such an oxidative coupling with vinyl acetate, a convenient and inexpensive vinyl source, to access *ortho*-substituted styrenes [76]. Noteworthy, the Rh complex plays a dual role catalyzing both the C–H activation and the C–O cleavage (elimination of the acetate moiety). Very recently, geminal-substituted vinyl acetates were also utilized as original coupling partners in combination with benzoic acids to build up 3-substituted isocoumarins [77].

An alternative coupling implying vinylic coupling partners was investigated by Li [78]. The author surmised that 2-vinyloxiranes could also be potent substrates for the alkene insertion into the Rh–C bond of the metallacyclic intermediate (Scheme 4b). Formation of the corresponding rhodium alkyl intermediate should thus induce an epoxide opening via β-O elimination (owing to the release of the ring strain) allowing effective isolation of allyl alcohol derivatives. This redox-neutral

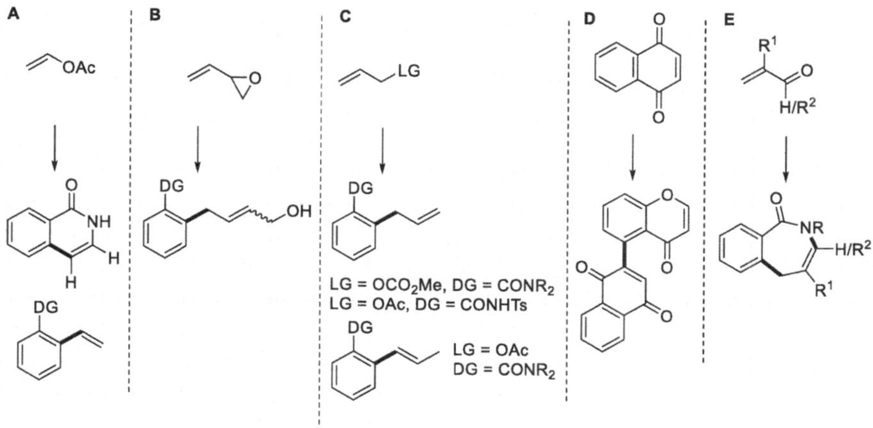

Scheme 4 Original alkene couplings partners used in [Cp*Rh(III)]-catalyzed C–H functionalization and the corresponding products. (**a**) Vinyl acetate; (**b**) 2-vinyloxiranes; (**c**) allylic compounds; (**d**) 1,4-naphthoquinone; (**e**) unsaturated aldehydes and ketones

transformation occurring at room temperature is rather general as both (hetero) aromatics and di- or trisubstituted olefins were well tolerated. A similar reactivity was also observed when 4-vinyl-1,3-dioxolan-2-ones were employed as coupling partners [79].

Otherwise, allyl-substituted arenes can also be synthesized when allylic electrophiles are used as coupling partners (Scheme 4c). This reactivity was firstly reported by Glorius [80]. Allyl carbonates were found to be the most potent coupling partners allowing direct functionalization of the aromatics bearing a tertiary amide or *N*-heterocyclic DGs. Comparable to the dehydrogenative coupling with the vinylic acetates, the products are probably generated via β-O elimination (β-H elimination might be prevented by the coordination between the carbonate moiety and the Rh atom). Intriguingly, when allyl acetates were used by Loh in combination with *N,N*-dimethylbenzamide C–H substrates at higher reaction temperature, olefination occurs and conjugated products were isolated [81]. This regioselectivity switch was attributed to the [Rh–H]-catalyzed migratory isomerization of the double bond of the initially formed nonconjugated product. Noteworthy, γ-selectivity of this C–H allylation with allyl acetate electrophiles was restored by using a weakly coordinating Ts-imide directing group [82]. Besides, the coupling between azobenzenes and allylic acetates allows the formation of alkylated products; β-O elimination seems to be disfavored in this specific case and a final saturated product is generated via protonolysis [83].

In addition, the use of Michael acceptors as potent coupling partners is attracting growing interests. Antonchick discovered that 1,4-naphthoquinones may be oxidatively coupled with chromones (Scheme 4d) [84]. [Cp*Ir(III)] complexes are also prompt to catalyze a related dehydrogenative coupling between benzoic acids and benzoquinone affording 2-hydroxy-6*H*-benzo[*c*]chromen-6-ones [85].

In 2013 Glorius hypothesized that the efficient insertion of the double bond of unsaturated aldehydes and ketones into the Rh–C bond of the metallacyclic intermediates should also be possible (Scheme 4e) [86]. Importantly, as these conjugated Michael acceptors can be considered as "three-carbon equivalents," their use in an annulative coupling with benzamides should open a window for the synthesis of challenging seven-membered *N*-containing rings.

Noteworthy, methylenecyclopropanes, an appealing building block, may also intercept rhodacyclic intermediates [87]. This C–H activation protocol enables an alternative synthesis of spiro dihydroisoquinolinones and furan-fused azepinones.

In addition, $[Cp^*RhCl_2]_2$-catalyzed dehydrogenative olefination enables the construction of diene scaffolds with excellent *Z*-control. The first step in this field using a Rh-based catalyst was achieved by Glorius in 2011 realizing an oxidative Heck reaction between vinylic substrates bearing various DGs (amide, ester, etc.) and activated olefins (acrylates and styrenes) (7) [88]. The desired products were obtained in good to moderate yields, but with encouraging *Z*-selectivity. Shortly afterward, Loh independently disclosed a closely related transformation [89]. In 2014, Glorius improved the scope of this transformation by employing enol-carbamate-substituted C–H substrates as novel "ketone equivalents" [90].

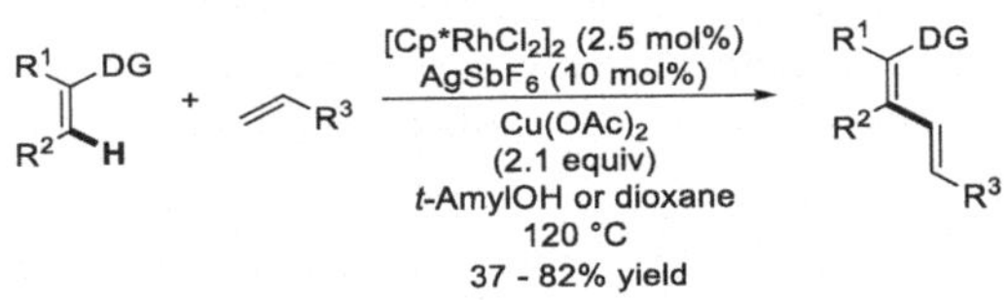

(7)

2.3 Couplings with Polyunsaturated Partners

Regarding the great potential of cyclometalated [Cp*Rh(III)] species to insert alkenes and alkynes into Rh–C bonds, it is not surprising that a design of closely related transformations implying polyunsaturated coupling partners has gathered interest of the scientific community. These transformations present several intrinsic difficulties like (1) chemo-selectivity (between two unsaturated moieties) and regiocontrol (selectivity for each migratory insertion step) issues and (2) selective mono- and diannulation. A pioneering example of such challenging transformations was reported in 2014 by Glorius et al. who disclosed an elegant catalytic system enabling a highly selective coupling between aromatic hydroxamic ester derivatives and 1,3-diynes (Scheme 5) [91]. This strategy enabled to build up selectively and under mild reaction conditions a large panel of bisisoquinolones, both in symmetrical and unsymmetrical fashion (when nonsymmetrical 1,3-diynes were used). Even more impressively a stepwise and/or one-pot diannulation with two different C–H activation substrates could also be performed, hence affording nonsymmetrical bisheterocycles bearing two different core structures.

Concomitantly, the same research group has shown that 1,3-dienes are also potent coupling partners for Rh(III)-catalyzed C–H functionalization [92]. A redox-neutral coupling between aromatic/vinylic oxime esters with these

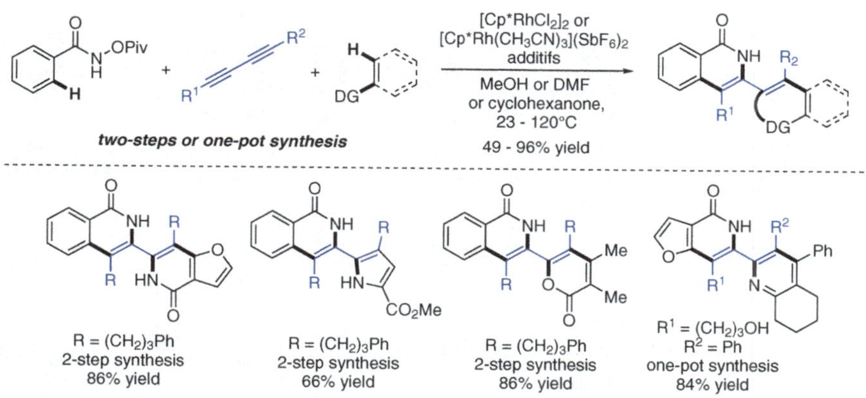

Scheme 5 Synthesis of bisheterocycles using 1,3-diynes as coupling partners

polyunsaturated coupling partners was astutely used to build up isoquinolines and pyridines.

Noteworthy, 1,6-enynes were also used as original coupling partners in a Rh (III)-catalyzed C–H activation cascade reaction by Tian and Lin [93]. Importantly, a slight modification of the directing group (OPiv- or OMe-substituted N-hydroxybenzamides) resulted in a different outcome, hence leading to the formation of either isoquinolone or hydrobenzofuran cores.

3 [Cp*M(C^X)] Carbometalation of Allenes

Allenes are important moieties in catalysis as their carbometalation provides access to highly versatile π-allyl species, which are key intermediates of the prominent Tsuji–Trost reaction. This is why the use of allenes as original coupling partners in the context of direct C–H activation has focused much attention. The pioneering work in [Cp*Rh(III)]-catalyzed C–H activation followed by annulation with allenes was disclosed by Glorius in 2012 [94]. The authors discovered that monosubstituted allenes, in the presence of Rh(III)-metallacyclic species, undergo rapid carborhodation resulting in a formation of a C–C bond involving the central carbon of the allene and the generation of a Rh(III)-allylic intermediate. Subsequently, reductive elimination at a less sterically hindered position is favored by the steric hindrance of the Cp* ligand, delivering a cyclized product with an exocyclic double bond (Scheme 6a). Importantly, as aromatic C–H substrates bearing an oxidizing directing group are used, this transformation occurs under very mild and external-oxidant-free conditions enabling original access to 3,4-dihydroisoquinolin-1(*2H*)-ones. Only 1 month later, Ma independently reported a closely related transformation based on the use of more substituted allenes [95]. Intriguingly, the Rh(III)-catalyzed coupling between *N*-methoxybenzamides

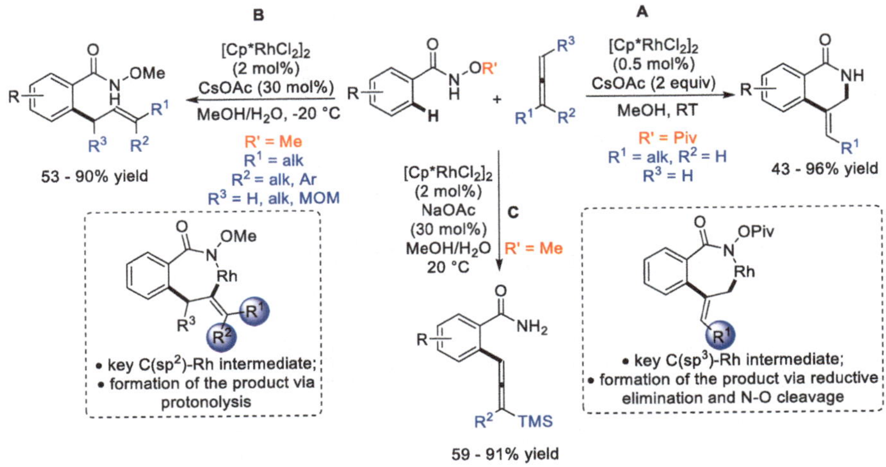

Scheme 6 Rh(III) catalyzed direct coupling with allenes. (**a**) Coupling with monosubstituted allenes; (**b**) coupling with di- and trisubstituted allenes; (**c**) coupling with sterically hindered allenylsilanes

and 1,1-disubstituted or internal allenes delivered non-cyclized monoallylated products (Scheme 6b). This different regioselectivity compared to the one observed by Glorius could be explained by the insertion of the less-substituted C=C bond into the Rh–C bond leading to the formation of a C(sp^2)–Rh intermediate. The final allylated product is liberated via protonolysis. Noteworthy, the stereoselective version of this transformation was developed by Cramer benefiting from a chiral [CpRh]-derived catalysts [96]. Pursuing his experimental work, Ma discovered also that the steric effect of the allene's substituent may drastically influence the outcome of this transformation [97]. Introduction of strongly sterically demanding substituents, like a trialkylsilyl group, prevents formation of the previously observed protonolysis product, and β-H elimination occurs preferentially, hence affording the extra-substituted allene product (Scheme 6c).

Subsequently, related couplings with allenes paved the way toward the innovative synthesis of valuable scaffolds. Glorius astutely used allenyl carbinol carbonate as a source of diene functionality to construct [3]dendralenes [98]. Cheng discovered that benzoic and vinylic carboxylic acids when reacted with a range of allenes are readily converted into phthalides cores [99]. In addition, 2,2-disubstituted 2*H*-chromenes could be build up from *ortho*-alkenylphenols and allenes via [Cp*RhCl$_2$]$_2$-catalyzed [5+1] annulation [100].

4 Formation of Rh–Carbene Species: Couplings with Diazo Compounds

Another exciting feature of [Cp*Rh(III)]- and [Cp*Ir(III)]-metallacyclic intermediates afforded via C–H activation was evidenced when the scientific community explored direct carbenoid functionalization. In 2012 Yu discovered that diazomalonate reacts smoothly, in the presence of a catalytic amount of $[Cp^*RhCl_2]_2$ and AgOAc, with aryl ketone oximes to deliver α-aryl malonate (Scheme 7) [101]. This unprecedented coupling is believed to occur via initial electrophilic C–H activation to form a rhodacyclic intermediate, prompt to coordinate the diazo coupling partner and hence yielding a diazonium intermediate. The following extrusion of N_2 leads to the formation of a Rh–carbene species, which after 1,2-migratory insertion can be converted into the targeted alkylated product.

This initial report clearly showcased the potential of diazo compounds as suitable "alkylating" agents for direct functionalization of $C(sp^2)$–H bonds. Subsequently, several classes of diazo compounds have been employed as one-, two-, or three-atom components. Rovis [102], Cui [103], and Yu [104] have applied this formal oxidative [4+1] cycloadditions to build up benzolactame scaffolds (Scheme 8a). Asymmetric version was subsequently reported by Cramer [105]. Besides, carbonyl-containing diazo precursors (Scheme 8b, c) or diazotized Meldrum's acid (Scheme 8d), used as "two-carbon components," enables tandem C–H activation/cyclization/condensation reaction sequence delivering valuable heterocyclic skeletons like isoquinolines and pyridine N-oxides [106], isocoumarins and α-pyrones [107], cinnolin-3(2*H*)-one derivatives [108], or *N*-methoxyisoquinolinediones [109]. Finally, C–H activation/[4+3] cycloaddition cascade transformations could be envisioned in the presence of vinylcarbenoids, hence enabling the expedient construction of generally difficult-to-prepare seven-membered azepinone scaffolds (Scheme 8e) [110].

Noteworthy, [Cp*Ir(III)] metallacycles may also undergo a closely related coupling with diazo compounds; however, a slightly higher reaction temperature is required (90°C vs. RT-60°C for the related Rh-catalyzed reaction) [111].

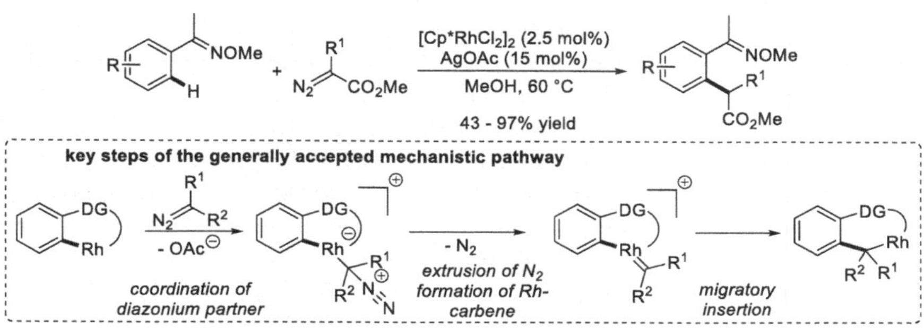

Scheme 7 Rh-catalyzed carbenoid functionalization

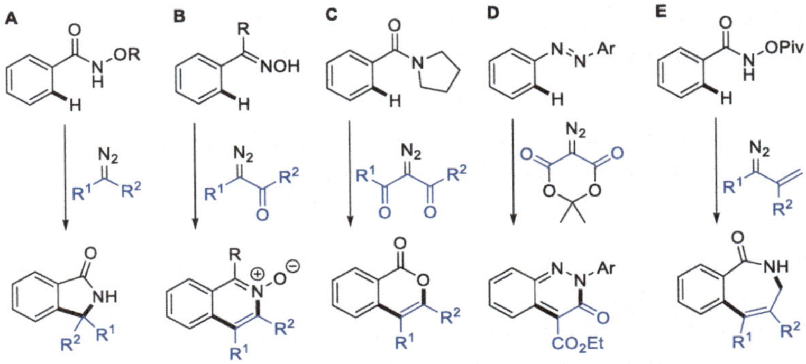

Scheme 8 Heterocyclic scaffolds constructed via C–H activation/cyclization with diazo compounds. (**a**) Coupling with diazo compounds as one-carbon components; (**b–d**) coupling with diazo compounds as two-carbon components; (**e**) coupling with diazo compounds as three-carbon components

5 Addition of Rh–C Bonds to C=X Polar π-Bonds

Nucleophilic addition of organometallic reagents to C=X (X=O, N, etc.) is one of the fundamental strategies to construct C–C bonds. In spite of the great importance of these couplings, tedious and waste-generating preparation of the organometallic nucleophiles (e.g., Grignard reagents or in situ generated lithium species) and their sometimes limited functional group tolerance strongly hamper the practical utility of this approach. The use of metallacyclic intermediates, obtained via C–H activation strategy, would greatly expand the potential of such transformations. Consequently, since the beginning of this decade, chemists have devoted considerable efforts to develop the addition reactions of C–Rh bonds ([Cp*Rh(III)]-metallacyclic intermediates) across polar C–N and C–O multiple bonds of imines and aldehydes [4, 112]. The regioselectivity issues, the weakly nucleophilic character of the C–Rh bonds, and the thermodynamic stability of metal alkoxides or metal alkylamide generated during the catalytic cycle were the challenges to be addressed.

This ambitious goal was reached independently by Bergman and Ellman [113] and Shi [114] who showed for the first time that phenylpyridine can undergo Rh(III)-catalyzed C–H activation and subsequent nucleophilic addition of the C–Rh bond to imines or aldimines delivering the expected amine products (Scheme 9). Further in-depth mechanistic investigations (intermediate isolation and kinetic studies) [115, 116] were conducted, and extensive experimental efforts revealed that several other classes of polar π-bonds containing compounds [112] such as isocyanates [117], aldehydes [118], or ketones [119] may be used (Scheme 10a–c). Importantly, when enantiopure *N*-perfluorobutanesulfinyl imines were used, the targeted addition proceeded with outstanding diastereoselectivity (>98:2 d.r.) providing, after removal of the sulfonyl group, enantiopure amine hydrochlorides [120].

Noteworthy, such nucleophilic reactivity of the C–Rh bond, when combined with a cyclative capture of the resulting intermediates, was also astutely used to construct valuable scaffolds. Furans and pyrroles [121] or indazoles [122] were

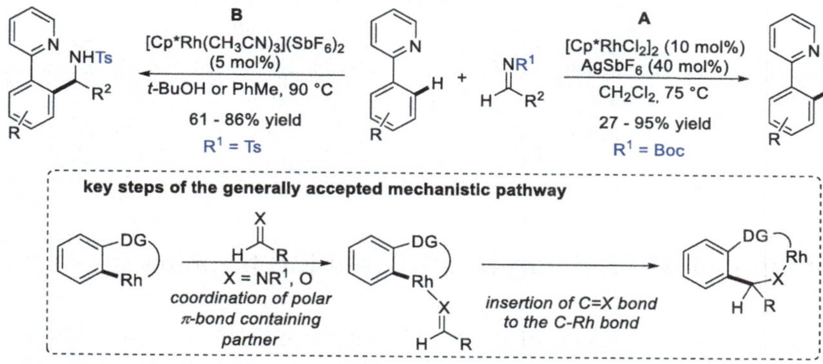

Scheme 9 Pioneering work on Rh(III)-catalyzed nucleophilic addition of non-acidic C–H bonds to C=X bonds. (**a**) Catalytic system developed by Bergman and Ellman; (**b**) catalytic system developed by Shi

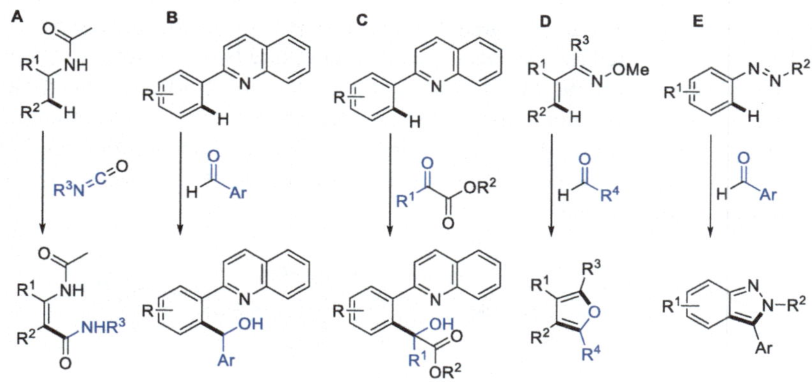

Scheme 10 Classes of coupling partners used in Rh-catalyzed additions to polar π-bonds. (**a**) Coupling with isocyanates; (**b**) coupling with aldehydes; (**c**) coupling ketones; (**d**) synthesis of furans; (**e**) synthesis of indazoles

accessed by coupling α,β-unsaturated oximes with ethyl glyoxylate or aldehydes/ azobenzene with aldehydes (Scheme 10d, e).

Migratory insertion of the carbonyl group of the cyclopropenones into the aryl group of the rhodacyclic intermediate also occurs smoothly (8) [123]. Subsequently, due to the ring strain, β-carbon elimination leading to a ring opening occurs yet affording chalcone derivatives at the end of the catalytic cycle.

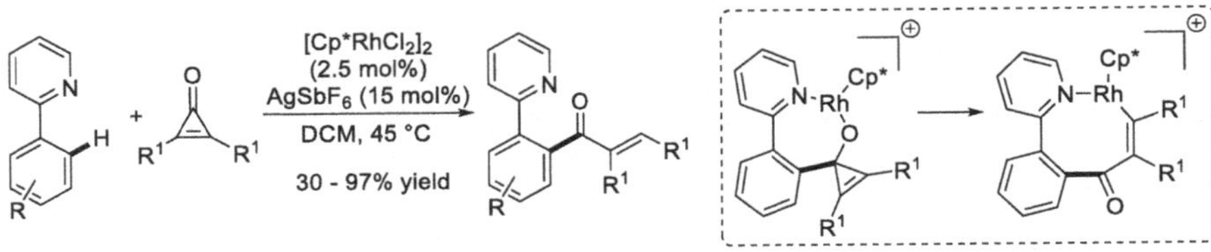

(8)

6 Addition of Rh–C Bonds to sp^3-Based Electrophiles

Regarding the rapid expansion of the Rh(III)-catalyzed C–H activation/condensation with electrophilic partners containing C=X bonds, it becomes quite obvious that the incipient Rh–C bond gives intrinsic reactivity that parallels Grignard reagents. Following this thought, Glorius et al. surmised that a related coupling with sp^3 electrophiles like α-(pseudo)halo ketones should also be viable even if several intrinsic difficulties such as regioselectivity issues and possible catalyst deactivation by an in situ liberated leaving group need to be overcome. This hypothesis was validated by discovering a direct coupling between aromatic amides and α-mesyloxyketones (9) [124]. Importantly, these preoxidized C(sp^3)-based electrophiles can be considered as a valuable alternative to the troublesome terminal alkynes. Accordingly, C–H activation/nucleophilic substitution/annulation reactional sequence allowed a redox-neutral synthesis of *N*-heterocycles like isoquinolones and pyridone.

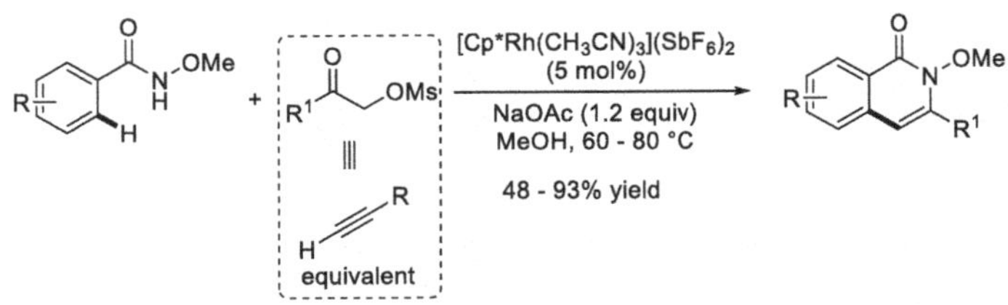

(9)

A somehow related reactivity of the Rh–C bond of the metallacyclic intermediate toward sp^3-hybridized electrophilic moieties was explored by Li and Wan when developing a dehydrogenative coupling between aromatic C–H substrates and aziridines (10) [125]. The authors discovered that a common rhodacyclic intermediate, in a presence of a phenyl-substituted aziridine, is readily converted into a novel coordinated complex in which a positive charge is built up at the benzylic position which enhances insertion of the Rh–C bond into the aziridine coupling partner.

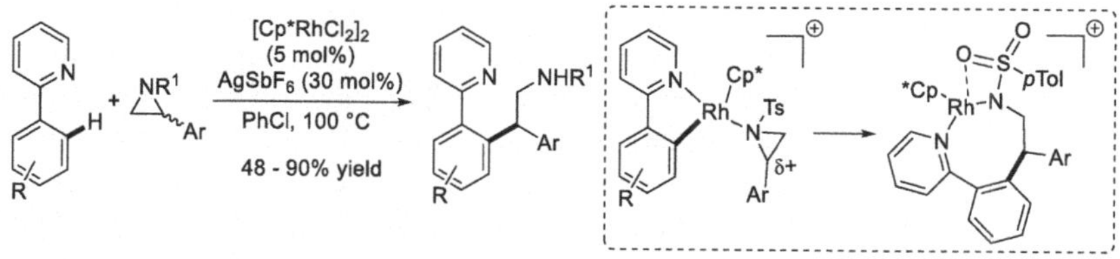

(10)

7 Oxidative Addition to Rh–C Bonds

7.1 Arylation with Haloarenes

With regard to the outstanding synthetic utility of the Suzuki–Miyaura reactions, their more sustainable alternative implying a C–H activation step has attracted great interest in the scientific community. Accordingly, numerous mainly Pd-based catalytic systems enabling couplings between Ar–H and Ar–X have been disclosed. In contrast, the direct arylation using Rh(III) or Ir(III) catalysts remained unexplored. Recently, the feasibility of such oxidative Ar–Ar coupling was proven by Cheng [126]. The *ortho*-arylation of anilides with aryl iodides was achieved using a cationic [Cp*Rh(III)] catalyst in combination with a silver salt additive (11). This reaction is believed to occur via oxidative addition of Ar–I to the rhodacyclic intermediate resulting from the C–H activation step. Thus, the existence of high-valent Rh(V) species is speculated. Iodine abstraction in the presence of the silver salt and final reductive elimination deliver the biaryl product and regenerate the Rh(III) catalyst. Importantly, KIE studies suggest that C–H cleavage is the rate-determining step and hence oxidative addition and generation of a rather rare Rh(V) species seem to be relatively fast and surprisingly straightforward.

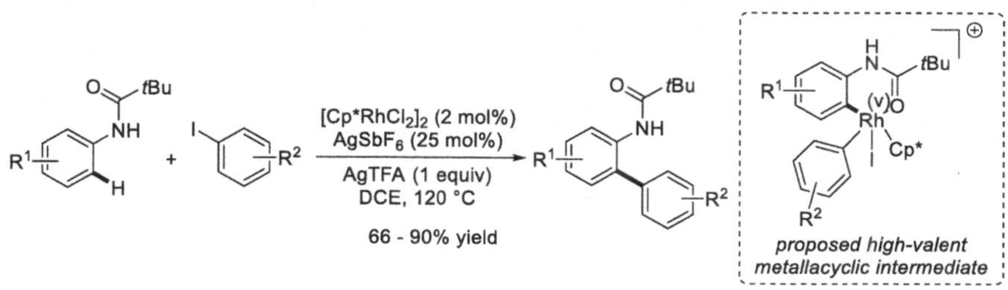

(11)

7.2 Alkynylation with the Hypervalent Alkynyl Iodine

Contrary to the largely explored C–H alkenylation reactions, the related alkynylations are more challenging due to an undesired homocoupling of terminal alkynes under commonly employed oxidative conditions. Moreover, the substrate scope of the rare C–H alkynylations is somehow limited to electronically activated arenes, and uncontrolled mono- and bis-functionalizations further hamper their synthetic utility. A valuable solution to these obstacles was found by discovering that hypervalent iodine reagents can be used as a powerful electrophilic alkyne source. The research group of Loh and slightly later the groups of Li and Glorius discovered independently that a large panel of aromatic and vinylic substrates undergoes an extremely mild alkynylation with 1-[(triisopropylsilyl)ethynyl]-1, 2-benziodoxol-3(1*H*)-one (TIPS-EBX) using [Cp*Rh(III)] catalyst (Scheme 11) [127–130]. Importantly, as this reaction does not require an addition of an external oxidant and occurs already at room temperature, almost equimolar amounts of both coupling partners may be used. Besides, the $[Cp^*IrCl_2]_2$ catalyst may also be used to perform this Sonogashira-type coupling [130]. Noteworthy, a complementary reactivity was observed when using the iridium congener; *N*-methoxybenzamides failed to undergo the desired alkynylation in the presence of $[Cp^*RhCl_2]_2$ catalyst, but the desired reactivity could be restored under the iridium-catalyzed conditions. As far as the mechanistic outcome of this transformation is concerned, several possible pathways have been evoked. Mechanistic investigations reported by Li (pathway a) suggest that a cyclometalated intermediate could oxidatively add to the

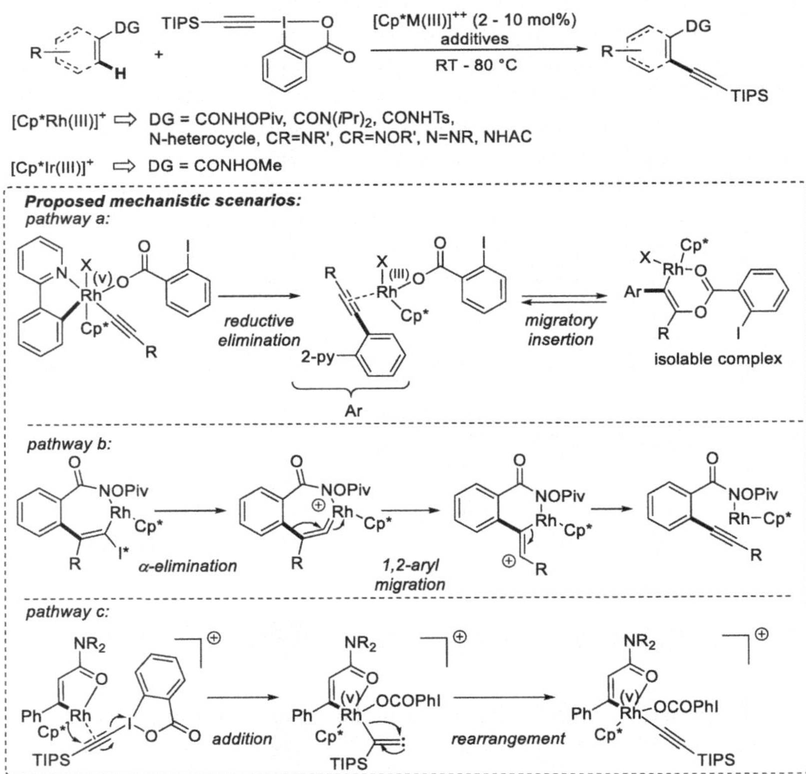

Scheme 11 Alkynylation with hypervalent iodine species

hypervalent iodine to give a Rh(V) alkynyl intermediate. Subsequent reductive elimination would lead to the formation of a Rh(III) alkyne benzoate species, which, after migratory insertion of the benzoate, would give an isolable Rh(III) vinyl complex [130]. Alternatively, a mechanistic scenario involving a regio-selective carborhodation by the metallacyclic intermediate, followed by α-elimination affording a rhodium vinylidene complex, intramolecular 1,2-aryl migration, and heterolytic C–Rh bond cleavage (pathway b), can be considered [127]. In addition, a pathway involving the addition of the rhodacycle to an alkyne with concomitant expulsion of 2-iodobenzoic acid affording carbene species was also proposed (pathway c) [129].

Notably, Rh(III)- and Ir(III)-catalyzed alkynylation using hypervalent iodine reagents was also used to selectively functionalize indoline cores at the C7 position [131, 132].

8 Transmetalation to Rhodacycles

Transmetalation is a fundamental step of classical transition metal-catalyzed cross coupling reactions. Surprisingly the potential of Rh(III)-based metallacyclic species to undergo transmetalation with diverse aromatic and aliphatic organometallic reagents has gathered attention of the scientific community only recently. In 2012, Cheng et al. observed that arylboronic acids may be coupled with *N*-methoxybenzamides in the presence of the $[Cp^{*}RhCl_2]_2$ catalyst (Scheme 12) [133]. This arylation reaction occurring via transmetalation of rhodacyclic intermediate is efficient at relatively low temperature (60°C) and shows excellent substrate scope. Importantly, as the secondary amide moiety is used as a DG, the newly generated biaryl skeletons undergo a second C–H activation event and subsequent C–N bond formation/reductive elimination hence delivering phenanthridinone skeletons.

A closely related strategy was also applied by Cui to selectively arylate *N*-amido-protected indoles at the C2 position [134]. Noteworthy, a slight modification

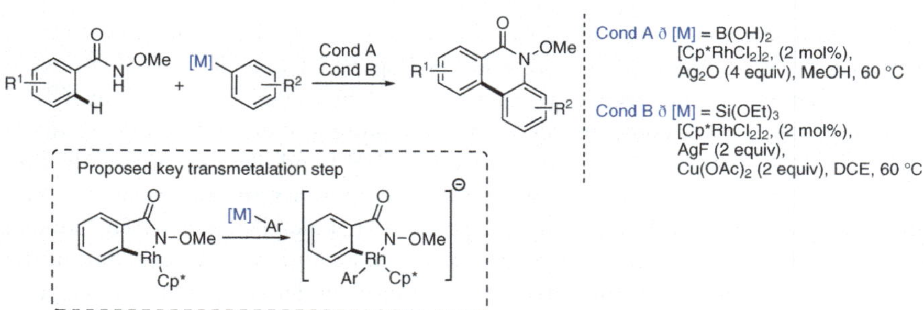

Scheme 12 Direct arylation occurring via transmetalation of arylboronic acids and arylsilanes

of the reaction conditions (choice of an oxidant and reaction temperature) alters the reaction's outcome from simple direct arylation to [4+2] and [4+1] cyclizations. Notably, mechanistically closely related direct arylations could also be achieved in the presence of arylsilanes [135, 136].

Impressively, the use of boroxines as an arylating agent turned out to be the key to success to perform an extremely challenging functionalization of homobenzylic sp^3 C–H bonds and secondary benzylic sp^3 C–H bonds, as disclosed by Glorius [137].

The use of boron reagents and the mechanistic scenario involving a transmetalation step provided a window of opportunity to perform challenging sp^2–sp^3 couplings. Indeed, Rh(III)-catalyzed alkylations are scarce due to an unfavorable reductive elimination. This ambitious target was achieved by Li who designed Rh (III)-catalyzed alkylation of various arenes in the presence of alkyl trifluoroborates as alkylating agents [138]. Although an excess of silver oxidant (2.8–4 equiv. of AgF) and a rather high reaction temperature (100°C) are essential to insure a satisfying efficiency of this transformation, this pioneering work showcases an underestimated potential of the boron reagents as coupling partners in the Rh(III)-catalyzed direct functionalizations. As in the case of the arylation reaction, a mechanistic pathway involving initial C–H activation followed by transmetalation and reductive elimination is proposed.

9 C–H Activation by Metallacyclic Intermediates

A powerful methodology to access biaryl scaffolds consists in the coupling of two non-prefunctionalized substrates by means of double C–H activation. Such cross couplings are however highly challenging as (1) two aryl partners need to be differentiated by the catalyst to avoid undesired homocouplings and (2) the metallacyclic intermediate generated via the first C–H activation should be able to perform direct C–H activation at the second coupling partner to generate a unique metallacyclic intermediate bearing both aromatic moieties. Alternatively, one could also envision the C–H activation of both substrates at different metal centers, which would then be united in a specific transmetalation event. Pioneering achievements in this field were reported in 2012 by Glorius et al. [139]. The authors discovered that tertiary benzamides, in the presence of $[Cp^*RhCl_2]_2$ and in a haloaromatic solvent such as bromobenzene, undergo a cross dehydrogenative coupling (CDC) affording biaryl compounds. Impressively, the coupling of bromobenzene occurs at *meta-* and *para-*positions implying a selective C–H bond cleavage rather than an oxidative addition into the C–Br bond. Extensive optimization studies showed that this dehydrogenative arylation is highly specific to halogenated arenes. The authors speculated that the catalytic cycle begins with the *ortho-*directed activation of the benzamide coupling partner delivering the common rhodacycle. This metallacyclic intermediate is supposed to catalyze a second, rate-determining C–H activation event on a haloarene. Theoretical studies supported the proposed mechanistic scenario involving a twofold C–H activation [140]. Further investigations enabled the discovery of a "second generation" of this CDC reaction. It was found that polyhalogenated aromatics, such as hexabromobenzene, can be used as additives hence enabling the desired coupling with non-halogenated simple arenes and heteroaromatics (12) [141]. A closely related twofold C–H activation between one aromatic partner bearing a directing group and a second, heteroaromatic partner was discovered independently by You [142] and Kambe [143].

Further major advances in CDCs involving Ar–H and (Het)Ar–H partners were reported independently by Su [144] and Lan and You [145]. Both groups astutely used carboxylic moieties as a traceless DG. In situ removal of this directing group at the final stage of the catalytic system results in the formation of simple Ar–(Het)Ar structures (13).

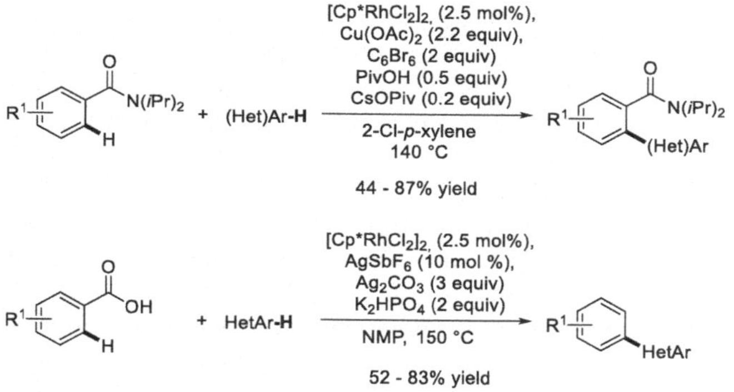

(12, 13)

As polyheteroaromatics are ubiquitous motifs in biologically active products and advanced materials, their construction by means of twofold C–H activation is particularly appealing. Following this ambitious goal, Glorius endeavored on designing such CDCs between two structurally similar heterocyclic scaffolds [146]. The desired cross coupling reaction could be achieved by carefully selecting both coupling partners (typically furans and (benzo)thiophenes), and after an intensive optimization study, the desired transformation delivered a panel of biheteroaryls (14).

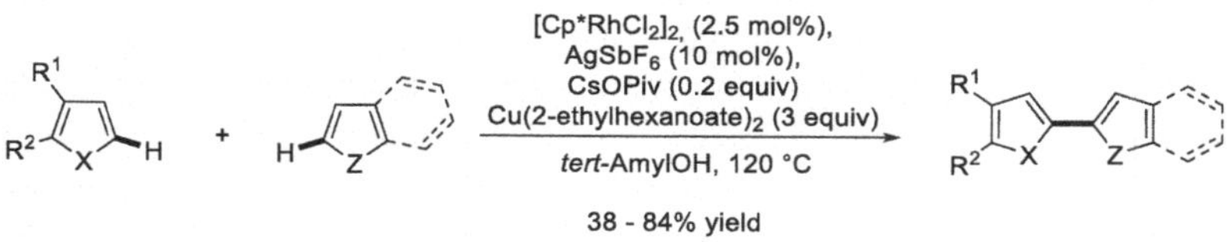

(14)

An additional interesting example of an intramolecular direct arylation occurring via double C–H activation was disclosed in 2013 by Miura [147]. The authors discovered that 2,2-diarylalkanoic acids undergo [Cp^ERh(III)]-catalyzed (Cp^E=1,3-bis(ethoxycarbonyl)-2,4,5-trimethylcyclopentadienyl) twofold C–H activation and subsequent decarboxylation yet delivering fluorene cores.

Interestingly, triarylmethanols failed to react under the optimized reaction conditions. However, in the presence of $[Cp^*IrCl_2]_2$ precursors, the reactivity was restored, and the triaryl substrates were smoothly converted into the fluoren-9-ol derivatives.

10 New Trends and Perspectives

Extraordinary advances achieved in the C–H activation field benefiting from the [Cp*M]-based complexes (M=Rh, Ir) propelled these half-sandwich organometallic complexes to the forefront of modern homogeneous organometallic catalysis. Indeed, an immense number of catalytic transformations initiated by the formation, via C–H activation, of the metallacyclic intermediates have been reported, enabling straightforward and waste- and step-economic synthesis of various compounds, mainly heterocyclic scaffolds. Originally, such C–H activation reactions required rather harsh reaction conditions, but recent developments have made those methods much milder, more tolerant to larger series of functional groups, and more user-friendly. The application of oxidizing directing groups was a real breakthrough as this approach enabled milder reaction conditions. Accordingly, redox-neutral transformations flourished by developing a large panel of oxidizing directing groups or by using preoxidized coupling partners.

A complementary strategy to design even more efficient transformations consists in elaborating modified Cp* ligands. Following this goal, extensive studies have been conducted by Tanaka who investigated the potential of electron-deficient Cp* ligands, bearing two ester substituents [148–150]. This complex showed an improved reactivity with regard to electron-rich C–H substrates like acetanilides and benzylic alcohols, enabling a C–H activation/annulation sequence at room temperature and benefiting from air as oxidant. In contrast, Rovis focused on the steric features of the Cp* ligand and synthesized a related Rh(III) complex with a cyclopentadienyl ligand bearing two *t*Bu substituents. The increased hindrance of this catalyst resulted in improved or complementary selectivity in the alkyne insertion event for the dehydrogenative coupling between vinylic amides or oximes and alkynes [39, 49]. Cp ligand design also recently enabled a diastereoselective [2 +1] annulation reaction [151]. A similar trend (in a context of the styrene incorporation into C–Rh bond) was recently described by Corminboeuf and Cramer [152]. The design of modified Cp ligands provides the opportunity to realize highly asymmetric C–H activation-based transformations. The pioneering works disclosed independently by Cramer [153] and Rovis and Ward [154] showcased that either stereogenic rhodacyclic intermediates (chirality introduces on the Cp ligand) or a chiral environment around the catalyst (enzyme) is prompt to transfer chiral information during the alkene insertion into the Rh–C bond. Enantioenriched dihydroisoquinolones could hence be accessed from very simple starting materials. Subsequently, the chiral [CpRh]-derived complexes were applied in several others asymmetric C–H activation reactions [155–157].

The design of advanced and even more powerful [CpIr(III)]- and [CpRh(III)]-derived catalysts will enable milder and unprecedented attractive transformations.

Many mechanistic studies combining experimental and theoretical investigations have been undertaken [140, 158–163]. Certainly, in a near future, even more efforts will be devoted to rationalize the observed exciting reactivity, and hopefully, in a near future, a more rational design of new transformations and more powerful catalytic systems will be made possible.

Another important obstacle of the herein presented chemistry to be truly synthetically useful at both laboratory and industrial scale is the high cost and limited natural abundance of these noble metals. Therefore, the development of related supported catalysts enabling their efficient recycling seems highly appealing. Although, to the best of our knowledge, immobilized [Cp*MX_2] complexes are still to be developed, some pioneering efforts concerning supported Rh- and Ir-based catalysts have been reported very recently. In particular, silica-supported Ir(III)-bipyridine complexes [164] and a Rh(I) complex coordinated by silica-supported triarylphosphine ligand [165] have been efficiently used as catalysts for C–H borylation. Besides, metal-organic frameworks have been used very recently to prepare heterogeneous Ir complexes which displayed interesting activity in direct C–H functionalizations [166]. An attempt to access recyclable [Cp*Rh] complexes by combining the [Cp*$RhCl_2$]$_2$ precatalyst with inorganic nanosheet-modified amino acid ligands was reported in 2014 [167]. Although this original catalytic system showed excellent reactivity and selectivity for the alkylation of *N*-(pivaloyloxy)benzamide, the recycling tests were disappointing. With regard to these recent reports, we believe that major advances will be achieved in these fields. Moreover, the use of cheaper first-row transition metals is also an important goal. Thus, the use of CoCp* instead of RhCp* is getting increasing attention since the first pioneering report by Kanai in 2013 [168–174].

In conclusion, Rh(III)- and Ir(III)-catalyzed direct transformations still have their best days ahead, and these transformations will continue to impose themselves as versatile synthetic tools for the rapid assembly of complex and relevant organic structures from simple starting materials.

References

1. Satoh T, Miura M (2010) Chem Eur J 16:11212–11222
2. Bouffard J, Itami K (2010) Top Curr Chem 292:231–280
3. Patureau FW, Wencel-Delord J, Glorius F (2012) Aldrichimica Acta 45:31–41
4. Colby DA, Tsai AS, Bergman RG, Ellman JA (2012) Acc Chem Res 45:814–825
5. Song G, Wang F, Li X (2012) Chem Soc Rev 41:3651–3678
6. Chiba S (2012) Chem Lett 41:1554–1559
7. Kuhl N, Schröder N, Glorius F (2014) Adv Synth Catal 356:1443–1460
8. Song G, Li X (2015) Acc Chem Res 48:1007–1020
9. Kang JW, Maitlis PM (1968) J Am Chem Soc 90:3259–3261
10. Kang JW, Moseley K, Maitlis PM (1969) J Am Chem Soc 91:5970–5977

11. Kisenyi JM, Sunley GJ, Cabeza JA, Smith AJ, Adams H, Salt NJ, Maitlis PM (1987) J Chem Soc Dalton Trans 2459–2466
12. Janowicz AH, Bergman RG (1983) J Am Chem Soc 105:3929–3939
13. Jones WD, Feher FJ (1985) J Am Chem Soc 107:620–631
14. Han YF, Jin GX (2014) Chem Soc Rev 43:2799–2823
15. Li L, Brennessel WW, Jones WD (2009) Organometallics 28:3492–3500
16. Matsumoto T, Periana RA, Taube DJ, Yoshida H (2002) J Catal 206:272–280
17. Ueura K, Satoh T, Miura M (2007) Org Lett 9:1407–1409
18. Ueura K, Satoh T, Miura M (2007) J Org Chem 72:5362–5367
19. Umeda N, Tsurugi H, Satoh T, Miura M (2008) Angew Chem Int Ed 47:4019–4022
20. Umeda N, Hirano K, Satoh T, Miura M (2009) J Org Chem 74:7094–7099
21. Boutadla Y, Davies DL, Jones RC, Singh K (2011) Chem Eur J 17:3438–3448
22. Choi J, Goldman AS (2011) Top Organomet Chem 34:139–168
23. Wang N, Li B, Song H, Xu S, Wang B (2013) Chem Eur J 19:358–364
24. Li L, Brennessel WW, Jones WD (2008) J Am Chem Soc 130:12414–12419
25. Boutadla Y, Davies DL, Al-Duaij O, Fawcett J, Jones RC, Singh K (2010) Dalton Trans 39: 10447–10457
26. Stuart DR, Bertrand-Laperle M, Burgess KMN, Fagnou K (2008) J Am Chem Soc 130: 16474–16475
27. Rakshit S, Patureau FW, Glorius F (2010) J Am Chem Soc 132:9585–9587
28. Zhang J, Qian H, Liu Z, Xiong C, Zhang Y (2014) Eur J Org Chem 8110–8118
29. Morimoto K, Hirano K, Satoh T, Miura M (2011) Chem Lett 40:600–602
30. Hyster TK, Rovis T (2010) J Am Chem Soc 132:10565–10569
31. Mochida S, Umeda N, Hirano K, Satoh T, Miura M (2010) Chem Lett 39:744–746
32. Guimond N, Gouliaras C, Fagnou K (2010) J Am Chem Soc 132:6908–6909
33. Luo CZ, Jayakumar J, Gandeepan P, Wu YC, Cheng CH (2015) Org Lett 17:924–927
34. Satoh T, Ueura K, Miura M (2008) Pure Appl Chem 80:1127–1134
35. Jia J, Shi J, Zhou J, Liu X, Song Y, Xu HE, Yi W (2015) Chem Commun 51:2925–2928
36. Morimoto K, Hirano K, Satoh T, Miura M (2010) Org Lett 12:2068–2071
37. Morimoto K, Hirano K, Satoh T, Miura M (2011) J Org Chem 76:9548–9551
38. Su Y, Zhao M, Han K, Song G, Li X (2010) Org Lett 12:5462–5465
39. Hyster TK, Rovis T (2011) Chem Sci 2:1606–1610
40. Mochida S, Hirano K, Satoh T, Miura M (2009) J Org Chem 74:6295–6298
41. Unoh Y, Hashimoto Y, Takeda D, Hirano K, Satoh T, Miura M (2013) Org Lett 15: 3258–3261
42. Qi Z, Wang M, Li X (2013) Org Lett 15:5440–5443
43. Umeda N, Hirano K, Satoh T, Shibata N, Sato H, Miura M (2011) J Org Chem 76:13–24
44. Song G, Gong X, Li X (2011) J Org Chem 76:7583–7589
45. Han YF, Li H, Hu P, Jin GX (2011) Organometallics 30:905–911
46. Frasco DA, Lilly CP, Boyle PD, Ison EA (2013) ACS Catal 3:2421–2429
47. Guimond N, Gorelsky SI, Fagnou K (2011) J Am Chem Soc 133:6449–6457
48. Rakshit S, Grohmann C, Besset T, Glorius F (2011) J Am Chem Soc 133:2350–2353
49. Hyster TK, Rovis T (2011) Chem Commun 47:11846–11848
50. Wang H, Grohmann C, Nimphius C, Glorius F (2012) J Am Chem Soc 134:19592–19595
51. Liu G, Shen Y, Zhou Z, Lu X (2013) Angew Chem Int Ed 52:6033–6037
52. Too PC, Wang YF, Chiba S (2010) Org Lett 12:5688–5691
53. Chuang SC, Gandeepan P, Cheng CH (2013) Org Lett 15:5750–5753
54. Muralirajan K, Cheng CH (2014) Adv Synth Catal 356:1571–1576
55. Xu X, Liu Y, Park CM (2012) Angew Chem Int Ed 51:9372–9376
56. Zhou B, Du J, Yang Y, Li Y (2014) Chem Eur J 20:12768–12772
57. Zhou B, Yang Y, Tang H, Du J, Feng H, Li Y (2014) Org Lett 16:3900–3903
58. Zhang X, Qi Z, Li X (2014) Angew Chem Int Ed 53:10794–10798
59. Sharma U, Park Y, Chang S (2014) J Org Chem 79:9899–9906

60. Dateer RB, Chang S (2015) J Am Chem Soc 137:4908–4911
61. Patureau FW, Besset T, Kuhl N, Glorius F (2011) J Am Chem Soc 133:2154–2156
62. Yang XF, Hu XH, Loh TP (2015) Org Lett 17:1481–1484
63. Chen Y, Wang F, Zhen W, Li X (2013) Adv Synth Catal 355:353–359
64. Wang F, Qi Z, Sun J, Zhang X, Li X (2013) Org Lett 15:6290–6293
65. Patureau FW, Besset T, Glorius F (2011) Angew Chem Int Ed 50:1064–1067
66. Patureau FW, Glorius F (2010) J Am Chem Soc 132:9982–9983
67. Wang F, Song G, Du Z, Li X (2011) J Org Chem 76:2926–2932
68. Wang F, Song G, Li X (2010) Org Lett 12:5430–5433
69. Zhao D, Vásquez-Céspedes S, Glorius F (2015) Angew Chem Int Ed 54:1657–1661
70. Tsai AS, Brasse M, Bergman RG, Ellman JA (2011) Org Lett 13:540–542
71. Li X, Gong X, Zhao M, Song G, Deng J, Li X (2011) Org Lett 13:5808–5811
72. Takahama Y, Shibata Y, Tanaka K (2015) Chem Eur J 21:9053–9056
73. Shi Z, Boultadakis-Arapinis M, Glorius F (2013) Chem Commun 49:6489–6491
74. Shi Z, Boultadakis-Arapinis M, Koester DC, Glorius F (2014) Chem Commun 50:2650–2652
75. Webb NJ, Marsden SP, Raw SA (2014) Org Lett 16:4718–4721
76. Otley KD, Ellman JA (2015) Org Lett 17:1332–1335
77. Zhang M, Zhang HJ, Han T, Ruan W, Wen TB (2015) J Org Chem 80:620–627
78. Yu S, Li X (2014) Org Lett 16:1200–1203
79. Zhang SS, Wu JQ, Lao YX, Liu XG, Liu Y, Lv WX, Tan DH, Zeng YF, Wang H (2014) Org Lett 16:6412–6415
80. Wang H, Schröder N, Glorius F (2013) Angew Chem Int Ed 52:5386–5389
81. Feng C, Feng D, Loh TP (2013) Org Lett 15:3670–3673
82. Feng C, Feng D, Loh TP (2015) Chem Commun 51:342–345
83. Deng H, Li H, Wang L (2015) Org Lett 17:2450–2453
84. Samanta R, Narayan R, Antonchick AP (2012) Org Lett 14:6108–6111
85. Engelman KL, Feng Y, Ison EA (2011) Organometallics 30:4572–4577
86. Shi Z, Grohmann C, Glorius F (2013) Angew Chem Int Ed 52:5393–5397
87. Cui S, Zhang Y, Wu Q (2013) Chem Sci 4:3421–3426
88. Besset T, Kuhl N, Patureau FW, Glorius F (2011) Chem Eur J 17:7167–7171
89. Zhang J, Loh TP (2012) Chem Commun 48:11232–11234
90. Boultadakis-Arapinis M, Hopkinson MN, Glorius F (2014) Org Lett 16:1630–1633
91. Yu DG, de Azambuja F, Gensch T, Daniliuc CG, Glorius F (2014) Angew Chem Int Ed 53: 9650–9654
92. Zhao D, Lied F, Glorius F (2014) Chem Sci 5:2869–2873
93. Fukui Y, Liu P, Liu Q, He ZT, Wu NY, Tian P, Lin GQ (2014) J Am Chem Soc 136: 15607–15614
94. Wang H, Glorius F (2012) Angew Chem Int Ed 51:7318–7322
95. Zeng R, Fu C, Ma S (2012) J Am Chem Soc 134:9597–9600
96. Ye B, Cramer N (2013) J Am Chem Soc 135:636–639
97. Zeng R, Wu S, Fu C, Ma S (2013) J Am Chem Soc 135:18284–18287
98. Wang H, Beiring B, Yu DG, Collins KD, Glorius F (2013) Angew Chem Int Ed 52: 12430–12434
99. Gandeepan P, Rajamalli P, Cheng CH (2015) Chem Eur J 21:9198–9203
100. Casanova N, Seoane A, Mascareñas JL, Gulías M (2015) Angew Chem Int Ed 54:2374–2377
101. Chan WW, Lo SF, Zhou Z, Yu WY (2012) J Am Chem Soc 134:13565–13568
102. Hyster TK, Ruhl KE, Rovis T (2013) J Am Chem Soc 135:5364–5367
103. Zhang Y, Wang D, Cui S (2015) Org Lett 17:2494–2497
104. Lam HW, Man KY, Chan WW, Zhou Z, Yu WY (2014) Org Biomol Chem 12:4112–4116
105. Ye B, Cramer N (2014) Angew Chem Int Ed 53:7896–7899
106. Shi Z, Koester DC, Boultadakis-Arapinis M, Glorius F (2013) J Am Chem Soc 135: 12204–12207
107. Li XG, Sun M, Liu K, Jin Q, Liu PN (2015) Chem Commun 51:2380–2383

108. Son JY, Kim S, Jeon WH, Lee PH (2015) Org Lett 17:2518–2521
109. Shi J, Zhou J, Yan Y, Jia J, Liu X, Song H, Xu HE, Yi W (2015) Chem Commun 51:668–671
110. Cui S, Zhang Y, Wang D, Wu Q (2013) Chem Sci 4:3912–3916
111. Xia Y, Liu Z, Feng S, Zhang Y, Wang J (2015) J Org Chem 80:223–236
112. Zhang XS, Chen K, Shi ZJ (2014) Chem Sci 5:2146–2159
113. Tsai AS, Tauchert ME, Bergman RG, Ellman JA (2011) J Am Chem Soc 133:1248–1250
114. Li Y, Li BJ, Wang WH, Huang WP, Zhang XS, Chen K, Shi ZJ (2011) Angew Chem Int Ed 50:2115–2119
115. Tauchert ME, Incarvito CD, Rheingold AL, Bergman RG, Ellman JA (2012) J Am Chem Soc 134:1482–1485
116. Li Y, Zhang XS, Li H, Wang WH, Chen K, Li BJ, Shi ZJ (2012) Chem Sci 3:1634–1639
117. Hesp KD, Bergman RG, Ellman JA (2011) J Am Chem Soc 133:11430–11433
118. Li Y, Zhang XS, Chen K, He KH, Pan F, Li BJ, Shi ZJ (2012) Org Lett 14:636–639
119. Zhang XS, Zhu QL, Luo FX, Chen G, Wang X, Shi ZJ (2013) Eur J Org Chem 6530–6534
120. Wangweerawong A, Bergman RG, Ellman JA (2014) J Am Chem Soc 136:8520–8523
121. Lian Y, Huber T, Hesp KD, Bergman RG, Ellman JA (2013) Angew Chem Int Ed 52:629–633
122. Lian Y, Bergman RG, Lavis LD, Ellman JA (2013) J Am Chem Soc 135:7122–7125
123. Yu S, Li X (2014) Org Lett 16:1220–1223
124. Yu DG, de Azambuja F, Glorius F (2014) Angew Chem Int Ed 53:2754–2758
125. Li X, Yu S, Wang F, Wan B, Yu X (2013) Angew Chem Int Ed 52:2577–2580
126. Haridharan R, Muralirajan K, Cheng CH (2015) Adv Synth Catal 357:366–370
127. Feng C, Loh TP (2014) Angew Chem Int Ed 53:2722–2726
128. Feng C, Feng D, Luo Y, Loh TP (2014) Org Lett 16:5956–5959
129. Collins KD, Lied F, Glorius F (2014) Chem Commun 50:4459–4461
130. Xie F, Qi Z, Yu S, Li X (2014) J Am Chem Soc 136:4780–4787
131. Yang XF, Hu XH, Feng C, Loh TP (2015) Chem Commun 51:2532–2535
132. Wu Y, Yang Y, Zhou B, Li Y (2015) J Org Chem 80:1946–1951
133. Karthikeyan J, Haridharan R, Cheng CH (2012) Angew Chem Int Ed 51:12343–12347
134. Zheng J, Zhang Y, Cui S (2014) Org Lett 16:3560–3563
135. Senthilkumar N, Parthasarathy K, Gandeepan P, Cheng CH (2013) Chem Asian J 8: 2175–2181
136. Lu MZ, Lu P, Xu YH, Loh TP (2014) Org Lett 16:2614–2617
137. Wang X, Yu DG, Glorius F (2015) Angew Chem Int Ed. DOI 10.1002/anie.201503888
138. Wang H, Yu S, Qi Z, Li X (2015) Org Lett 17:2812–2815
139. Wencel-Delord J, Nimphius C, Patureau FW, Glorius F (2012) Angew Chem Int Ed 51: 2247–2251
140. Zhao D, Li X, Han K, Li X, Wang Y (2015) J Phys Chem A 119:2989–2997
141. Wencel-Delord J, Nimphius C, Wang H, Glorius F (2012) Angew Chem Int Ed 51: 13001–13005
142. Dong J, Long Z, Song F, Wu N, Guo Q, Lan J, You J (2013) Angew Chem Int Ed 52:580–584
143. Reddy VP, Qiu R, Iwasaki T, Kambe N (2013) Org Lett 15:1290–1293
144. Zhang Y, Zhao H, Zhang M, Su W (2015) Angew Chem Int Ed 54:3817–3821
145. Qin X, Sun D, You Q, Cheng Y, Lan J, You J (2015) Org Lett 17:1762–1765
146. Kuhl N, Hopkinson MN, Glorius F (2012) Angew Chem Int Ed 51:8230–8234
147. Itoh M, Hirano K, Satoh T, Shibata Y, Tanaka K, Miura M (2013) J Org Chem 78:1365–1370
148. Shibata Y, Tanaka K (2011) Angew Chem Int Ed 50:10917–10921
149. Hoshino Y, Shibata Y, Tanaka K (2014) Adv Synth Catal 356:1577–1585
150. Fukui M, Hoshino Y, Satoh T, Miura M, Tanaka K (2014) Adv Synth Catal 356:1638–1644
151. Piou T, Rovis T (2014) J Am Chem Soc 136:11292–11295
152. Wodrich MD, Ye B, Gonthier JF, Corminboeuf C, Cramer N (2014) Chem Eur J 20: 15409–15418
153. Ye B, Cramer N (2012) Science 338:504–506

154. Hyster T, Knörr L, Ward TR, Rovis T (2012) Science 338:500–503
155. Ye B, Cramer N (2015) Acc Chem Res 48:1308–1318
156. Wencel-Delord J, Colobert F (2013) Chem Eur J 19:14010–14017
157. Zheng C, You SL (2014) RSC Adv 4:6173–6214
158. Xu L, Zhu Q, Huang G, Cheng B, Xia Y (2012) J Org Chem 77:3017–3024
159. Quiñones N, Seoane A, García-Fandiño R, Mascareñas JL, Gulías M (2013) Chem Sci 4: 2874–2879
160. Liu L, Wu Y, Wang T, Gao X, Zhu J, Zhao Y (2014) J Org Chem 79:5074–5081
161. Algarra AG, Davies DL, Khamker Q, Macgregor SA, McMullin CL, Singh K, Villa-Marcos B (2015) Chem Eur J 21:3087–3096
162. Xing Z, Huang F, Sun C, Zhao X, Liu J, Chen D (2015) Inorg Chem 54:3958–3969
163. Yu S, Liu S, Lan Y, Wan B, Li X (2015) J Am Chem Soc 137:1623–1631
164. Wu F, Feng Y, Jones CW (2014) ACS Catal 4:1365–1375
165. Kawamorita S, Miyazaki T, Iwai T, Ohmiya H, Sawamura M (2012) J Am Chem Soc 134: 12924–12927
166. Manna K, Zhang T, Greene FX, Lin W (2015) J Am Chem Soc 137:2665–2673
167. Liu H, An Z, He J (2014) ACS Catal 4:3543–3550
168. Yoshino T, Ikemoto H, Matsunaga S, Kanai M (2013) Angew Chem Int Ed 52:2207–2211
169. Ikemoto H, Yoshino T, Sakata K, Matsunaga S, Kanai M (2014) J Am Chem Soc 136: 5424–5431
170. Yu DG, Gensch T, de Azambuja F, Vasquez-Cespedes S, Glorius F (2014) J Am Chem Soc 136:17722–17725
171. Li J, Ackermann L (2015) Angew Chem Int Ed 54:3635–3638
172. Hummel JR, Ellman JA (2015) J Am Chem Soc 137:490–498
173. Pawar AB, Chang S (2015) Org Lett 17:660–663
174. Zhao D, Kim JH, Stegemann L, Strassert CA, Glorius F (2015) Angew Chem Int Ed 54: 4508–4511

Top Organomet Chem (2016) 55: 29–52
DOI: 10.1007/3418_2015_123

Published online: 10 June 2015

Rh(III)- and Ir(III)-Catalyzed Direct C–H Bond Transformations to Carbon–Heteroatom Bonds

Jeung Gon Kim, Kwangmin Shin, and Sukbok Chang

Abstract The direct manipulation of C–H bonds is now a powerful tool in chemical synthesis. In achieving the current high standard of research progresses, Rh(III) and Ir(III) complexes played an important role to understand the nature of C–H bond activation. While numerous stoichiometric reactions of hydrocarbons with Rh (III) or Ir(III) complexes were scrutinized, their use in catalytic transformations has been relatively undeveloped until recently. Given their outstanding reactivity in C–H activation, they are highly promising candidates for inducing mild C–H functionalizations. In spite of a short development history, numerous contributions from leading research groups made big strides in highly efficient and selective C–H bond transformations for the C–C and C–heteroatom bond formation. In this report, we specifically focus on the Rh(III)- or Ir(III)-mediated direct C–H functionalizations for the C–heteroatom bond formation that is now a rapidly growing area. This report presents the current status of such catalytic systems including scope of substrates and coupling partners as well as brief mechanistic descriptions.

Keywords C–H bond activation · C–heteroatom bond formation · Iridium(III) catalyst · Rhodium(III) catalyst

Contents

J.G. Kim, K. Shin, and S. Chang (✉)
Center for Catalytic Hydrocarbon Functionalizations, Institute for Basic Science (IBS), Daejeon 305-701, Republic of Korea

Department of Chemistry, Korea Advanced Institute of Science and Technology (KAIST), Daejeon 305-701, Republic of Korea
e-mail: sbchang@kaist.ac.kr

1 Introduction

The direct functionalization of a carbon–hydrogen bond was not previously considered as a logical retrosynthetic disconnection mainly due to their high dissociation energy and omnipresence in organic molecules, thus resulting in a low level of efficiency and selectivity. Mainly due to the better understanding of the activation pathway, highly efficient catalytic systems have been developed during the last decades. Because of these tremendous research efforts on the direct C–H transformation, this strategy is now widely utilized in making functionalized molecules in various areas including organic synthesis, medicinal chemistry, and materials science. Direct use of abundant carbon–hydrogen connections could eliminate prefunctionalization steps, thus leading to atom and step economy in chemical synthesis. Therefore, transition metal-mediated C–H functionalization presents a high promise to chemo-, regio-, and stereoselective synthetic approaches, desirably under mild conditions [1–3]. In this regard, while significant achievements have been made by employing a wide range of transition metal catalysts, the group 9 metals of Co, Rh, and Ir have also played an important role in advancing the C–H functionalization routes [4–7]. Sufficiently high reactivity of Rh(III) and Ir(III) complexes made them suitable for studying the nature of C–H activation process since stoichiometric manipulations with hydrocarbons give an easy access to cyclometalated intermediates [8]. As the C–H bond cleavage of hydrocarbons can readily be achievable by tuning the ligated Rh and Ir species, the resultant mild catalytic C–H functionalization procedures are expected to have a high impact in broad research areas. Surprisingly, it was not until recently that their catalytic efficiency of group 9 metal complexes has successfully applied to the useful organic transformations. In this context, impressive achievements have been made recently to broaden the scope of Rh(III)- and Ir(III)-catalyzed construction of carbon–heteroatom bonds as well as carbon–carbon bonds.

The formation of carbon–heteroatom bonds is of special interest due to the fact that the introduction of heteroatom into molecules brings functions in addition to the structural modification [9–11]. In this direct C–H functionalization, although a wide range of catalytic systems have been developed, described herein are the recent advances achieved by Rh(III)- and Ir(III)-based catalytic systems. Since excellent reviews on the direct C–H borylation reactions are available due to their extensive research efforts [12–14], this chapter does not cover the borylation chemistry.

2 Rh(III)-Catalyzed Reactions

In 2003, Davies and coworkers demonstrated that a Rh(III) complex generated from $[RhCp^*Cl_2]_2$ is capable of activating a C–H bond to form a metallacycle with an imine directing group (1) [15]. Derivatives of Cp*Rh(III) species turned out to be highly reactive enough to undergo the C–H activation process of various hydrocarbons under mild conditions, sometimes even at room temperature, which caught a special attention [5, 16]. After the Miura group's seminal report on the Rh(III)-based catalytic C–H bond functionalization to form C–C bonds [17–19], significant advances have been made in the Rh(III)-mediated C–H bond activation strategy for the C–C and C–heteroatom bond formations including B, N, S, Cl, Br, and I [4–6].

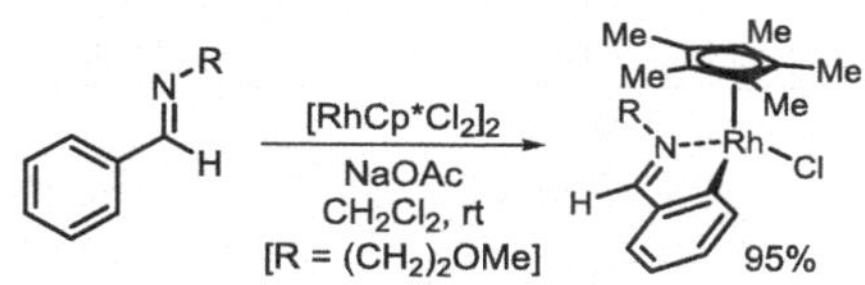

(1)

2.1 C–N Bond Formation

The direct amination of C–H bonds has been studied as an environmentally benign alternative to the Ullmann–Goldberg and Buchwald–Hartwig aminations, which employ (hetero)aryl halides or their equivalents to react with amines [20–26]. Significant research efforts are in progress for an ideal C–H amination system: coupling of unactivated hydrocarbons with readily available amino sources under mild conditions while producing minimal amounts of wastes [27, 28]. In 2012, Chang, Glorius, and Yu simultaneously reported Cp*Rh(III)-catalyzed direct C–H amination protocols with azides or *N*-chloroamines, operating under mild conditions in the absence of external oxidants. In contrast to the Rh(II)-catalyzed C–H aminations which undergo an initial metal-nitrenoid formation and subsequent insertion into C–H bonds (outer sphere mechanism) [29–32], Cp*Rh(III)-mediated C–H aminations are believed to proceed through sequential C–H activation and C–N bond formation (inner sphere pathway) [33].

Chang and coworkers presented the use of organic azides as a unique nitrogen source [34]. A cationic Cp*Rh(III) catalyst generated in situ from $[Cp^*RhCl_2]_2$ and silver additive was discovered to turnover C–N bond formation between 2-phenylpyridines and sulfonyl azides. This protocol does not require external oxidants and releases only nitrogen gas as sole by-product, enabling an environmentally benign catalytic process. This method displayed a high reactivity over a wide range of substrates and azides. Assorted types of directing groups [34–39]

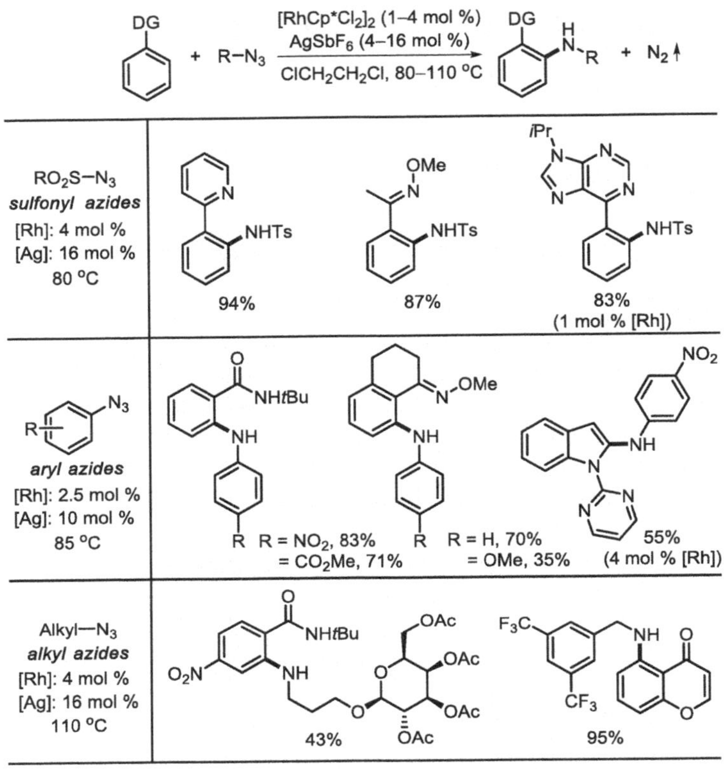

Scheme 1 Cp*Rh-catalyzed direct C–N bond-forming reactions using organic azides

including pyridyl, azobenzene, imidate, amide, oxime, and ketone effectively underwent the direct C–H aminations with sulfonyl, alkyl [37], and aryl azides [38] (Scheme 1).

Detailed mechanistic studies and theoretical calculations were carried out by the Chang research group, especially clearing the mist in the C–N bond-forming step (Scheme 2) [40]. While two mechanisms, redox-active nitrenoid pathway and redox-neutral concerted pathway, were considered, DFT calculations pointed out that the stepwise pathway involving a Rh(V)-imido intermediate is energetically more feasible.

Wang and coworkers extended the scope of Cp*Rh(III)-catalyzed C–H amination protocol to sp^3 C–H bonds. A cationic Cp*Rh(III) species successfully activated benzylic C–H bonds of 8-methylquinolines and the resultant rhodacycles reacted with sulfonyl azides in moderate efficiency (2) [41].

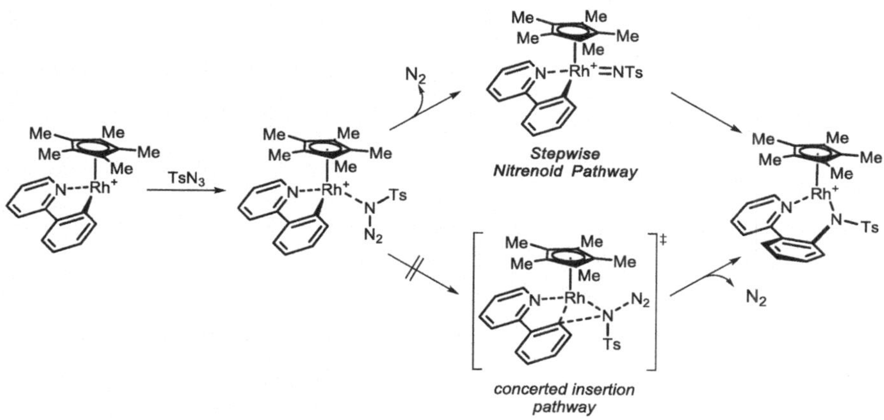

Scheme 2 Reaction mechanism of C–N bond formation step using organic azides

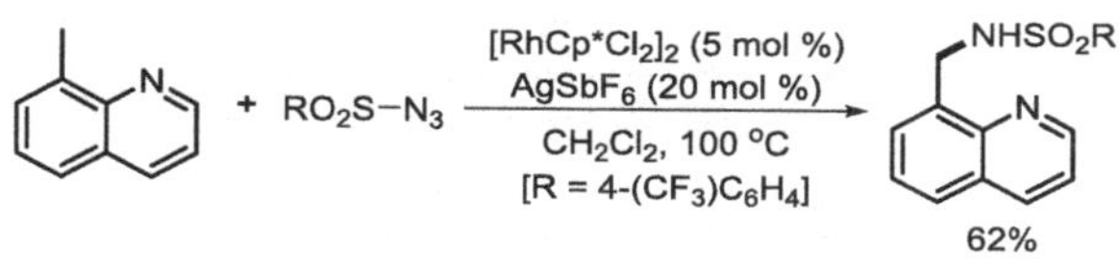

(2)

Aminated products obtained from the reaction with azides were shown to be a valuable motif for the preparation of many synthetic building blocks (Scheme 3). Syntheses of 1*H*-indazoles (Glorius, Scheme 3a) [42], acridines, phenazines (Bergman/Ellman, Scheme 3b) [43], 2-aryl-2*H*-Benzotriazoles (Lee, Scheme 3c) [35], and carbazoles (Chang) [38] showcased the synthetic versatility of aminated products. The Chang group also applied this method to 6-arylpurine functionalizations, an important unit in medicinal chemistry (Scheme 3d) [44].

While *N*-chloroamines have been used for the electrophilic C–H amination of hydrocarbons by the action of various metal catalysts [45–47], it was proven to be especially well suited with the Rh(III) catalytic system (Scheme 4). In 2012, Glorius and Yu groups independently demonstrated that *N*-chloroamines would undergo a S_N-type reaction by utilizing chloride as a leaving group [48–50]. Oximes (Scheme 4a), hydroxamic acids (Scheme 4b), and benzoic acid derivatives (Scheme 4c) smoothly proceeded the desired C–H aminations under mild conditions. Whereas secondary *N*-chloroamines were used most efficiently, Yu and coworkers devised a method to tame reactive and unstable primary *N*-chloroamines. The presence of an external base or in situ generation of primary *N*-chloroamines led to the efficient production of secondary amine compounds [49]. Additionally, their recent works widened the synthetic scope by employing *N*-chlorocarbamates [50]. Both Glorius and Yu postulated that the extra Ag or Cs salts, which are required for high turnovers, facilitate chloride removal at the stage of C–N bond formation (Scheme 4d). Initial mechanistic studies revealed the electrophilic nature of this amination protocol although a Rh(V)-imido pathway could not be completely ruled out at the present stage.

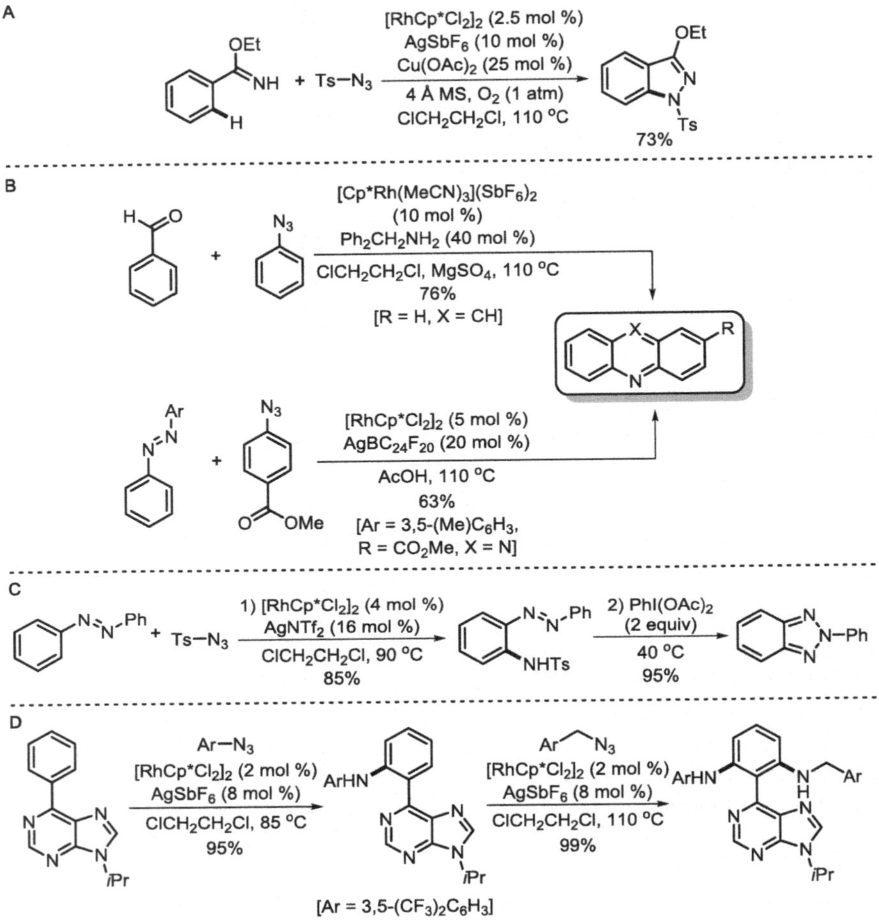

Scheme 3 Synthetic applications of C-H aminations using organic azides: 1*H*-indazoles (**a**), acridines and phenazines (**b**), 2-aryl-2*H*-benzotriazoles (**c**), and functionalized 6-arylpurines (**d**)

A number of hydroxylamine derivatives were also shown to be effective in the Cp*Rh(III)-catalyzed amination reactions (Scheme 5) [51–55]. This C–H amination takes place upon the cleavage of N–O bond of aminating reagents to serve as an internal oxidant, thus making the procedure free from the use of stoichiometric amounts of external oxidants. An initial study was reported by Glorius and coworkers, where they hypothesized that an electron-deficient carboxylate would act as a leaving group (Scheme 5a) [51]. However, the exact reaction pathway is unclear especially in regard to whether a Rh(III) migratory insertion or a Rh(V)-nitrenoid pathway would be involved.

An analogous type of amidating source of *N*-arenesulfonated imides was developed by Li and coworkers, and it was successfully applied to the amidation of

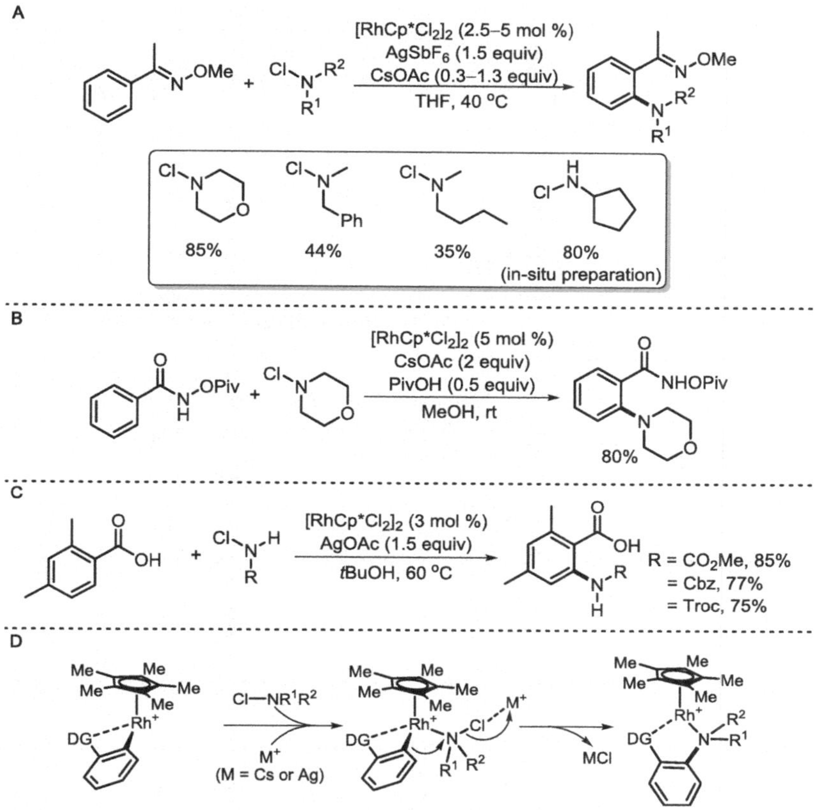

Scheme 4 Cp*Rh(III)-mediated C-H amination with *N*-chloroamines of oxime (**a**), hydroxamic acid (**b**), and benzoic acid (**c**) derivatives and its mechanism (**d**)

substrates containing pyridine or oxime directing groups (Scheme 5b) [52]. Zhang and coworkers utilized *N*-benzoylated alkylamines for the functionalization of phenidones and edaravone (neuroprotective drugs) analogues (Scheme 5c) [53, 54]. In 2015, Lu and coworkers developed a unique procedure of the Cp*Rh(III)-catalyzed amidation on water using *tert*-butyl 2,4-dinitrophenoxycarbamate as an amidating reagent [55]. $[RhCp^*Cl_2]_2$ was found to catalyze the directed C–N bond formation in water without additives. The authors proposed that an active cationic Rh(III) species would be formed by the hydroxyl group on water surface, thus replacing a silver salt that was used in most Cp*Rh(III)-catalyzed activations.

An interesting usage of nitroso compounds as an amino source was shown by Li and coworkers (3) [56–58]. An electrophilic nitrogen of nitrosobenzene was able to react with a rhodacycle species to afford *N*-diarylhydroxyamines at room temperature [56]. Later they established one-pot protocol of C–N bond formation and subsequent dehydroxylation to form diarylamines [57]. After the formation of *N*-diarylhydroxyamines, Rh(III) catalyst continues to promote dehydroxylation with

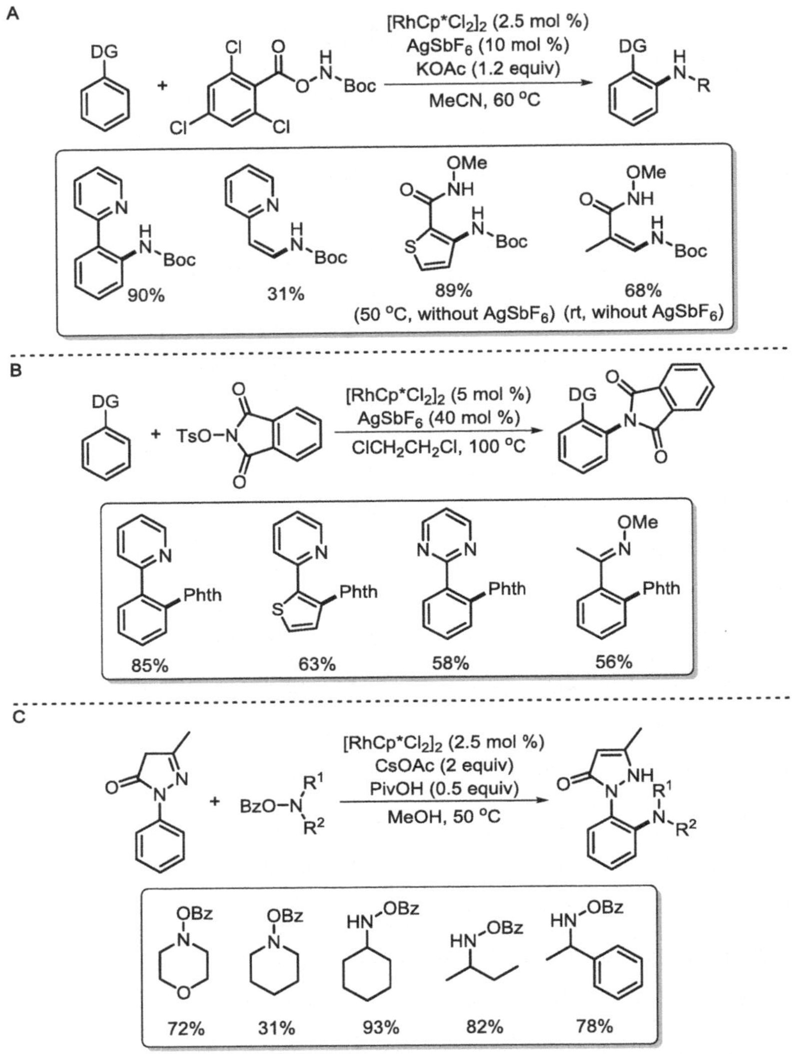

Scheme 5 Cp*Rh(III)-mediated C-H amination with *N*-substituted hydroxylamines of Aroyloxycarbamates (**a**) and *N*-arenesulfonated imides (**b**), and its application to edaravone functionalization (**c**)

an aid of molecular oxygen and carboxylic acid at elaborated temperature. The same research group extended this tandem protocol to in situ nitroso group formation. Three-step sequence of oxidation of *N*-hydroxylcarbamates to nitroso compounds, C–N bond formation, and dehydroxylation was realized in one pot to produce *N*-Boc protected arylamines [58].

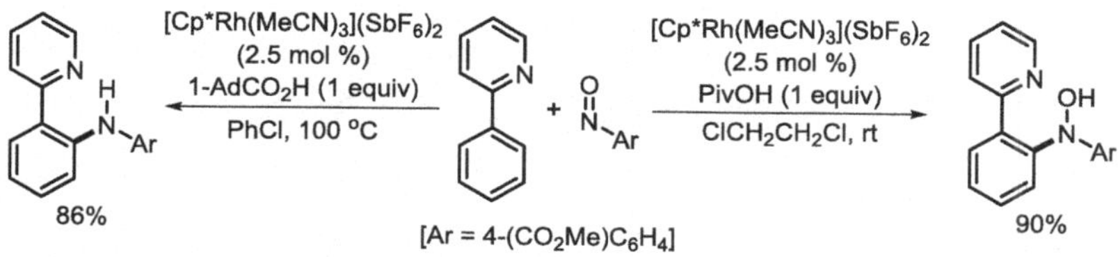

(3)

N-Fluorobenzenesulfonimide (NFSI), a well-known fluorination reagent, also participated in Rh(III)-catalyzed C–H functionalization as an amino source (4) [59, 60]. Yang and Li utilized NFSI as an electrophilic amidating reagent [61], exhibiting reasonable efficiency with 2-pyridyl-directed arenes.

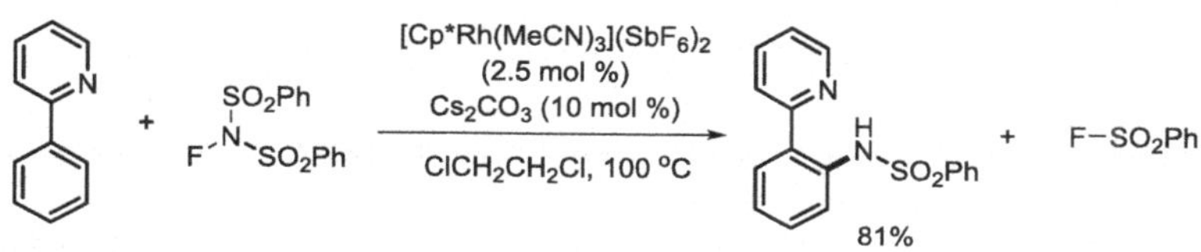

(4)

Chang and coworkers recently added another class of a highly reactive C–H amidating reagent, 1,4,2-dioxazol-5-one [62, 63]. During the mechanistic investigations on the Cp*Rh(III)-catalyzed direct C–H amination reaction, they revealed that an amino source with high coordination ability to metal center would increase the reaction rate. An acyl nitrene precursor, 1,4,2-dioxazol-5-one, showed high coordination affinity to the metal center, but with low activation energy in an imido-insertion process, which resulted in highly efficient amidation system, operating with as low as 0.5 mol% catalyst loading (5) [64].

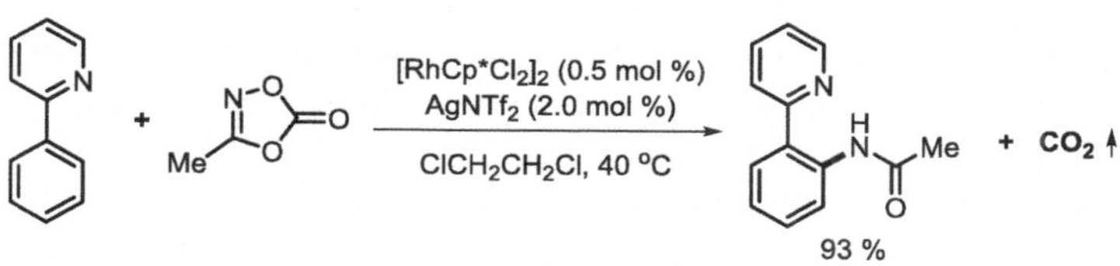

(5)

While various types of activated amine surrogates are in use for C–H aminations, the use of unmodified parent amines would be ultimately desirable for the step- and atom-efficient synthesis. Su and coworkers successfully applied the dehydrogenative coupling conditions by employing electron-deficient sulfonamides and acyl amides as amidating reagents (Scheme 6) [65]. Although stoichiometric amounts of oxidant such as $PhI(OAc)_2$ are required, this approach has an obvious advantage in that the reaction procedure does avoid an additional step of parent amine derivatization.

Along with the amination, C–H azidation and nitration were also developed [66]. Li and coworkers succeeded in the in situ formation of electrophilic azide source from NaN_3, which was then reacted with a rhodacycle to produce aryl azide products. An electrophilic azidation reagent of $PhI(N_3)OTs$ was obtained from a mixture of $PhI(OAc)_2$, TsOH, and NaN_3. Similar to azidation, the formation of an

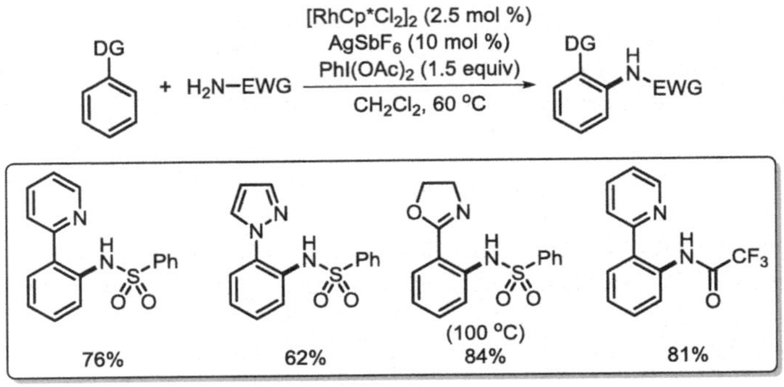

Scheme 6 Rh(III)-catalyzed dehydrogenative C–H amidation

electrophilic nitration source was achieved from the reaction of $NaNO_2$ with hyperiodine reagent. For both azidation and nitration, high conversion was observed over various substrates bearing rather strong directing groups such as pyridine and its analogues (6).

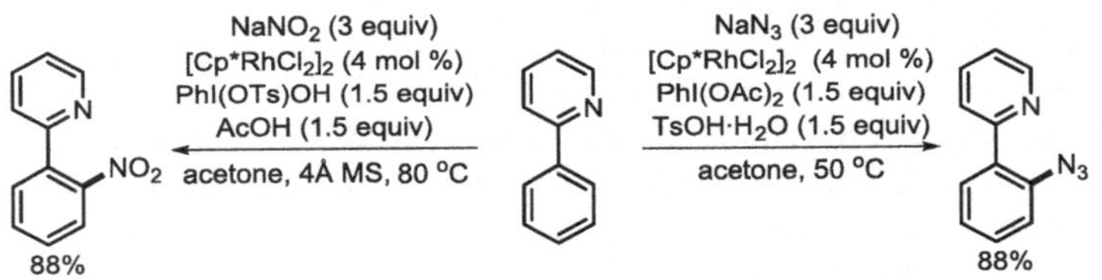

(6)

While the majority of known examples have focused on the (hetero)arene C–H bond activation and subsequent functionalization, several groups paid attention to other types of C–H bonds such as aldehydes [67–69]. Direct transformation of carbonyl C–H bonds into other functional groups such as ketones, amides, or esters would be highly valuable in synthetic chemistry. In this regard, Li group successfully developed the Cp*Rh(III)-catalyzed amide synthesis starting from aldehydes (7) [70, 71]. It was proposed that a cationic Cp*Rh(III) species activates a carbonyl C–H bond of 8-quinolinecarbaldehyes or 2-(methylthio)benzaldehydes to form a rhodium acyl intermediate that then reacts with amino sources including *N*-chloroamines and azides to deliver the desired amide products.

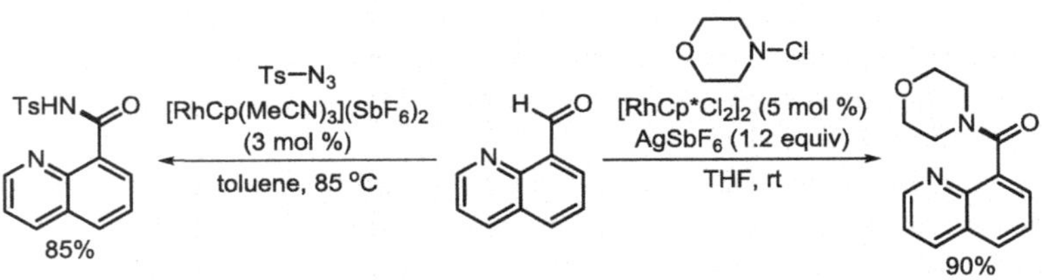

(7)

2.2 C–X (X = Cl, Br, and I) Bond Formation

With the aim of site selective halogenation, electrophilic halide sources such as *N*-bromosuccinimide (NBS) and *N*-iodosuccinimide (NIS) were successfully added to the list of reactive coupling partners. The Glorius group found that a cationic Cp*Rh(III) species is capable of coupling NBS and NIS with a broad range of arene substrates to produce *ortho*-halogenated arenes, which are otherwise difficult to obtain regioselectively via conventional halogenation methods (Scheme 7) [72–74]. It is notable that this halogenation reaction is facile with substrates containing ketone, ester, and carbamate directing groups where Cp*Rh(III)-based catalysis does not work efficiently for other types of C–H functionalizations [72]. The replacement of an additive from PivOH to $Cu(OAc)_2$ was key to success for substrates bearing enolizable ketones and esters as a directing group. In addition to aromatic C–H bonds, olefinic C–H bonds were also shown to be feasible for this halogenation protocol to give synthetically valuable (*Z*)-halogenated acrylamides [73]. In the presence of Cp*Rh(III) catalyst, non-catalyzed halogenation (background reaction) was effectively suppressed. Glorius and coworkers extended this method to the halogenation of important heterocycles such as furans, thiophenes,

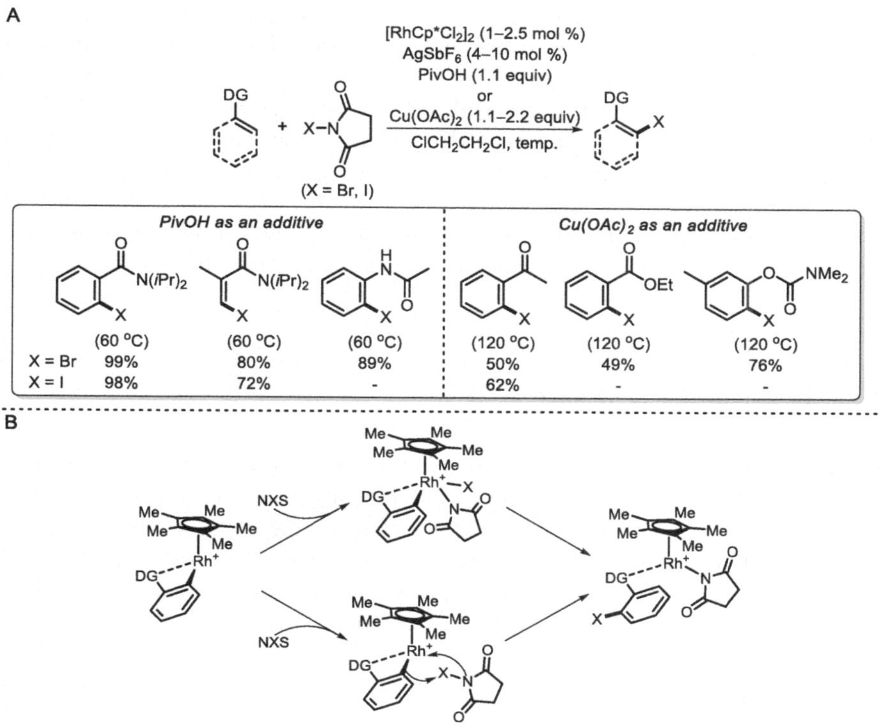

Scheme 7 Rh(III)-catalyzed C-H halogenation using NBS and NIS: scope (**a**) and mechanism (**b**)

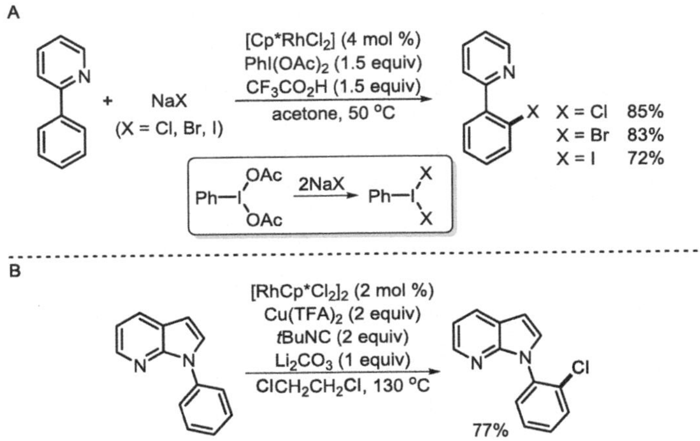

Scheme 8 Cp*Rh(III)-catalyzed halogenation using sodium halides (**a**) and 1,2-dichloroethane (**b**)

pyrazoles, quinolines, and chromones [74]. Recently, Chang and coworkers developed the selective halogenation at the C-8 position of quinoline *N*-oxides [75]. Wang and coworkers improved the synthetic utility of this approach by using readily available sodium halide salts as a halogen source (Scheme 8a) [76]. They proposed that the in situ formation of an active iodobenzene dihalide from $PhI(OAc)_2$ and sodium halide is a key to success. The achievement of chlorination is an additional merit, while NCS is not reported yet to be effective with Cp*Rh(III) catalysis. In addition, Xu and coworkers revealed an interesting reactivity of 1,2-dichloroethane as a chlorinating agent (Scheme 8b) [77]. In the functionalization of 7-azaindoles, 1,2-dichloroethane, which is a conventional solvent for numerous C–H functionalization reactions, was proven to be a facile chlorine source although its mechanism is unclear at the present stage.

2.3 C–S Bond Formation

Li and coworkers reported that the formation of a C–S bond could be achieved by using a Cp*Rh catalyst (8) [78]. Given that the sulfur atom can strongly bind to metals and, as a result, deactivate metal catalysts, the development of metal-catalyzed sulfenylation has been challenging. In the presence of extra oxidants, both mono- and di-thiolation were optimized to give thioethers with directing groups of pyridine, pyrimidine, pyrazole, and oxime.

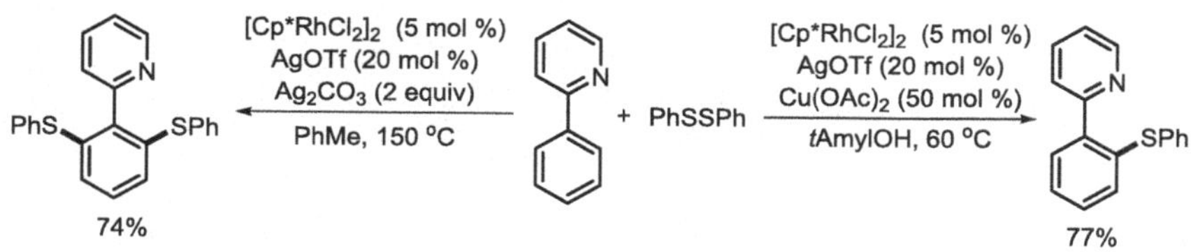

(8)

3 Ir(III)-Catalyzed Reactions

Whereas iridium complexes have been shown to be highly active for the *stoichiometric* C–H bond activation, their use in catalytic reactions was challenging [16, 79, 80]. The high stability of iridacycles, especially of half-sandwich complexes, hampered their catalytic turnovers. Thus, while their group 9 neighbor Cp*Rh (III) system has been widely applied to various C–H functionalizations, only a handful of Cp*Ir(III)-based catalytic methods have been developed in the area of C–C bond formation and H/D exchange [81–84]. As the difference in reactivity and mechanistic pathway between Ir(III) and its congeners is not fully understood yet [85], a number of notable achievements have been made in catalytic C–N and C–O bond formation in recent years.

3.1 *C–N Bond Formation*

Iridium(III)-catalyzed C–N bond formation was previously investigated for an intramolecular asymmetric amination. Katsuki and coworkers prepared a series of chiral Ir(III)–salen catalysts for the intramolecular C–H amination to produce optically active benzosultams (Scheme 9) [86]. High level of enantioselectivity was achieved with catalyst loading as low as 3 mol% under mild conditions.

In case of intermolecular amination, Cp*Ir(III)-based catalytic systems showed excellent reactivity with various amino sources such as organic azides,

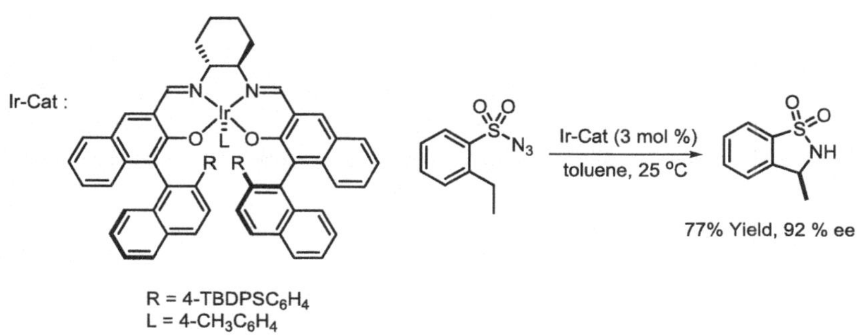

Scheme 9 (Salen)Ir(III)-catalyzed asymmetric intramolecular C–H amidation

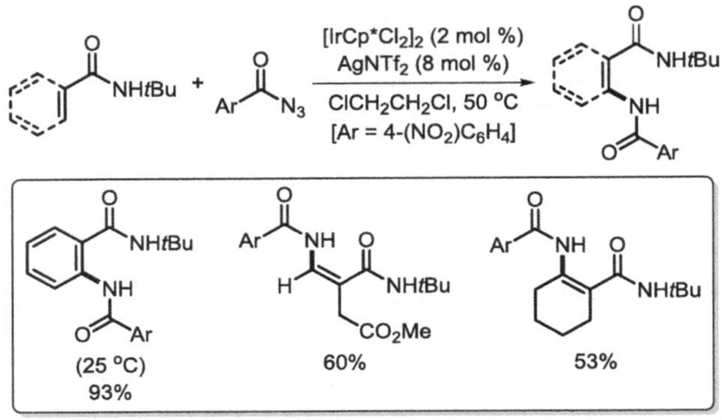

Scheme 10 Ir(III)-catalyzed C–H amidation using acyl azides

acetoxycarbamates, and anilines. As described above, the use of organic azides is environmentally benign due to the fact that no external oxidants are required and that N_2 is released as a single by-product. In this context, while Cp*Ir and Cp*Rh are both facile, Cp*Ir(III) system displayed a distinctive reactivity. Indeed, whereas Cp*Rh(III)-mediated aminations with azides require moderate- to high-temperature conditions (>80°C), the reactivity of Cp*Ir(III) catalyst is high enough to allow the reaction to run even at room temperature. In 2013, Chang and coworkers reported the first example of Ir(III)-catalyzed amidation of aryl and olefinic C–H bonds with acyl azides as the amino source [87, 88]. The fact that the mild reaction conditions permitted the use of this labile type of azides for the C–H amidation is significant because acyl azides easily undergo thermal Curtius rearrangement producing isocyanates (Scheme 10).

Another distinct feature of the Cp*Ir(III) catalytic system is in its high reactivity for substrates bearing weak directing groups [89]. As many C–H functionalizations require relatively strong coordination groups to facilitate the formation of a metallacycle intermediate, the use of weakly coordinating groups for the direct C–H activation has been a challenging task. While Cp*Rh(III) catalyst allowed the direct C–H amination of benzamides and aryl ketones at high temperature (>80°C), Cp*Ir(III) could form the C–N bonds by the action of ester or ketone directing groups at ambient temperature with the help of AcOH/Li_2CO_3 additives (Scheme 11).

In 2015, the research team of Genentech reported an Ir(III)-catalyzed C–H sulfonamidation of arenes by using benzylic amine as a directing group (9) [90]. The Cp*Ir(III)-derived catalyst was found to be more effective over Rh (III) and Ru(II) with tertiary benzylic amines.

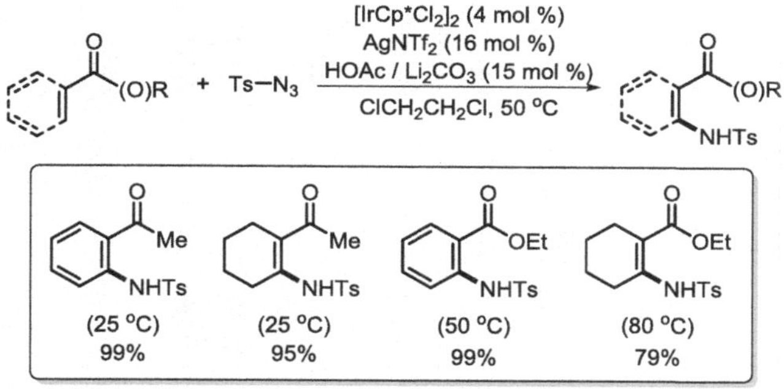

Scheme 11 Ir(III)-catalyzed C–H amidation with weakly coordinating carbonyl directing groups under mild conditions

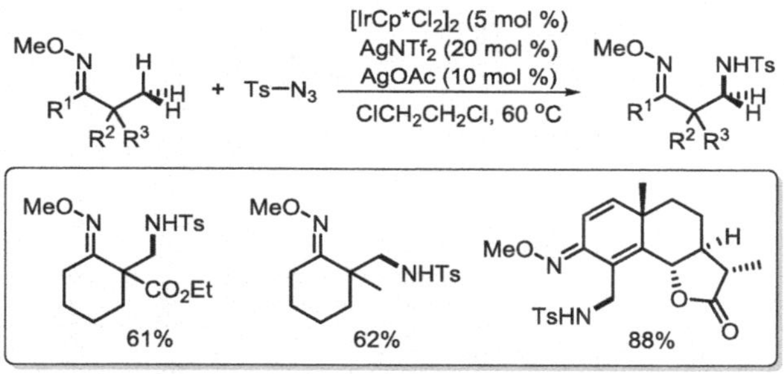

Scheme 12 Ir(III)-catalyzed amidation of unactivated sp^3 methyl C–H bonds

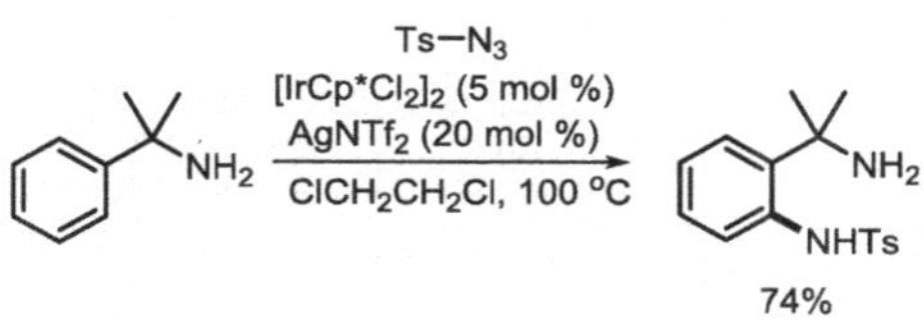 (9)

Excellent reactivity of Cp*Ir(III) complexes in the C–H activation process allowed for the functionalization of unactivated sp^3 C–H bonds [91, 92]. While Cp*Rh(III) catalyzed only activated benzylic C–H amidation of 8-methylquinolines, Cp*Ir(III) promoted the amidation of unactivated sp^3 methyl C–H bonds with the help of acetate additive. With a ketoxime directing group, their β-position was selectively amidated with various sulfonyl azides under mild conditions (Scheme 12) [91]. The same research group of Chang also devised a facile synthetic route accessing 1,2-amino alcohols by using direct C–H amidation reactions (Scheme 13) [92].

Asymmetric induction via C–H functionalization is a highly desirable but a challenging goal [93–95]. Chang and coworkers investigated a diastereoselective C–H amidation of diarylphosphoryl substrates bearing a chiral auxiliary (Scheme 14a) [96]. With a C_2-symmetric chiral pyrrolidine moiety, *P*-stereogenic center was newly generated by the Cp*Ir(III)-mediated diastereoselective

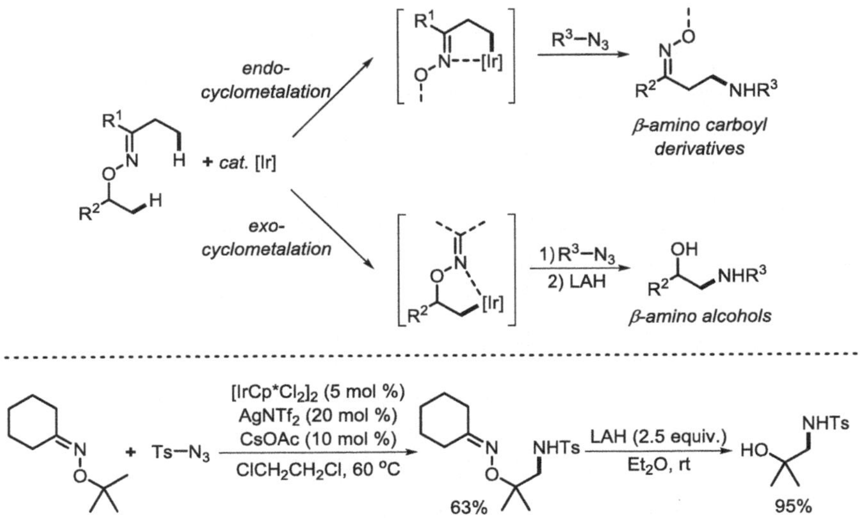

Scheme 13 Synthesis of 1,2-aminoalcohols via *exo*-directed sp^3 activation

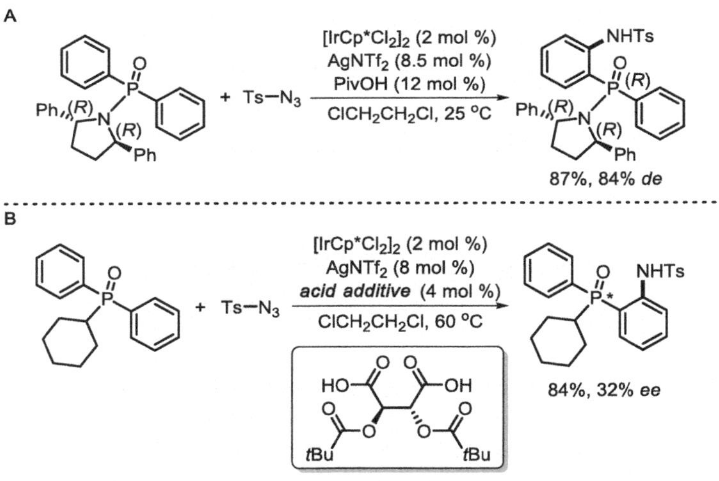

Scheme 14 Ir(III)-catalyzed diastereo-(**a**) and enantio-selective (**b**) C-H amidations of diarylphosphoryl compounds

amidation with notable selectivity. In the course of diarylphosphoryl amidation, detailed mechanistic studies revealed that Cp*Ir(III)–monoacetate is a reactive species, thus leading the authors to postulate that an introduction of chiral carboxylic acids would form a chiral Cp*Ir(III) intermediate for the asymmetric catalysis. Based on this assumption, they explored an enantioselective C–H amidation approach (Scheme 14b) [97]. Although various types of chiral carboxylic acids were examined for this purpose, the level of enantioselectivity is remained still low to moderate.

High reactivity of Cp*Ir(III) catalyst in the direct C–H amination with organic azides was successfully utilized in the synthesis of valuable products. Chang and coworkers proved that 8-aminoquinoline *N*-oxides could readily be obtained by a Cp*Ir(III)-catalyzed reaction of quinoline *N*-oxides with sulfonyl or acyl azides (Scheme 15) [75]. As most other metal systems were known to provide functionalization at the 2-position, the unique activation mode of the Cp*Ir(III) system at the C-8 position significantly improved quinoline derivatization. Later, they also demonstrated that a one-pot protocol of C8 functionalization of quinolines would be feasible and, therefore, a series of tandem processes was optimized consisting of *N*-oxide formation from quinolines, C–H functionalization at the C-8 position, and then deoxygenation of functionalized *N*-oxides [98].

The selective C-7 amination of indolines was achieved with Cp*Ir(III) catalyst. Chang and Li independently scrutinized the selectivity as well as reactivity of the indoline moiety that is an important structural core in pharmaceutical chemistry (Scheme 16) [99, 100]. Cp*Ir(III)-catalyzed C-7 amination proceeded efficiently

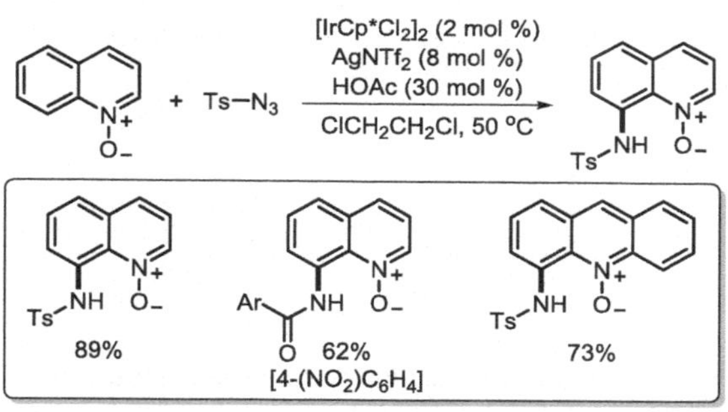

Scheme 15 Synthesis of C-8 functionalized quinoline derivatives

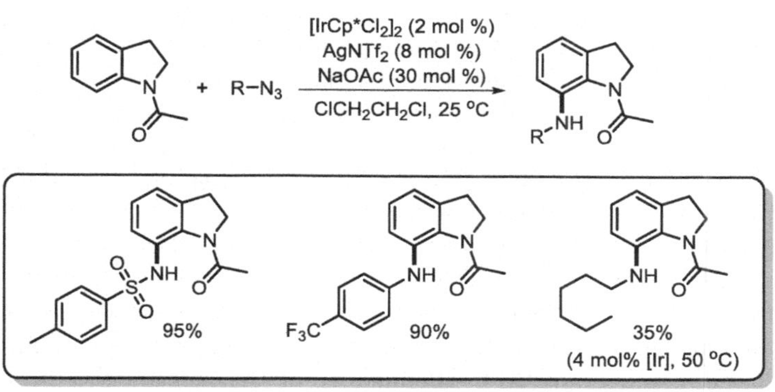

Scheme 16 Ir(III) catalyzed C–H amination with *N*-acetylindoline

under mild conditions and the scope of azides was broad including aryl, alkyl, and sulfonyl groups. In addition, a new synthetic route to phosphoramidates was developed using an intermolecular C–H amidation by the action of cationic Cp*Ir (III) catalyst with phosphoryl azides (10). As a result, this carbon–nitrogen bond formation strategy allows a quick access to a library of biologically important compounds [101, 102].

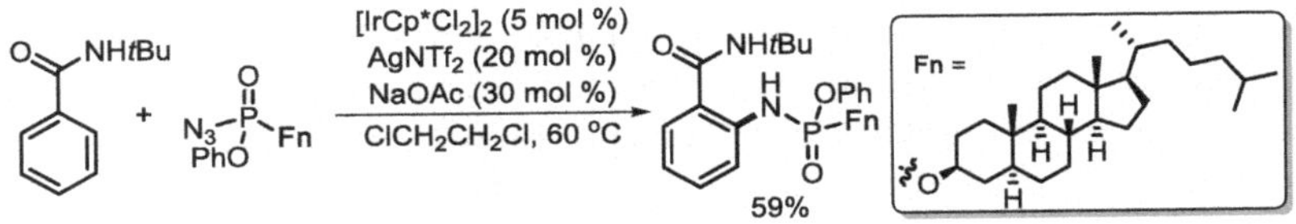

(10)

The Chang group recently developed a method installing an amino group at the *meta*- or *para*-position of arenes via the direct C–H amidation of substituted benzoic acids followed by tandem decarboxylation [103]. Traceless directing group approach consisting of sequential amidation by Cp*Ir(III)-based catalyst and one-pot Pd or Cu-promoted decarboxylation procedure provided a novel route to *meta*- or *para*-aniline products [104–107]. Remote steric effects by preexisting *meta*-substituents in benzoic acids made this method unique to produce *para*-amidated compounds exclusively, which are inaccessible by other C–H activation approaches (11).

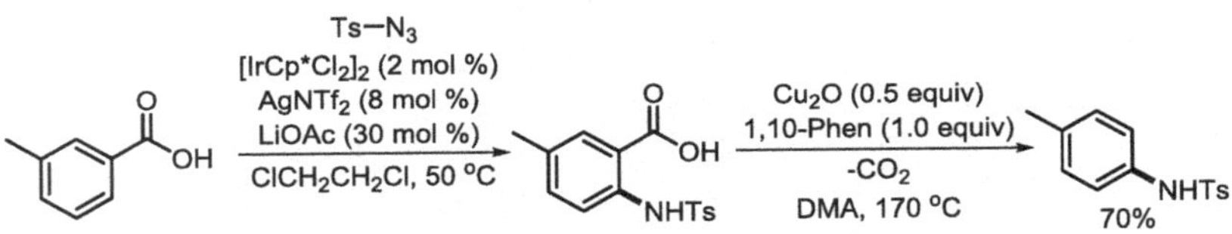

(11)

Amino sources other than organic azides were also examined for the direct C–H amidation reactions. Chang and coworker successfully utilized *N*-substituted hydroxylamines as highly efficient amidating reagents with a Cp*Ir(III) catalyst system. Aryloxy- and acryloxycarbamates were found to undergo the C–H amidation over an assorted array of arene substrates at room temperature [108].

The high catalytic activity of Cp*Ir(III) species toward direct C–N bond-forming reactions enabled the use of anilines as an amino source in the inter- and intramolecular C–H amination reactions. In 2014, Chang group reported the first example of iridium-catalyzed oxidative C–H amination of benzamides with anilines (Scheme 17a) [109]. In a stark contrast to the conventional cross-dehydrogenative coupling reactions, this amination proceeds even at room temperature to prove the excellent performance of the iridium(III) catalyst toward C–H amination. Although the exact mechanism still needs to be explored at the present stage, the authors suggested the intermediacy of high-valent Ir(V) species formed by the assistance of an external oxidant AgNTf$_2$. Later, Miura and coworkers reported an intramolecular version of this Ir-catalyzed C–H amination of arenes leading to the synthesis of *N*–H carbazoles (Scheme 17b) [110]. Although, in contrast to the intermolecular

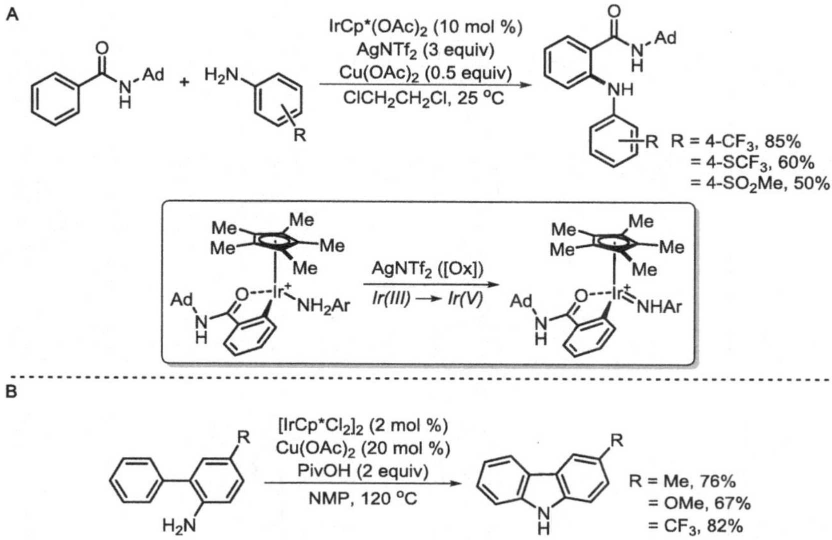

Scheme 17 Cp*Ir(III)-catalyzed dehydrogenative coupling between arenes and anilines: intermolecular (**a**) and intramolecular (**b**) C-H aminations

C–H amination with anilines, this procedure requires elevated temperature, the intramolecular reaction operates efficiently without the need of external oxidants, implying that this reaction may proceed via Ir(III)/Ir(I) catalytic cycle.

3.2 C–O Bond Formation

Cp*Ir(III)-based complexes have gained intense interests for their high catalytic activity in water oxidation [111–113]. In 2010, Crabtree and coworkers found that an iridacycle species could promote C–H hydroxylation with ceric ammonium nitrate (CAN) as the terminal oxidant and water as an oxygen source (12) [114]. Detailed mechanistic studies indicated that an alkane substrate is oxidized by an Cp*Ir(V)-oxo species, which is formed upon the oxidation of Cp*Ir(III)–H_2O complex with CAN [115]. While a radical pathway is ruled out, theoretical and experimental investigations support an oxygen insertion pathway, which is consistent with the observed retention of stereochemistry of *cis*-decaline and 1,4-dimethylcyclohexane.

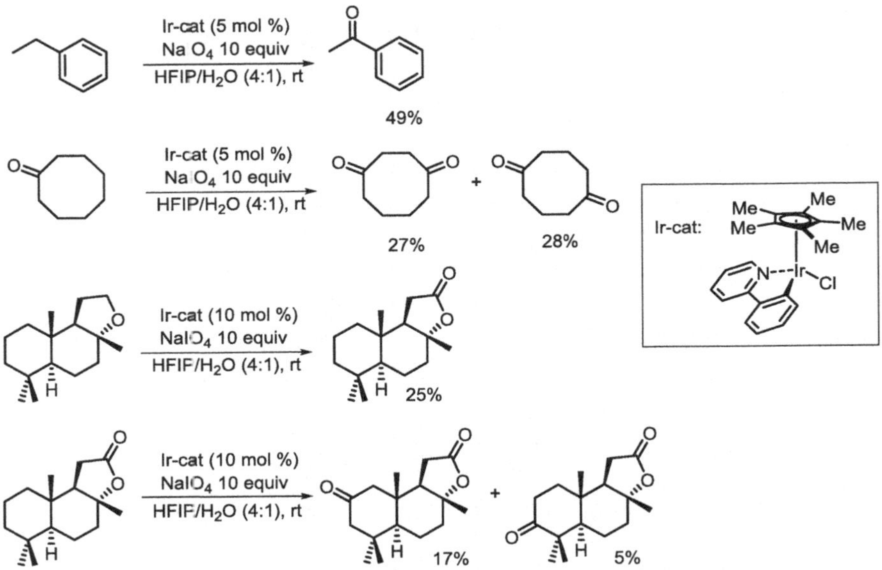

Scheme 18 Cp*Ir (III)-catalyzed C–H oxidations

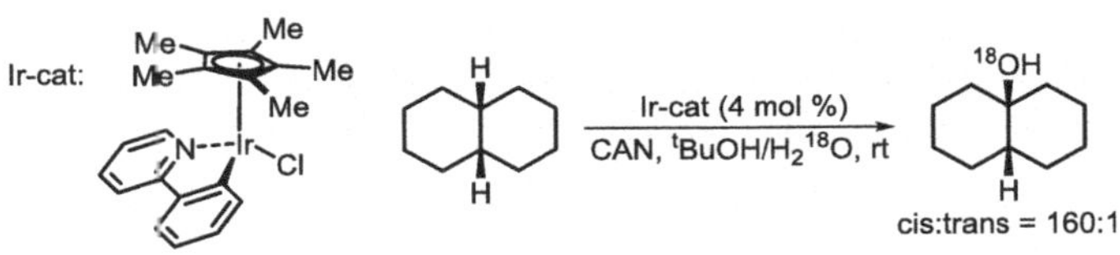

(12)

Later, $NaIO_4$ was introduced as a milder, but more efficient oxidant than CAN [116, 117] (Scheme 18). An array of functional groups was found to be compatible with the employed oxidative reaction conditions. In this context, functionalization of biologically relevant molecules such as (−)-ambroxide and sclareolide were successfully demonstrated.

In addition, molecular oxygen was shown to be an effective oxidant for the C–H oxidation with the Ir(III)-based catalyst system. In 2013, Yan and coworkers developed an aerobic oxidation system under solvent-free conditions [118]. The catalyst generated in situ from $[Cp^*IrCl_2]_2$, picolinic acid, and iodobenzene efficiently performed the reaction by using atmospheric O_2 with high turnover numbers up to 20,000 to furnish benzoic acids and aryl ketones via the selective benzylic C–H oxidation.

4 Outlook

Highly robust catalytic systems of Cp*Rh(III) and Cp*Ir(III) are expected to continue broadening their scope and improving their efficiency and selectivity in the direct C–H functionalizations and related transformations. Considering that many reactions presented in this report have not been fully understood, more detailed mechanistic elucidations of the C–H bond activation and functional group transfer steps would be crucial for the next level advance in terms of scope, reactivity, and selectivity. Particularly, a better understanding of the difference in reactivity between rhodium and iridium catalyst systems would provide a driving force to develop more practical C–H functionalization reactions even including their congener: cobalt catalyst system.

References

1. Bergman RG (2007) Nature 446:391
2. Crabtree RH (2004) J Organomet Chem 689:4083
3. Goldman AS, Goldberg KI (2004) In: Goldberg KI, Goldman AS (eds) Activation and functionalization of C–H bonds, vol 885, ACS symposium series. American Chemical Society, Washington, p 1
4. Satoh T, Miura M (2010) Chem Eur J 16:11212
5. Song G, Wang F, Li X (2012) Chem Soc Rev 41:3651
6. Kuhl N, Schröder N, Glorius F (2014) Adv Synth Catal 356:1443
7. Pan S, Shibata T (2013) ACS Catal 3:704
8. Han Y-F, Jin G-X (2014) Chem Soc Rev 43:2799
9. Muñiz K (2010) Top Organomet Chem 31:1
10. Glueck DS (2010) Top Organomet Chem 31:65
11. Eisen MS (2010) Top Organomet Chem 31:157
12. Mkhalid IAI, Barnard JH, Marder TB, Murphy JM, Hartwig JF (2010) Chem Rev 110:890
13. Hartwig JF (2011) Chem Soc Rev 40:1992
14. Ros A, Fernández R, Lassaletta JM (2014) Chem Soc Rev 43:3229
15. Davies DL, Al-Duaij O, Fawcett J, Giardiello M, Hilton ST, Russell DR (2003) Dalton Trans 4132
16. Li L, Brennessel WW, Jones WD (2008) J Am Chem Soc 130:12414
17. Ueura K, Satoh T, Miura M (2007) Org Lett 9:1407
18. Umeda N, Tsurugi H, Satoh T, Miura M (2008) Angew Chem Int Ed 47:4019
19. Stuart DR, Bertrand-Laperle M, Burgess KMN, Fagnou K (2008) J Am Chem Soc 130:16474
20. Ullmann F (1903) Ber Dtsch Chem Ges 36:2382
21. Monnier F, Taillefer M (2009) Angew Chem Int Ed 48:6954
22. Evano G, Blanchard N, Toumi M (2008) Chem Rev 108:3054
23. Paul F, Patt J, Hartwig JF (1994) J Am Chem Soc 116:5969
24. Hartwig JF (2008) Acc Chem Res 41:1534
25. Guram AS, Buchwald SL (1994) J Am Chem Soc 116:7901
26. Surry DS, Buchwald SL (2008) Angew Chem Int Ed 47:6338
27. Collet F, Dodd RH, Dauban P (2009) Chem Commun 14(34):5061
28. Louillat M-L, Patureau FW (2014) Chem Soc Rev 43:901
29. Espino CG, Wehn PM, Chow J, Du Bois J (2001) J Am Chem Soc 123:6935
30. Espino CG, Du Bois J (2001) Angew Chem Int Ed 40:598

31. Fleming JJ, McReynolds MD, Du Bois J (2007) J Am Chem Soc 129:9964
32. Lebel H, Huard K, Lectard S (2005) J Am Chem Soc 127:14198
33. Shin K, Kim H, Chang S (2015) Acc Chem Res 48:1040
34. Kim JY, Park SH, Ryu J, Cho SH, Kim SH, Chang S (2012) J Am Chem Soc 134:9110
35. Ryu T, Min J, Choi W, Jeon WH, Lee PH (2014) Org Lett 16:2810
36. Jia X, Han J (2014) J Org Chem 79:4180
37. Shin K, Baek Y, Chang S (2013) Angew Chem Int Ed 52:8031
38. Ryu J, Shin K, Park SH, Kim JY, Chang S (2012) Angew Chem Int Ed 51:9904
39. Zhang C, Zhou Y, Deng Z, Chen X, Peng Y (2015) Eur J Org Chem 2015(8):1735
40. Park SH, Kwak J, Shin K, Ryu J, Park Y, Chang S (2014) J Am Chem Soc 136:2492
41. Wang N, Li R, Li L, Xu S, Song H, Wang B (2014) J Org Chem 79:5379
42. Yu D-G, Suri M, Glorius F (2013) J Am Chem Soc 135:8802
43. Lian Y, Hummel JR, Bergman RG, Ellman JA (2013) J Am Chem Soc 135:12548
44. Kim HJ, Ajitha MJ, Lee Y, Ryu J, Kim J, Lee Y, Jung Y, Chang S (2014) J Am Chem Soc 136:1132
45. Kawano T, Hirano K, Satoh T, Miura M (2010) J Am Chem Soc 132:6900
46. Barker TJ, Jarvo ER (2009) J Am Chem Soc 131:15598
47. Ng K-H, Zhou Z, Yu W-Y (2012) Org Lett 14:272
48. Grohmann C, Wang H, Glorius F (2012) Org Lett 14:656
49. Ng K-H, Zhou Z, Yu W-Y (2013) Chem Commun 49:7031
50. Ng F-N, Zhou Z, Yu W-Y (2014) Chem Eur J 20:4474
51. Grohmann C, Wang H, Glorius F (2013) Org Lett 15:3014
52. Yu S, Wan B, Li X (2013) Org Lett 15:3706
53. Xue Y, Fan Z, Jiang X, Wu K, Wang M, Ding C, Yao Q, Zhang A (2014) Eur J Org Chem 2014(33):7481
54. Wu K, Fan Z, Xue Y, Yao Q, Zhang A (2014) Org Lett 16:42
55. Ali MA, Yao X, Sun H, Lu H (2015) Org Lett 17:1513
56. Zhou B, Du J, Yang Y, Feng H, Li Y (2013) Org Lett 15:6302
57. Du J, Yang Y, Feng H, Li Y, Zhou B (2014) Chem Eur J 20:5727
58. Zhou B, Du J, Yang Y, Feng H, Li Y (2014) Org Lett 16:592
59. Sun K, Li Y, Xiong T, Zhang J, Zhang Q (2011) J Am Chem Soc 133:1694
60. Iglesias Á, Álvarez R, de Leca ÁR, Muñiz K (2012) Angew Chem Int Ed 51:2225
61. Tang R-J, Luo C-P, Yang L, Li C-J (2013) Adv Synth Catal 355:869
62. Sauer J, Mayer KK (1968) Tetrahedron Lett 9:319
63. Bizet V, Buglioni L, Bolm C (2014) Angew Chem Int Ed 53:5639
64. Park Y, Park KT, Kim JG, Chang S (2015) J Am Chem Soc 137:4534
65. Zhao H, Shang Y, Su W (2013) Org Lett 15:5106
66. Xie F, Qi Z, Li X (2013) Angew Chem Int Ed 52:11862
67. Murphy SK, Bruch A, Dong VM (2015) Chem Sci 6:174
68. Jun C-H, Lee H, Hong J-B (1997) J Org Chem 62:1200
69. Ko S, Kang B, Chang S (2005) Angew Chem Int Ed 44:455
70. Zhou B, Du J, Yang Y, Li Y (2013) Org Lett 15:2934
71. Zhou B, Yang Y, Shi J, Feng H, Li Y (2013) Chem Eur J 19:10511
72. Schröder N, Wencel-Delord J, Glorius F (2012) J Am Chem Soc 134:8298
73. Kuhl N, Schröder N, Glorius F (2013) Org Lett 15:3860
74. Schröder N, Lied F, Glorius F (2015) J Am Chem Soc 137:1448
75. Hwang H, Kim J, Jeong J, Chang S (2014) J Am Chem Soc 136:10770
76. Zhang P, Hong L, Li G, Wang R (2015) Adv Synth Catal 357:345
77. Qian G, Hong X, Liu B, Mao H, Xu B (2014) Org Lett 16:5294
78. Yang Y, Hou W, Qin L, Du J, Feng H, Zhou B, Li Y (2014) Chem Eur J 20:416
79. Liu J, Wu X, Iggo JA, Xiao J (2008) Coord Chem Rev 252:782
80. Kohl G, Pritzkow H, Enders M (2008) Eur J Inorg Chem 2008(27):4230–35

81. Klei SR, Tan KL, Golden JT, Yung CM, Thalji RK, Ahrendt KA, Ellman JA, Tilley TD, Bergman RG (2004) In: Goldberg KI, Goldman AS (eds) Activation and functionalization of C–H bonds, vol 885, ACS symposium series. American Chemical Society, Washington, p 46
82. Choi J, Goldman AS (2011) Top Organomet Chem 34:139
83. Atzrodt J, Derdau V, Fey T, Zimmermann J (2007) Angew Chem Int Ed 46:7744
84. Lehman MC, Gary JB, Boyle PD, Sanford MS, Ison EA (2013) ACS Catal 3:2304
85. Figg TM, Park S, Park J, Chang S, Musaev DG (2014) Organometallics 33:4076
86. Ichinose M, Suematsu H, Yasutomi Y, Nishioka Y, Uchida T, Katsuki T (2011) Angew Chem Int Ed 50:9884
87. Ryu J, Kwak J, Shin K, Lee D, Chang S (2013) J Am Chem Soc 135:12861
88. Lee D, Kim Y, Chang S (2013) J Org Chem 78:11102
89. Kim J, Chang S (2014) Angew Chem Int Ed 53:2203
90. Chen H, Huestis MP (2015) ChemCatChem 7:743
91. Kang T, Kim Y, Lee D, Wang Z, Chang S (2014) J Am Chem Soc 136:4141
92. Kang T, Kim H, Kim JG, Chang S (2014) Chem Commun 50:12073
93. Chu L, Wang X-C, Moore CE, Rheingold AL, Yu J-Q (2013) J Am Chem Soc 135:16344
94. Giri R, Chen X, Yu J-Q (2005) Angew Chem Int Ed 44:2112
95. Collet F, Lescot C, Dauban P (2011) Chem Soc Rev 40:1926
96. Gwon D, Lee D, Kim J, Park S, Chang S (2014) Chem Eur J 20:12421
97. Gwon D, Park S, Chang S (2015) Tetrahedron 71:4504
98. Jeong J, Lee D, Chang S (2015) Chem Commun 51:7035
99. Shin K, Chang S (2014) J Org Chem 79:12197
100. Hou W, Yang Y, Ai W, Wu Y, Wang X, Zhou B, Li Y (2015) Eur J Org Chem 2015 (2):395–400
101. Kim H, Park J, Kim JG, Chang S (2014) Org Lett 16:5466
102. Pan C, Jin N, Zhang H, Han J, Zhu C (2014) J Org Chem 79:9427
103. Lee D, Chang S (2015) Chem Eur J 21:5364
104. Rousseau G, Breit B (2011) Angew Chem Int Ed 50:2450
105. Maehara A, Tsurugi H, Satoh T, Miura M (2008) Org Lett 10:1159
106. Mochida S, Hirano K, Satoh T, Miura M (2010) Org Lett 12:5776
107. Mochida S, Hirano K, Satoh T, Miura M (2011) J Org Chem 76:3024
108. Patel P, Chang S (2014) Org Lett 16:3328
109. Kim H, Shin K, Chang S (2014) J Am Chem Soc 136:5904
110. Suzuki C, Hirano K, Satoh T, Miura M (2015) Org Lett 17:1597
111. Hull JF, Balcells D, Blakemore JD, Incarvito CD, Eisenstein O, Brudvig GW, Crabtree RH (2009) J Am Chem Soc 131:8730
112. McDaniel ND, Coughlin FJ, Tinker LL, Bernhard S (2008) J Am Chem Soc 130:210
113. Hintermair U, Sheehan SW, Parent AR, Ess DH, Richens DT, Vaccaro PH, Brudvig GW, Crabtree RH (2013) J Am Chem Soc 135:10837
114. Zhou M, Schley ND, Crabtree RH (2010) J Am Chem Soc 132:12550
115. Zhou M, Balcells D, Parent AR, Crabtree RH, Eisenstein O (2012) ACS Catal 2:208
116. Zhou M, Hintermair U, Hashiguchi BG, Parent AR, Hashmi SM, Elimelech M, Periana RA, Brudvig GW, Crabtree RH (2013) Organometallics 32:957
117. Hohloch S, Kaiser S, Duecker FL, Bolje A, Maity R, Košmrlj J, Sarkar B (2015) Dalton Trans 44:686
118. Yan Y, Chen Y, Yan M, Li X, Zeng W (2013) Catal Commun 35:64

Top Organomet Chem (2016) 55: 53–76
DOI: 10.1007/3418_2015_125

Published online: 17 July 2015

Computational Studies on Heteroatom-Assisted C–H Activation and Functionalisation at Group 8 and 9 Metal Centres

Kevin J.T. Carr, Stuart A. Macgregor, and Claire L. McMullin

Abstract This chapter surveys computational studies on heteroatom-assisted C–H activation at group 8 and 9 metal centres and will cover the literature since 2009. The chapter first considers work where the mechanism of the C–H activation step is the primary concern and categorizes these into intramolecular (with directing groups) and intermolecular processes. Studies on C–H activation and functionalization will be presented, classified in terms of the nature of the functionalization process (oxidative coupling to form heterocycles, alkenylation and amination).

Keywords Catalysis · C–H activation · Computation · DFT · Gp 8 · Gp 9 · Mechanism

Contents

K.J.T. Carr, S.A. Macgregor (✉), and C.L. McMullin
Institute of Chemical Sciences, Heriot-Watt University, Edinburgh EH14 4AS, UK
e-mail: S.A.Macgregor@hw.ac.uk

1 Introduction

Computational chemistry has played a leading role in providing insight into the mechanistic complexity of C–H bond activation. The ability to characterise short-lived intermediates and transition states provides an ideal complement to experiment where such information is often extremely difficult if not impossible to obtain. Computational studies of C–H activation were the subject of several reviews towards the end of the last decade [1–5] and this body of work has catalogued the diversity of the C–H activation mechanistic landscape. For late transition metals, mechanisms based on the broad classes of oxidative addition (OA), σ-bond metathesis (SBM) and electrophilic activation (EA) have all been identified. Vastine and Hall [4] have proposed that the first two of these represent extremes on a continuum where the degree of M···H interaction in the C–H activation transition state progresses from a maximum for OA to a minimum for SBM, with a plethora of named mechanistic possibilities lying in between. The position on this continuum can depend on the metal centre, the steric and electronic properties of the coordination environment and the substrate involved in the C–H activation process.

The third broad class, EA, has been joined in the last decade by the concepts of concerted metalation deprotonation (CMD) proposed by Fagnou and co-workers [5] and ambiphilic metal ligand assisted (AMLA) C–H activation put forward by Davies and Macgregor [2]. These two terms describe a process by which an electron-deficient metal centre and a metal-bound heteroatom base combine to facilitate C–H bond cleavage via deprotonation. Computationally, this type of process was first characterised in 2000 by Sakaki and co-workers who studied the C–H activation of benzene and methane at $Pd(O_2CH)_2$ [6]. In 2005, Davies, Macgregor and co-workers then showed that the cyclometalation of dimethylbenzylamine at $Pd(OAc)_2$ involves an agostic intermediate from which C–H cleavage proceeds via facile deprotonation [7]. Here we shall use the term AMLA to describe this type of heteroatom-assisted C–H activation and reserve the term 'electrophilic activation' for instances where no base assistance is in place [8, 9].

As well as progress in mechanistic understanding, the last 10 years have seen important developments in the computational methodologies available to model transition metal reactivity. While density functional theory (DFT) remains the core method of choice, the ability to model larger systems that more closely reflect experiment has highlighted the known shortcomings of DFT in describing dispersion interactions. These long-range, stabilising interactions are individually weak, but their cumulative effect in large systems can be significant. Methods to incorporate this component include its separate calculation (e.g. with Grimme's D3 parameter set [10]), use of functionals that include a treatment of dispersion (e.g. B97D [11], ωB97X-D [12]) or use of a Minnesota functional (e.g. M06 or its variants [13]) where the functional parameterisation (e.g. to reproduce molecular structures from crystallographic data) captures dispersion effects without explicitly identifying them. The validity of such approaches is seen, for example, in marked improvements in calculated L_nM-PR_3 dissociation energies [14]. These developments make the choice of functional for the study of C–H activation and functionalisation especially critical when larger models are employed in calculations.[1]

This chapter will describe computational work on heteroatom-assisted C–H activation at group 8 and 9 metal centres and will survey the literature since the previous review of 2009 [2] up to April 2015. The chapter will first cover studies where the C–H activation step is the primary concern and will consider both intramolecular (with directing groups) and intermolecular processes. A significant trend has been a shift in focus away from the actual C–H activation and towards the subsequent functionalisation, and this will be presented in terms of the nature of the functionalisation process (oxidative coupling to form heterocycles, alkenylation and amination). The major functional employed in each study will be highlighted, along with any solvent used in brackets. The notation DFT2//DFT1 will be used to indicate cases where a second functional (DFT2) has been used to recompute the energy of a system optimised with an initial functional (DFT1). The original papers should be consulted for full computational details.

2 Intramolecular C–H Activation

2.1 *Activation of C(sp^2)–H Bonds*

The importance of treating dispersion effects was highlighted in a study by Davies, Macgregor and co-workers on the relative reactivity of C(sp^2)–H bonds in *N*-alkylimines (**H-L$_{1–5}$**) and 2-phenylpyridine (**H-L$_6$**) at $[MCl_2Cp^*]_2$ dimers (M = Rh, Ir) [15]. In the presence of NaOAc, these substrates give well-defined cyclometalated species as their chloride adducts, indicating that these reactions are all thermodynamically accessible (Scheme 1). Moreover, H/D exchange experiments indicated irreversible C–H activation at Ir (i.e. kinetic control), while this process was reversible at Rh (thermodynamic control). Competition experiments then indicated different trends in relative reactivity at each metal. These various aspects were studied with BP86-D3 calculations, corrected for either MeOH (Rh) or CH_2Cl_2 (Ir) solvent.

The computed profiles for the reactions of the 2-thiophenyl derivative **H-L$_3$** are shown in Fig. 1 which displays several features common to many of the C–H activation reactions described in this chapter. The precursor to C–H activation, **IV$_M$**, is formed via OAc-induced opening of the $[MCl_2Cp^*]_2$ dimer and ligand substitution to introduce the substrate. For Rh, a two-step C–H activation is

[1] High-level wave function methods promise the accurate treatment of these long-range electron correlation effects; however, such methods are still too computationally demanding for routine use on large model systems.

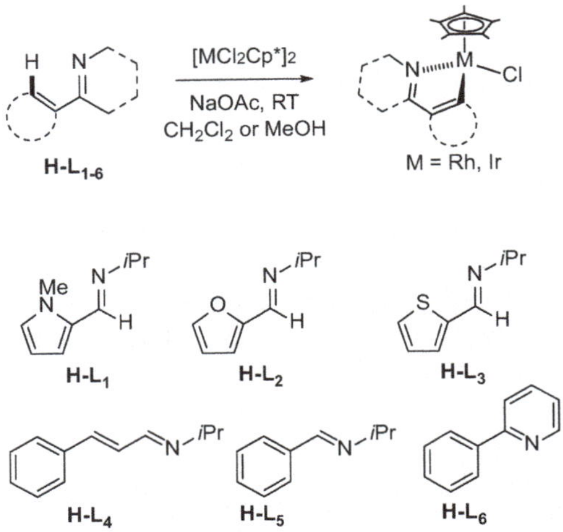

Scheme 1 Cyclometalation of *N*-alkylimines (**H-L$_{1-5}$**) and 2-phenylpyridine (**H-L$_6$**) at $[MCl_2Cp^*]_2$

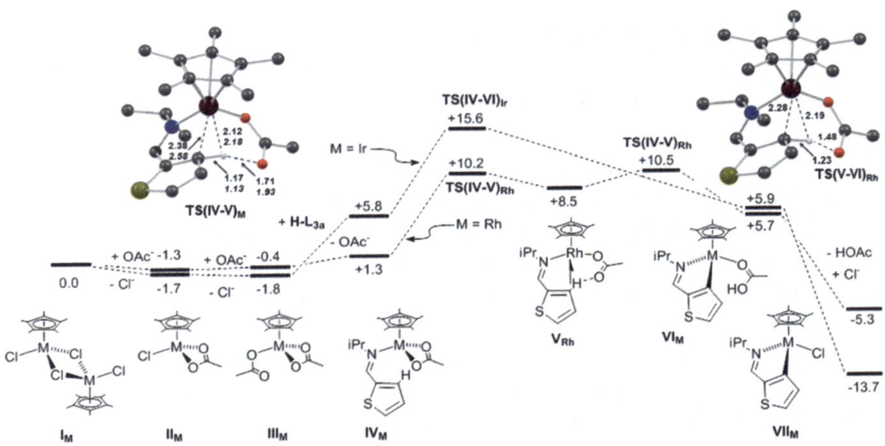

Fig. 1 Computed free energy reaction profiles (kcal/mol) for C–H activation of **H-L$_3$** at $[MCl_2Cp^*]_2$ for M = Ir (in CH_2Cl_2) and M = Rh (in MeOH). Computed C–H activation transition states are shown with key distances in Å (Rh: *plain text*; Ir: *italics*) and non-participating H atoms omitted for clarity

characterised with the initial κ^2–κ^1 displacement of OAc leading to an agostic intermediate, **V$_{Rh}$**. This renders the C–H bond more acidic, permitting facile intramolecular deprotonation by the intramolecular OAc base. These are the characteristic features of AMLA C–H activation, specifically an AMLA-6, as it involves the Rh, the C–H bond and the acetate {CO_2} group in a six-membered transition state. HOAc/Cl substitution then leads to the observed product **VII$_{Rh}$**. The energetics of this process ($\Delta G_{calc} = -4.0$ kcal/mol; $\Delta G^{\ddagger}_{calc} = 11.8$ kcal/mol) indicate reversible C–H activation, consistent with the observed H/D exchange. It is also worth noting that the energy surface associated with C–H activation is very flat with, in this case, the C–H bond cleavage transition state (C···H = 1.23 Å) being slightly higher in energy than that for κ^2-κ^1 displacement of OAc. For Ir, the reaction is both more exergonic ($\Delta G_{calc} = -14.2$ kcal/mol) and has a much higher barrier ($\Delta G^{\ddagger}_{calc} = 18.4$ kcal/mol), consistent with an irreversible reaction under kinetic control. Moreover, C–H activation proceeds in a single step, corresponding to κ^2–κ^1 displacement of OAc, in which minimal lengthening of the C–H bond has occurred (1.13 Å). Experimentally the reaction of the Rh system is far less efficient and takes much longer to achieve reasonable yields. The lower barrier for Rh indicates that this is not a kinetic effect but rather reflects a slow approach to equilibrium in favour of product formation.

It is important to note that use of an extended basis set (with diffuse functions) and the D3 correction were both essential to reproduce the favourable thermodynamics expected for all metal/substrate combinations. This reflects the stabilisation derived from having the bulky substrate adjacent to the {MCp*} fragment, which in turn stresses the importance of using the full experimental system in the calculations. With this approach, the more subtle trends in the relative reactivity of the six substrates were also reasonably captured. Thus for Rh (thermodynamic control) the trend in ΔG_{calc} largely follows the increase in substrate reactivity (Fig. 2i), whereas for Ir (kinetic control) increased substrate reactivity is mirrored in a decrease in $\Delta G^{\ddagger}_{calc}$ (Fig. 2ii). One exception was **H-L₄**, the reactivity of which was overestimated for both metals. This discrepancy is thought to relate to the different shape of this substrate (which features a vinyl CH = CHPh substituent,

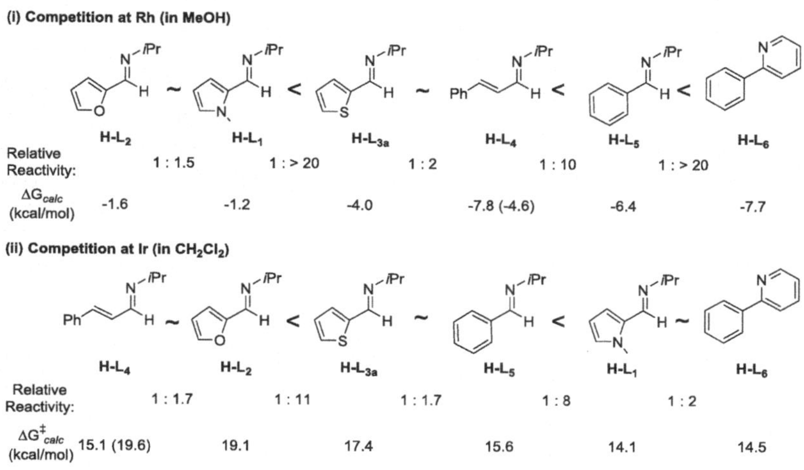

Fig. 2 Relative experimental and computed reactivities for substrates **H-L$_{1-6}$** at $[MCl_2Cp^*]_2$ (M = Rh and Ir). Computed data (kcal/mol) give the overall free energy changes, ΔG_{calc}, for M = Rh and calculated activation barriers, and $\Delta G^{\ddagger}_{calc}$, for M = Ir

rather than the aromatic substituents in **H-L$_{1\text{–}3}$** and **H-L$_5$**) and the propensity of the dispersion correction to emphasise intramolecular stabilisation over interaction with the environment (e.g. solvent). The different degree of dispersion stabilisation therefore occurs upon binding **H-L$_4$** to the metal fragment. To test this, model substrate **H-L$_{4'}$** featuring a smaller $CH{=}CH_2$ substituent, was computed (values in parenthesis in Fig. 2), and this gave both a reduced ΔG_{calc} at Rh and an increased $\Delta G^{\ddagger}_{calc}$ at Ir.

Around the same time, Zheng and co-workers [16] modelled the acetate-assisted cyclometalation of *N*-phenylbenzaldimines at $[MCl_2Cp^*]_2$ dimers (M = Rh, Ir). B3LYP(MeOH) calculations found cyclometalation at Ir to be favoured thermodynamically (Rh: $\Delta G_{calc} = -0.5$ kcal/mol; Ir: $\Delta G_{calc} = -8.4$ kcal/mol), but, contrary to the above study, the C–H activation barrier was higher for Rh (Rh : $\Delta G^{\ddagger}_{calc}$ = 26.4 kcal/mol; Ir : $\Delta G^{\ddagger}_{calc}$ = 19.7 kcal/mol). Experimentally the reaction is slower with Rh [17]. The effect of phenyl substituents, R, positioned *meta* to the directing group, was also assessed. When R = CF_3 or Me, C–H activation is computed to be kinetically favoured at the less hindered *ortho*-C–H bond, as seen experimentally. With R = OMe the two pathways are computed to be in competition, while for R = F activation of the adjacent *ortho*-C–H bond is favoured (perhaps due to the '*ortho*-effect' of fluorine [18]). Experimentally a mixture of products is seen when R = OMe and F.

Matsunaga, Kanai and co-workers described the first computational study of acetate-assisted C–H activation at Co, in their work on the {Co(III)Cp*}-catalysed annulation of *N*-dimethylcarbamoyl-indole (**1**) with alkynes to form pyrroloindolones [19]. With B3LYP, overall barriers of 20.5 kcal/mol (singlet) and 20.9 kcal/mol (triplet) were computed for C–H activation at the 2-position, relative to singlet $[Co(OAc)Cp^*(\kappa\text{-}O\text{-}\mathbf{1})]^+$ at 0.0 kcal/mol. On the triplet surface, C–H activation proceeds in a single step via **^{3}2** which features significant C–H bond elongation (1.35 Å, Scheme 2). In contrast, three transition states were located on a very flat singlet surface with the highest point, **^{1}2**, having a C–H distance of 1.13 Å. AMLA-4 processes (i.e. involving Rh, the C–H bond and the Rh-bound acetate oxygen in a 4-membered transition state) were much higher in energy. C–H activation was endergonic with ΔG_{calc} = +19.2 kcal/mol (singlet) or + 16.8 kcal/mol (triplet), consistent with experimental H/D exchange without the observation of cyclometalated intermediates. The higher efficiency of the {CoCp*} catalysts

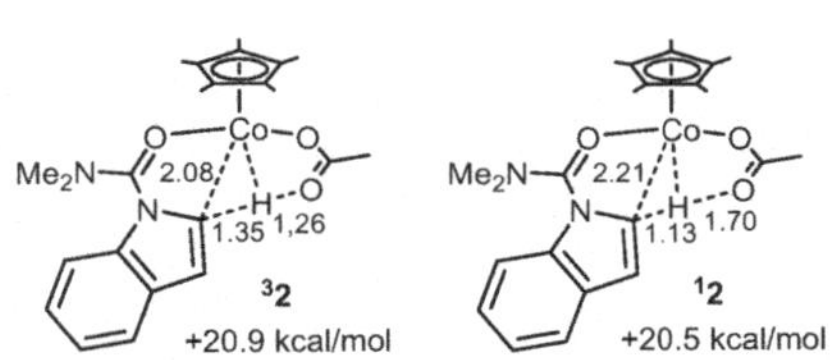

Scheme 2 Key distances (Å) and relative free energies of transition states for C2–H activation of *N*-dimethylcarbamoylindole at $Co(OAc)_2Cp^*$, computed in the triplet (**^{3}2**) and singlet (**^{1}2**) states

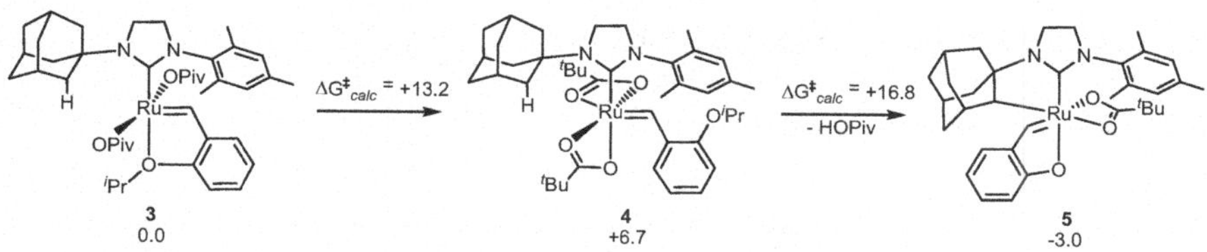

Scheme 3 Pivalate-induced C(sp^3)–H activation in a Grubb's catalyst; free energies in kcal/mol

over {RhCp*} species was attributed to a greater polarity of the Co–C bond that might aid the final reductive annulation step.

2.2 *Activation of C(sp^3)–H Bonds*

Houk, Grubbs and co-workers have studied the OPiv-induced (OPiv = pivalate, $^tBuCO_2^-$) intramolecular C(sp^3)–H activation in a Grubbs' catalyst, featuring an *N*-heterocyclic carbene (NHC) with an *N*-adamantyl and an *N*-mesityl substituent using M06(THF)//B3LYP calculations [20]. C–H activation requires two Cl/OPiv substitutions (**3**, Scheme 3), as well as displacement of the axial benzylidene oxygen to give **4**. C–H activation then proceeds with an overall $\Delta G^{\ddagger}_{\text{calc}}$ of 23.5 kcal/mol, in good agreement with experiment (22.2 ± 0.1 kcal/mol) and consistent with a significant k_H/k_D kinetic isotope effect (KIE) (computed: 6.5; observed: 8.1 ± 1.7 at 25 °C). Alternative processes involving C–H activation by an equatorial OPiv or with OPiv acting as an external base proved less accessible, while the presence of two OPiv ligands allows one of these to act as a κ^2-spectator that stabilises the Ru centre (see **4**). The axial position of the active OPiv accounts for the diastereoselectivity of the C(sp^3)–H bond activation, while the alternative C(sp^3)–H activation at mesityl is disfavoured as rotation of this group is hindered by the NHC backbone. In contrast, C(sp^3)–H activation of an *N*-^{t}Bu substituent is predicted to be accessible ($\Delta G^{\ddagger}_{\text{calc}}$ = 15.2 kcal/mol). This system compares with cyclometalation of Ru(I^{i}Pr$_2$Me$_2$)(PPh$_3$)$_2$(H)(CO) where deprotonation of a non-agostic C–H bond is effected by a bulky NHC acting as an external base [21].

Rh(C^C)I$_2$(OAc) complexes (where C^C = bidentate normal and abnormal NHCs linked by a –(CH$_2$)$_3$– unit) undergo C(sp^3)–H activation of the central CH$_2$ of the linker moiety. BP86(MeCN) calculations suggest that the greater reactivity seen with the abnormal NHC is due to an easier dissociation of iodide that is required, rather than the subsequent AMLA-6 C–H activation step [22].

3 Intermolecular C–H Activation

3.1 C(sp^2)–H Activation

Goddard, Periana and co-workers have investigated the intermolecular C–H activation of benzene at the $Ir(OH)_2(C_5H_5N)(NNC^R)$ system, **6**, with B3LYP(H_2O) and M06(H_2O) calculations (R = H) [23]. Experimentally (R = ^{t}Bu), H/D exchange occurs in the presence of KOH base with a TOF of ca. $6 \times 10^{-3}\ s^{-1}$ at 190 °C ($\Delta H^{\ddagger}_{exp} = 33 \pm 2$ kcal/mol). The computed mechanism involved pyridine/C_6H_6 substitution followed by C–H activation with a *cis*-OH ligand, assisted by the electrophilic Ir. This AMLA-4 process is termed 'internal substitution' by these authors. With B3LYP, $\Delta G^{\ddagger}_{calc} = 46.7$ kcal/mol, and this drops to 38.5 kcal/mol with M06. Alternative processes involving water assistance or deprotonation of the Ir–OH by external hydroxide and C–H activation over an Ir = O bond proved higher in energy.

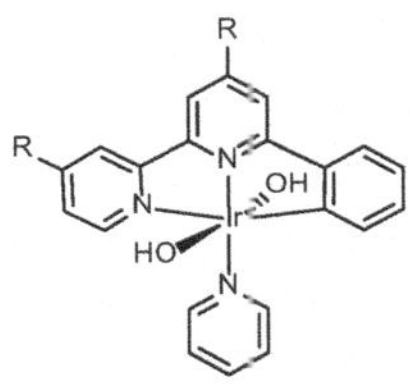

6

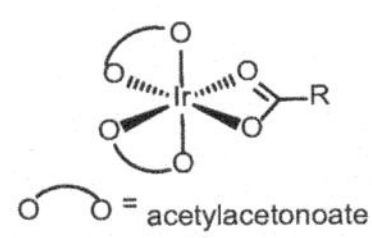

7

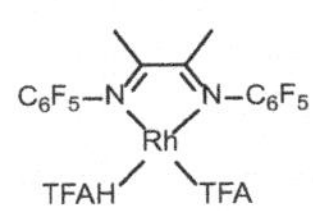

8a

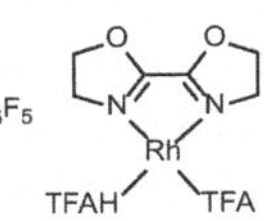

8b

A subsequent study considered the different behaviour seen in the H/D exchange reactions of benzene at *cis*-$Ir(acac)_2(\kappa^2\text{-}O_2CR)$ complexes, **7**, depending on whether R = CH_3 (multiple exchange) or CF_3 (single exchange) [24]. With R = CH_3, B3LYP($MeCO_2H$) calculations indicate rate-limiting dissociative substitution to form *cis*-$Ir(acac)_2(\kappa^1\text{-}O_2CR)(\eta^2\text{-}C_6H_6)$ ($\Delta H^{\ddagger}_{calc} = 23$ kcal/mol). This is then followed by an AMLA-6 C–H activation transition state at +21 kcal/mol. H/D exchange at the bound acetic acid followed by the microscopic reverse process would then give C_6H_5D. However, as further C–H activation is more facile than arene dissociation, several H/D exchanges can proceed, giving higher $C_6H_{6\text{-}n}D_n$ species before the arene is eventually lost. In contrast with R = CF_3, the initial substitution has a much lower barrier of only 5 kcal/mol and C–H activation proceeds through a transition state at 10 kcal/mol. Only one H/D exchange event is therefore possible before the arene dissociates as C_6H_5D. The difference of 13 kcal/mol between the values of $\Delta H^{\ddagger}_{calc}$ is in excellent agreement with experiment, although both barriers are underestimated by ca. 5 kcal/mol. The preferential formation of *cis*-$Ir(acac)_2(\kappa^2\text{-}OAc)$ in acetic acid is also proposed to reduce the efficiency of ethene hydroarylation.

Benzene H/D exchange has also been observed by Gunnoe and co-workers at [(N,N)Rh(TFA)(TFA–H)] complexes, where N,N = bidentate diazobutadiene (**8a**)

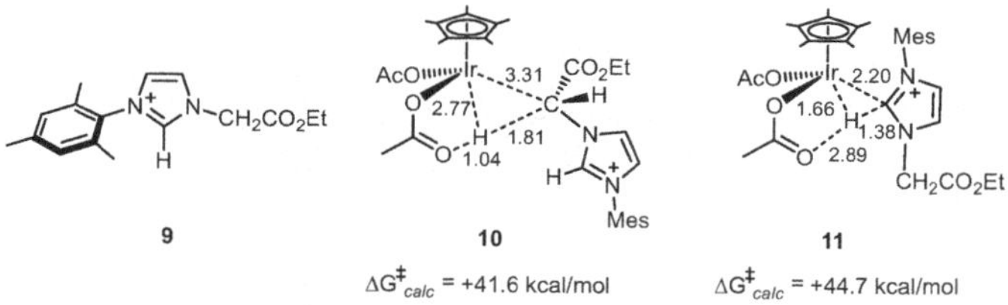

Scheme 4 Key distances (Å) and relative free energies (kcal/mol) in transition states for C(sp^3)–H activation (**10**) and C(sp^2)–H activation (**11**) of an α-imidazolium ester, **9**, at Ir(OAc)$_2$Cp*

or bis(2-oxazolin-2-yl) ligands (**8b**); TFA = $CF_3CO_2^-$ and TFA–H = CF_3CO_2H [25]. Computed pathways (M06(TFA–H)//B3LYP) based on initial oxidative addition or TFA–H/benzene substitution and an AMLA-6 C–H activation have similar overall barriers. For **8a** (the most active species experimentally), $\Delta G^{\ddagger}_{\text{calc}} = 30.9$ kcal/mol at 498 K, and this is in reasonable agreement with experimental turnover numbers of 450 after 2 h at 423 K (implying $\Delta G^{\ddagger}_{\text{exp}} \approx 27.4$ kcal/mol).

3.2 *C(sp³)–H Activation*

Competing C(sp^2)–H and C(sp^3)–H bond activation processes of α-pyridinium and α-imidazolium esters (Scheme 4, **9**) at [IrCl$_2$Cp*]$_2$/NaOAc were reported by Cross and co-workers. M06(CH$_2$Cl$_2$)//B3LYP calculations showed the observed C(sp^3)–H activation to be kinetically favoured over the thermodynamically preferred C(sp^2)–H activation [26]. Both processes involve a nondirected C–H activation and initiate from H-bonded adducts formed at Ir(κ^1-OAc)$_2$Cp*. Although the computed barriers of > 40 kcal/mol are rather high for room temperature processes, the different transition state geometries show some interesting features: C(sp^3)–H activation is dominated by interaction with the base (**10**: C···H = 1.81 Å, O···H = 1.04 Å Ir···H = 2.77 Å), while the metal plays a more prominent role in the C(sp^2)–H activation (**11**, C···H = 1.38 Å, O···H = 2.89 Å Ir···H = 1.66 Å). This is consistent with the AMLA concept, with complementary involvement of base and metal combining to accommodate the activation of a wide range of different C–H bonds [27].

Jones and co-workers have shown that [M(bdmpza)Cl$_3$]$^-$ complexes (M = Rh, Ir; bdmpza = *bis*-(3,5-dimethylpyrazol-1-yl)acetate) promote H/D exchange of both arene C(sp^2)–H bonds and the β-alkyl-C(sp^3)–H bonds of appropriate alkylbenzenes in TFA–H [28]. B3LYP(HOAc) calculations modelled H/D exchange in *iso*-propylbenzene, starting from [Rh(bdmpza)(κ^2-O$_2$CCF$_3$)(C$_6$H$_5$^{i}Pr)]$^+$, **12** (Scheme 5), in which the arene binds in a η^2-C–H fashion through one of the *ortho*-C–H bonds. **12** is a common precursor to both *ortho*-C(sp^2)–H activation ($\Delta G^{\ddagger}_{\text{calc}} = 23.2$ kcal/mol) and β-alkyl-C(sp^3)–H activation

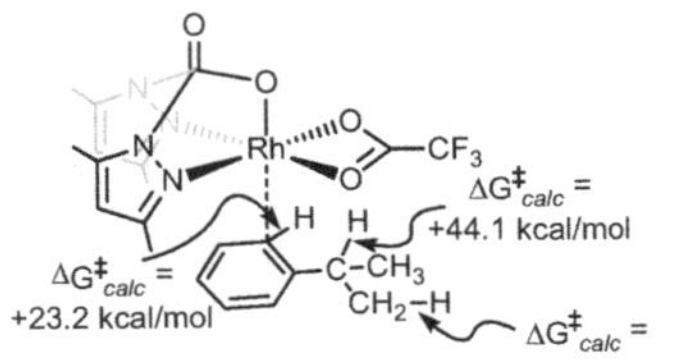

12

Scheme 5 $[Rh(bdmpza)(\kappa^2\text{-}O_2CCF_3)(C_6H_5{}^iPr)]^+$, intermediate **12**, with computed free energies barriers to *ortho*-C(sp^2)–H, β-alkyl-C(sp^3)–H and α-alkyl-C(sp^3)–H activation indicated in kcal/mol

($\Delta G^{\ddagger}_{\text{calc}} = +32.3$ kcal/mol) and these calculated barriers agree well with experiment. Activation of the methine C–H bond has a higher barrier of 44.1 kcal/mol, consistent with no H/D exchange at that position. Similar AMLA-6 transition states are computed for all three processes, the major difference being in the Rh···C distances (C(sp^2)–H: 2.16 Å; β-C(sp^3)–H: 2.29 Å; α-C(sp^3)–H: 2.39 Å).

As part of their screening for new catalysts for the selective conversion of CH_4 to CH_3OH, Gunnoe, Goddard and co-workers have computed energetics for H_3C–H bond activation at a range of Rh complexes bearing bi- and tridentate ligands. Their first study [29] considered reaction in both water and TFA–H and adopted an M06//B3LYP protocol. The combination of an aminidate ligand bearing three C_6F_5 substituents (as in **13**) and TFA–H solvent gave the lowest C–H activation barrier of 27.6 kcal/mol (relative to **13** and a free proton) at 298 K. The C–H activation process comprised TFA–H/CH_4 substitution followed by an AMLA-6 C–H activation transition state involving one of the TFA ligands. This process is promoted by the electron-withdrawing aminidate ligand which facilities σ-donation from the methane C–H bond. The computed Rh–Me bond strength was a good indicator of the relative size of the C–H activation barrier. The subsequent functionalisation of the Rh–Me complex (via S_N2 attack of TFA at the methyl group) gave a barrier of 36.8 kcal/mol. Interestingly, the accessibility of C–H activation and nucleophilic attack swapped when computed with a model formed in water, transition states being located at +35.0 and 33.3 kcal/mol, respectively.

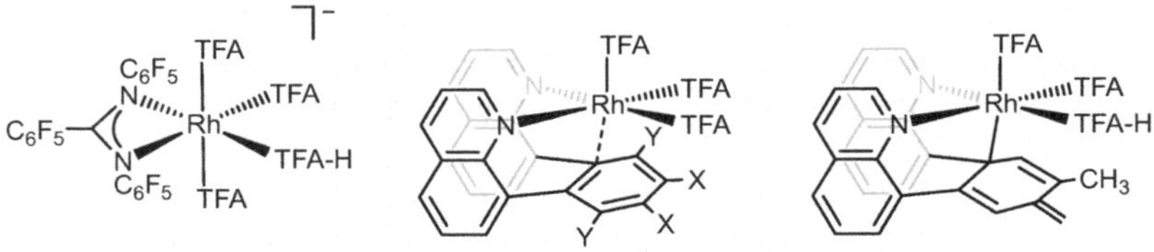

13 14 15

In a subsequent study, a series of $Rh(TFA)_3$(bis(quinolinyl)benzene) systems, **14**, were assessed with the parent system (X = Y = H) providing the lowest barriers of 33.4 kcal/mol for H_3C–H bond activation and 35.8 kcal/mol for functionalisation with TFA [30]. In parallel experimental studies, H/D exchange was noted in the Me substituents of an *o*-xylene-based ligand (Y = H, X = Me) [31]. B3LYP (HOAc) calculations characterised this unusual 'remote' C(sp^3)–H bond activation which proceeds via deprotonation by a Rh-bound TFA ligand. This induces dearomatisation of the xylyl moiety and metalation at the *para* position to give intermediate **15** at +4.8 kcal/mol. The value of $\Delta H^{\ddagger}_{calc}$ (22.4 kcal/mol) is in excellent agreement with experiment (21 ± 2 kcal/mol).

4 C–H Activation and Functionalisation

A range of studies have now appeared modelling complete catalytic cycles based on an initial C–H bond activation and the subsequent functionalisation steps. The functionalisation processes involve reactions with an insertion partner (an alkyne, an alkene or a carbene or nitrene source) and lead to the formation of new heterocyclic rings or direct replacement of the C–H bond with a new C–R group. In many cases, the focus in the computational studies is on these functionalisation processes, as well as the reoxidation steps (particularly with internal oxidants). The C–H activation in these catalytic processes is typically of the intramolecular AMLA-6 type, although exceptions do exist, as will be detailed below. Input from parallel experimental studies, including the determination of k_H/k_D KIEs, the observation of H/D exchange, competition experiments and the isolation of key intermediates both complement and provide benchmark data for the computational modelling. In the following, figures and schemes will show the model systems used in the calculations and relevant experimental results and conditions will be given in the text.

4.1 Heterocycle Formation with Internal Oxidants

In 2011, Guimond and co-workers published the first computational study of Rh (III)-catalysed heterocycle formation based on initial C–H activation and subsequent functionalisation [32]. The coupling of benzamides with alkynes to give isoquinolones was considered, with an OR group (OR = OMe, OPiv) installed on the benzamide nitrogen to act as an internal oxidant. B3LYP(MeOH) calculations were performed for the model reaction of *N*-acetoxybenzamide with acetylene at a $Rh(OAc)_2Cp$ catalyst (see Fig. 3). After *N*-deprotonation and loss of HOAc, directed C–H activation proceeds with an overall barrier of 20.5 kcal/mol. The alternative C–H activation of the neutral (i.e. *N*-protonated) substrate was considerably less accessible. Facile HOAc/HCCH substitution, insertion and N(sp^3)–C (sp^2) reductive coupling then led to intermediate **V** in which the isoquinolone is bound to Rh (Rh–N = 2.13 Å) and the *N*-acetate group bridges the Rh–N bond

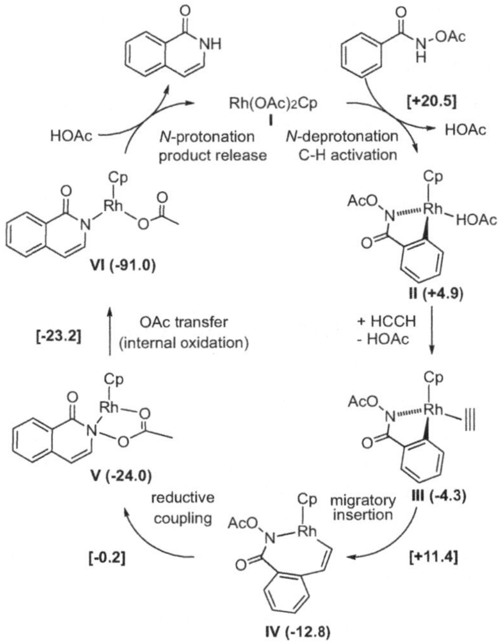

Fig. 3 Computed catalytic cycle for the coupling of *N*-acetoxybenzamide with acetylene at $Rh(OAc)_2Cp$ [32]. Computed free energies of intermediates and transition states are given in kcal/mol, with the latter indicated in square brackets, and are quoted relative to the reactants at 0.0 kcal/mol

(N–O = 1.49 Å; Rh–O = 2.21 Å). N–O bond cleavage with transfer of acetate to Rh then occurs with a minimal barrier of 0.8 kcal/mol with concomitant formal Rh(I) to Rh(III) oxidation. Protonolysis by HOAc then releases the isoquinolone product and completes the catalytic cycle. The calculations suggest rate-limiting C–H activation, consistent with the observation of a large k_H/k_D KIE when an *N*-(pivaloyloxy)benzamide substrate was used experimentally. However, as no KIE is observed with *N*-methoxybenzamide, the nature of the rate limiting step is clearly substrate dependent.

An important complement to this study was published by Xia and co-workers in 2012 who considered the mechanism of 3,4-dihydroquinolone formation via the coupling of alkenes with PhC(O)NH(OR), where the OR group dictates whether N(sp^3)-C(sp^3) coupling (OR = OPiv) or β-H elimination occurs (OR = OMe; see Fig. 4a for the case of ethene) [33]. M06(MeOH) calculations compared profiles for the reactions of both benzamides with ethene at a $Rh(OAc)_2Cp^*$ catalyst. Very similar N–H and C–H activation profiles are computed in each case, with overall barriers of ca. 18 kcal/mol and the endergonic formation of the HOAc adducts ($\Delta G_{calc} \approx +4$ kcal/mol) indicating reversible C–H activation. After HOAc/ethene substitution, insertion forms a conformationally flexible 7-membered rhodacycle, ($\mathbf{I_{OR}}$, Fig. 4b) from which a low-energy isomer can be accessed when OR = OPiv ($\mathbf{I'_{OR}}$, Fig. 4c) which features an interaction between the pendant oxygen and the metal centre.

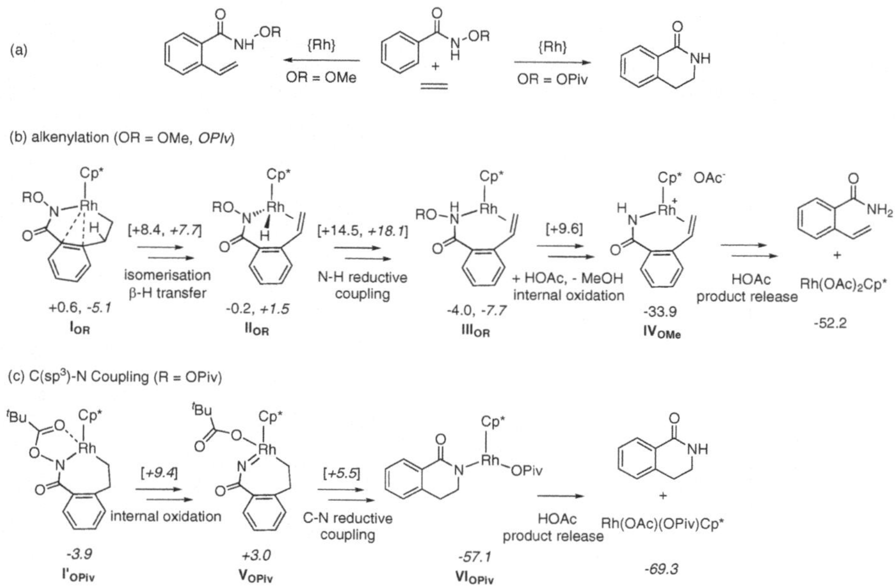

Fig. 4 (**a**) Reactions of PhC(O)NH(OR) (OR = OMe, OPiv) with ethene at $Rh(OAc)_2Cp^*$ [33]. Key stationary points (kcal/mol; free energies quoted relative to the reactants at 0.0 kcal/mol) for (**b**) alkenylation for OR = OMe and OPiv (in *italics*); data for the onward reaction of $\mathbf{III_{OR}}$ for OMe only; and (**c**) $C(sp^3)$–N coupling for OR = OPiv. Double arrows indicate that several steps are involved with the energy of the highest transition state between the two minima indicated in square brackets

From these 7-membered rhodacycles, the mechanisms diverge. For OR = OMe (Fig. 4b), facile β-H transfer followed by formal N–H reductive coupling produces the Rh(I) intermediate $\mathbf{III_{OMe}}$. Reaction with HOAc then facilitates the cleavage of the N–O bond ($\Delta G^{\ddagger}_{calc}$ = 13.6 kcal/mol), forming MeOH and the Rh(III) species $\mathbf{IV_{OMe}}$. A transition state for reoxidation via OMe transfer onto Rh in $\mathbf{III_{OMe}}$ is 22.4 kcal/mol higher in energy. A second HOAc then protonates the amide nitrogen in $\mathbf{IV_{OMe}}$, releasing the product and regenerating the catalyst. A transition state for alternative direct C–N coupling from $\mathbf{I_{OMe}}$ was located, but at + 43.8 kcal/mol it is too high to be competitive. In contrast, N–O bond cleavage in $\mathbf{I'_{OPiv}}$ (Fig. 4c) with transfer of the OPiv group onto Rh ($\Delta G^{\ddagger}_{calc}$ = 14.3 kcal/mol) is now favoured over β-H transfer and N–H reductive coupling ($\Delta G^{\ddagger}_{calc}$ = 23.2 kcal/mol, $\mathbf{I_{OR}}$ to $\mathbf{III_{OPiv}}$, Fig. 4b). The OPiv group is thought to play several roles in this mechanistic swap: the extra Rh–O interaction in the 7-membered rhodacycle $\mathbf{I'_{OPiv}}$ disfavours the initial β-H transfer step, while the greater bulk of the OPiv unit also destabilises the alkene intermediate, $\mathbf{II_{OPiv}}$, making β-H transfer endergonic. The Rh–O interaction in $\mathbf{I'_{OPiv}}$ also facilitates N–O bond cleavage and effects the internal oxidation to give a Rh(V) nitrene intermediate ($\mathbf{V_{OPiv}}$, Rh–N = 1.871 Å). This species then undergoes facile C–N reductive coupling and protonolysis to complete the cycle.

Scheme 6 Regioselective dihydroisoquinolone formation with Rh-cyclopentadienyl catalysts [34]

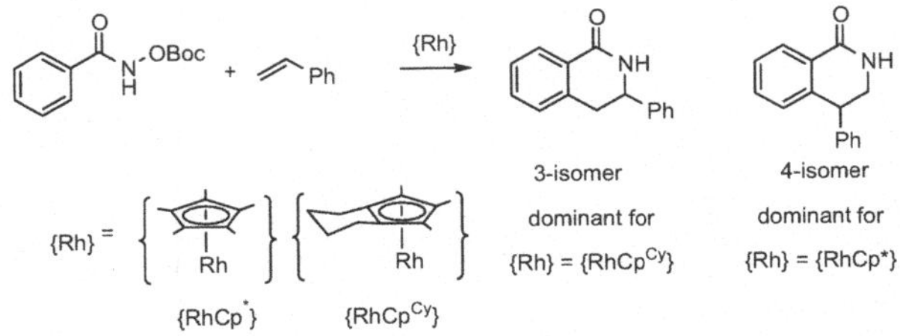

Scheme 7 Rh-catalysed lactam formation from *N*-acetoxybenzamide and 3-methylhexa-1,2,5-triene [35]

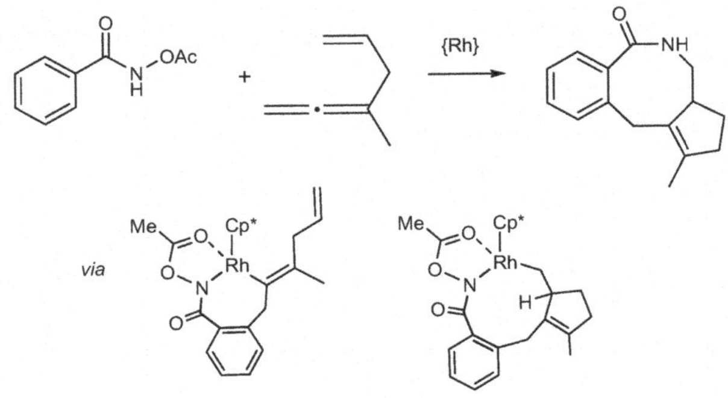

The regioselectivity of dihydroisoquinoline formation between PhC(O)NHOBoc (OBoc = OC(O)^{t}Bu) and styrene has been studied with ωB97X-D (MeOH) calculations (Scheme 6) [34]. Experimentally, {RhCp*} catalysts favour the 4-isomer and {RhCp^{Cy}} catalysts favour the 3-isomer. Calculations based on the pathway defined by Xia above indicate very similar transition state energies for the insertion and internal oxidation steps. This made quantitative predictions about selectivity difficult and so the computed data were used as a basis for kinetics simulations. These predicted a 76:24 ratio in favour of the 3-isomer with {RhCp^{Cy}}, in excellent agreement with experiment. For {RhCp*}, the computed ratio is 49:51, indicating that the 4-isomer does become relatively more favoured, although this is underestimated compared to experiment where a 16:84 ratio is seen. The authors point out that a 0.5 kcal/mol shift in energy would reproduce this result, highlighting the challenge to theory when attempting to model such subtle selectivity effects.

Ma and co-workers have considered the related coupling of *N*-acetoxybenzamide with allene-enes to construct 8-membered lactam rings (Scheme 7) [35]. Using a $Rh(CO_3)Cp^*$ catalyst, M06(MeOH) calculations suggest that the carbonate can effect both the *N*-deprotonation and (as HCO_3^-) the subsequent C–H activation. The C–H activation barrier of 20.8 kcal/mol is the highest along the computed profile, consistent with efficient catalysis at room temperature, and a significant k_H/k_D KIE. C–H activation was computed to be endergonic ($\Delta G_{calc} = +5.4$ kcal/mol), although experimentally no H/D exchange was seen in the absence of the coupling partner. Coordination of the allene-ene permits

Scheme 8 Rh-catalysed benzocyclopentanone formation from benzophenone ammonium salts and α-diazoesters ($R = CO_2{}^iPr$) [37]

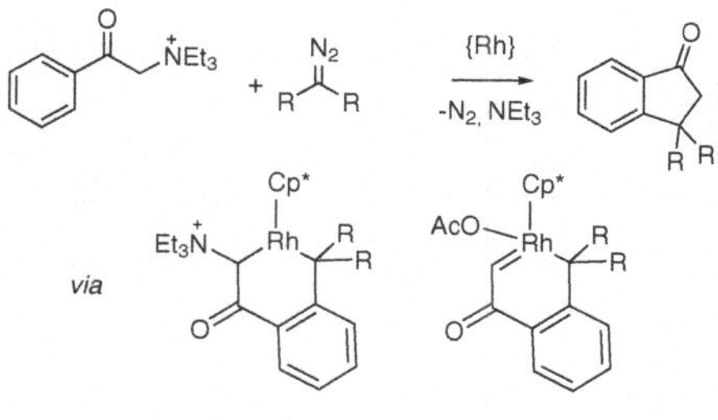

insertion of first the distal allenic double bond and then the terminal alkene to form 7- and then 9-membered rhodacyclic intermediates (**16** and **17,** respectively). The profile proceeds as in Fig. 4c, with the key OAc transfer to a Rh(V) nitrene again shown to be favoured over β-H transfer/N–H reductive coupling. A related study by Huang and Chen on the $Rh(OAc)_2Cp^*$-catalysed coupling of PhC(O)NHOR (OR = OPiv, OMe) with both cyclohexylallene and 1,1-dimethylallene was recently reported [36]. M06(MeOH)//B3LYP calculations confirm the observed regioselectivities: with cyclohexylallene, the 4-isomer (cf. Scheme 6) is favoured, but with 1,1-dimethylallene, greater steric hindrance in the insertion transition state drives the formation of the 3-isomer.

Ammonium groups can also act as internal oxidants, with C–N bond cleavage resulting in loss of an amine and formation of carbocyclic products. Lan, Wan, Li and co-workers used M06(MeCN) and B3LYP-D3(MeCN) calculations to model $Rh(OAc)_2Cp^*$-catalysed benzocyclopentanone formation from benzophenone ammonium salts and α-diazoesters (Scheme 8) [37]. Initial deprotonation and loss of HOAc forms an enolate intermediate from which C–H activation proceeds preferentially from the C-bound form ($\Delta G^{\ddagger}_{calc}$: M06 = 27.2 kcal/mol, B3LYP-D3 = 26.3 kcal/mol); barriers via the O-bound isomer were 2–3 kcal/mol higher. The subsequent steps involve diazoester coordination and N_2 loss followed by carbene insertion to give **18.** NEt_3 loss and coordination of OAc then gives **19** via a transition state at 25.2 kcal/mol (M06) or 20.7 kcal/mol (B3LYP-D3). **19** is a Rh (V) carbene complex that is isoelectronic with the previous Rh(V)-nitrene species. Carbene insertion and protonolysis then complete the cycle. Overall the initial C–H activation is rate limiting, consistent with an observed k_H/k_D KIE of ca. 4. C–H activation is, however, computed to be endergonic, which implies a reversibility not seen in attempted H/D exchange reactions. The authors postulate that this discrepancy may be due to the presence of CsOAc in the catalytic system that slows the reverse protonolysis.

N–N bond cleavage as a means of internal oxidation has been exploited in the coupling of 2-acetyl-1-arylhydrazines with alkynes to give indoles (Fig. 5). This Rh $(OAc)_2Cp^*$-catalysed reaction was studied by Chen and Lin who reported M06 (DCE) energetics with an additional D3 correction [38]. After deprotonation, the terminal NAc position acts as a directing group for a two-step C–H activation via

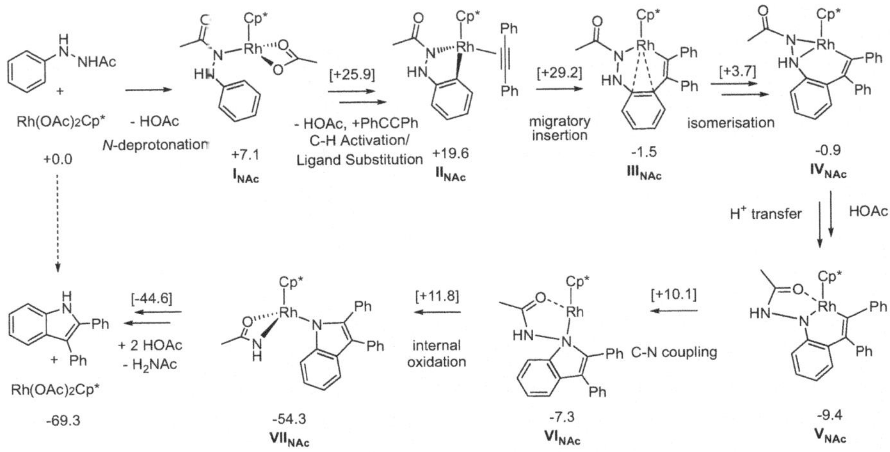

Fig. 5 Key stationary points on free energy profiles (kcal/mol) for the $Rh(OAc)_2Cp^*$-catalysed reaction of 2-acetyl-1-arylhydrazines with diphenylacetylene to give indoles [38]. Double arrows indicate that several steps are involved, with the energy of the highest transition state between the two minima indicated in square brackets

$\mathbf{I_{NAc}}$. Unusually, C–H bond cleavage from the agostic intermediate has a significant additional barrier of 10.5 kcal/mol, giving an overall C–H activation barrier of 25.9 kcal/mol. HOAc/PhCCPh substitution gives $\mathbf{II_{NAc}}$, from which alkyne insertion (via a transition state at 29.2 kcal/mol) gives $\mathbf{III_{NAc}}$ at −1.5 kcal/mol. This isomerises first via a facile 1,2-Rh shift to $\mathbf{IV_{NAc}}$ and then by an HOAc-assisted H^+ transfer to $\mathbf{V_{NAc}}$. C–N reductive coupling constructs the indole moiety ($\mathbf{VI_{NAc}}$) which is bound to a formally Rh(I) centre and an NHC(O)Me group. The latter then transfers onto Rh to generate a Rh(III) species ($\mathbf{VII_{NAc}}$). Reaction with HOAc induces loss of H_2NAc, protonolysis of the Rh–N bond and release of the indole.

In contrast to previous pathways, no Rh(V) species are invoked in this process, but rather a Rh(III)-Rh(I)-Rh(III) pathway is proposed. Experimentally the reaction proceeds at 70 °C [39] and a k_H/k_D KIE of 2.3 is observed. The rate-limiting alkyne insertion implied by the calculations therefore appears inconsistent with this; however an endothermic (i.e. reversible) initial C–H activation step would also be expected to affect the overall rate [40]. The use of kinetic simulations to quantify these effects would be of interest. Calculations comparing the -NH(NHAc) and -NH(OAc) groups suggest the former is a weaker oxidant that can promote C–N bond coupling with alkynes but not with alkenes.

4.2 *Heterocycle Formation without Internal Oxidants*

Davies, Macgregor and co-workers have studied the $Rh(OAc)_2Cp^*$-catalysed oxidative coupling of 5-methyl-3-phenylpyrazole and 4-octyne with BP86(DCE) and

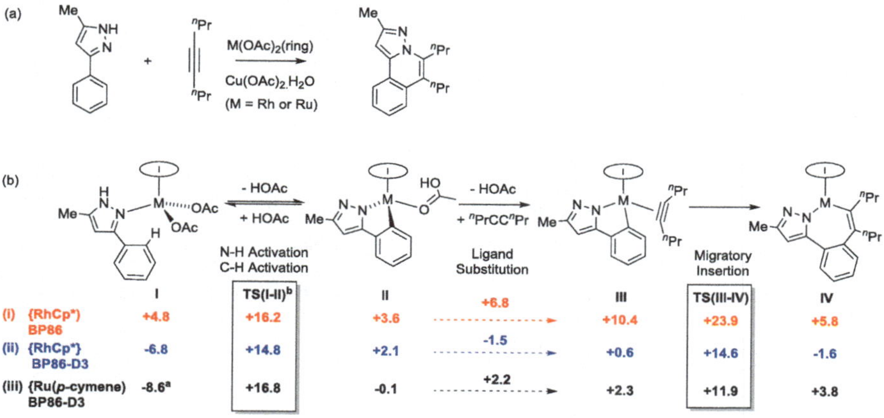

Fig. 6 (**a**) Rh- and Ru-catalysed oxidative coupling of 3-phenylpyrazole and 4-octyne to form an pyrazoloisoquinoline. (**b**) Key stationary points on the free energy profiles (kcal/mol) for (i) {RhCp*} (BP86(DCE), *red*), (ii) {RhCp*} (BP86-D3(DCE), *blue*) and (iii) {Ru(*p*-cymene)} (BP86-D3(MeOH), *black*), quoted relative to reactants set to 0.0 kcal/mol in each case [41]. [a] −8.6 kcal/mol corresponds to the lowest point on the profile and is the N–H-activated form of **I**.[b] The formation of **II** from **I** involves several steps and **TS(I–II)** is the highest point in this process

BP86-D3(DCE) calculations (Fig. 6) [41]. Experimental H/D exchange studies indicate reversible C–H activation both in the absence (40% *ortho*-deuteration) and in the presence of the alkyne (7% incorporation). This indicates close competition between protonolysis of the initial cyclometalated intermediate, **II** (via **TS(I–II)**), and the onward reaction involving HOAc/alkyne substitution and migratory insertion (via **TS(III–IV)**). This competition was only correctly reproduced when the D3 dispersion correction was employed and primarily reflects changes in the HOAc/^{n}PrCC^{n}Pr substitution step (ΔG_{calc}: BP86 = +6.8 kcal/mol; BP86-D3 = −1.5 kcal/mol). This is particularly sensitive to dispersion effects because of the different bulk of the two ligands involved. Alternative methods that include a treatment of dispersion effects (e.g. M06 or B97D or the addition of the D3 correction to other pure and hybrid functionals) also performed well in this regard.

Comparison with the $Ru(OAc)_2$(*p*-cymene) catalyst proved instructive. For both metals, rate-limiting C–H activation was computed, with a higher barrier for Ru (25.4 kcal/mol) compared to Rh (21.6 kcal/mol). This is consistent with the higher temperature required for Ru experimentally (2-methylbutanol, 120 °C; cf. DCE at 83 °C for Rh). For Rh, an observed k_H/k_D KIE of 2.7 ± 0.5 is consistent with the computed two-step C–H activation where the higher transition state, $\mathbf{TS(I–II)_{Rh}}$ (Fig. 7), corresponds to C–H bond cleavage occurring after an agostic intermediate (a calculated KIE based on $\mathbf{TS(I–II)_{Rh}}$ was 5.5). In contrast, the observed k_H/k_D KIE at Ru was small (1.1 ± 0.2). In this case, however, calculations indicated a one-step C–H activation via $\mathbf{TS(I–II)_{Ru}}$, which equates to the κ^2–κ^1 displacement of OAc. $\mathbf{TS(I–II)_{Ru}}$ therefore features minimal C–H bond elongation and the

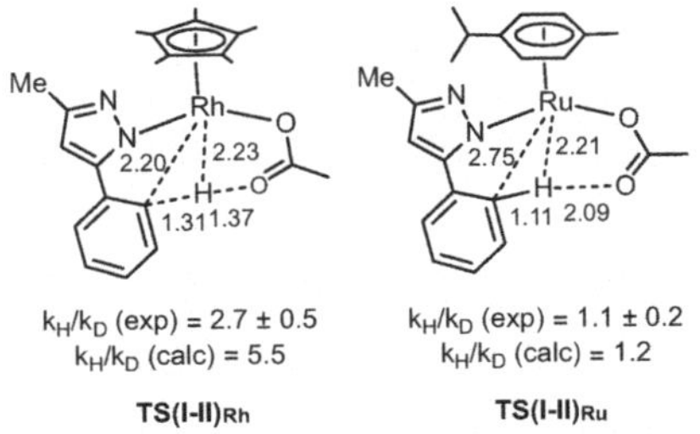

Fig. 7 Rate-limiting transition states with key distances (Å) for the C–H activation of 5-methyl-3-phenylpyrazole at {Rh(OAc)Cp*} (**TS(I–II)$_{Rh}$**) and {Ru(OAc)(*p*-cymene)} (**TS(I–II)$_{Ru}$**). Associated experimental and computed k_H/k_D KIE data are also shown [41]

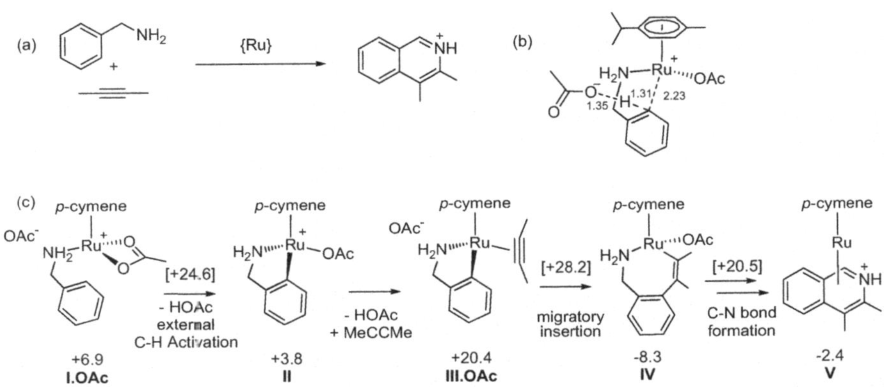

Fig. 8 (**a**) Ru-catalysed oxidative coupling of benzylamine with 2-butyne to form an isoquinoline. (**b**) Transition state for C–H activation via external deprotonation with key distances (Å). (**c**) Key stationary points on the free energy profile (kcal/mol; energies quoted relative to the reactants at 0.0 kcal/mol). Double arrows indicate that several steps are involved, with the energy of the highest transition state between the two minima indicated in square brackets [42]

computed KIE is only 1.2, in good agreement with experiment. An important conclusion from this study is that the lack of a significant k_H/k_D KIE does not necessarily mean that C–H activation is not rate limiting; it may simply be that the C–H activation has a very early transition state geometry.

Lledós, Urriolabeitia and co-workers have used M06(MeOH) calculations to model the oxidative coupling of benzylamines with 2-butyne at [RuCl$_2$(*p*-cymene)]$_2$ (Fig. 8) [42]. The presence of protic N–H bonds in benzylamine promotes the formation of H-bonded ion pairs upon OAc$^-$ dissociation, and these proved more reactive than non-ion-paired species where OAc$^-$ was modelled as fully dissociated in solution (see Fig. 8). Thus after OAc$^-$/amine substitution at Ru(OAc)$_2$(*p*-cymene) to give **I•OAc**, an outer-sphere C–H activation is computed via a transition state at 24.6 kcal/mol, 2.5 kcal/mol below a more conventional

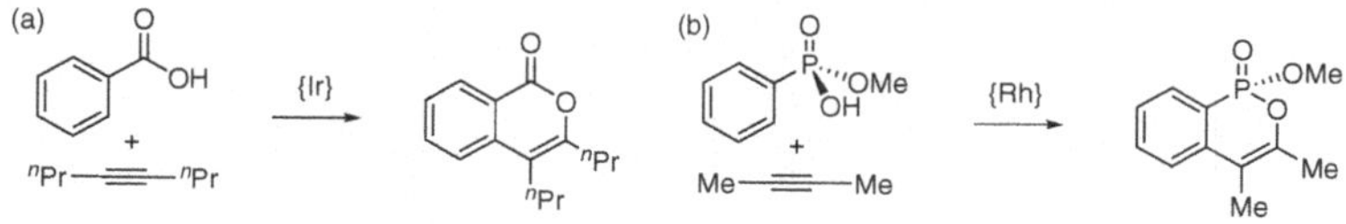

Scheme 9 (**a**) Ir-catalysed isocoumarin formation [43]; (**b**) Rh-catalysed phosphaisocoumarin formation [44]

intramolecular process. Although this external deprotonation cannot be strictly be termed 'AMLA' (as this implies a coordinated ligand acting as a base), it does display some common features, namely, the interaction of the C–H bond with an electron-deficient metal centre which activates it to deprotonation by a base. 'CMD' would describe this concerted process appropriately. After loss of HOAc, OAc^-/MeCCMe substitution gives a further ion-pair, **III•OAc**, that undergoes insertion to a 7-membered ruthenacycle, which is then trapped by OAc^- to form a saturated intermediate, **IV**. Of several routes considered for C–N bond formation, a process involving *N*-decoordination and nucleophilic attack of the free amine arm at the bound Ru-alkenyl moiety was found to be most accessible, with an overall barrier relative to **IV** of 28.8 kcal/mol; OAc^- is again H-bonded to the amine during this process. The protonated 1,2-dihydroisoquinoline is thus formed bound to a Ru(0) metal centre (**V**). Overall, no clear rate-determining process could be identified, with insertion and C–N bond formation both having similar barriers of ca. 29 kcal/mol, both slightly higher than C–H activation. Experimentally temperatures of ca. 100 °C are required. Related profiles were computed for 1-naphthylmethyl, 2-methylallylamine and 2-thiophenemethylamine.

The $Ir(OAc)_2Cp$*-catalysed oxidative coupling of benzoic acids with alkynes has been studied with BP86(MeOH) calculations by Ison and co-workers (Scheme 9a) [43]. These confirmed the previously reported result [45] that C–H activation proceeds via acetate assistance and not oxidative addition. The subsequent insertion of 4-octyne insertion is thought to be rate limiting and has a transition state at +29.6 kcal/mol. This is perhaps rather high for a reaction that completes after 1 h at 60 °C, and it would be interesting to see how this barrier was affected by a treatment of dispersion effects. After insertion, C–O reductive coupling then gives the isocoumarin product bound as an η^4-ligand at an Ir(I) centre. Oxidation via reaction with AgOAc to give an Ir(II)OAc intermediate and a Ag atom is proposed to be necessary to release the organic product, further reaction with a second equivalent of AgOAc then regenerating the $Ir(III)(OAc)_2$ catalyst.

The related synthesis of phosphaisocoumarins via the Rh-catalysed coupling of phenylphosphonates and alkynes (Scheme 9b) was modelled by Zhao and co-workers with B3LYP-D3(^{t}BuOH) calculations [44]. Using a model system ($Rh(OAc)_2Cp$, PhP(O)(OMe)(OH) and 2-butyne), the authors defined a pathway with an energy span [46] of 36.8 kcal/mol, which increases to 50.2 kcal/mol if a low-energy carbonate intermediate, $Rh(CO_3)Cp${PhP(O)(OMe)(OH)}, is taken into account. The computed rate-limiting transition state involves C–C coupling. It is difficult to reconcile these computed results with a process that operates efficiently

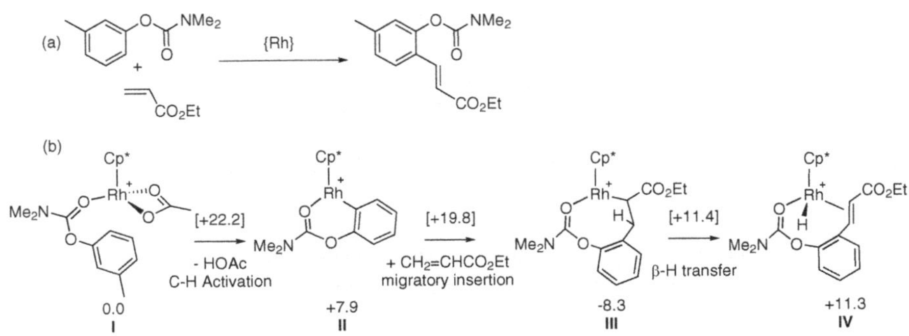

Fig. 9 (**a**) Rh-catalysed alkenylation of *m*-tolyldimethylcarbamate with ethyl acrylate. (**b**) Key stationary points on the free energy profile (kcal/mol; energies quoted relative to the reactants at 0.0 kcal/mol) [49]

at 90 °C and has an observed k_H/k_D KIE of 5.3 [47, 48]. Use of a truncated model was addressed with tests comparing Cp and Cp* and gave similar results for the rate-limiting C–C coupling. However, these were performed with 2-butyne, an alkyne that is not used experimentally. The importance of combining the full steric bulk of all substrates and catalysts to construct a reasonable model has already been stressed [41].

4.3 Alkenylation and Amination

The first computational contribution to this area came from Huang, Fu and co-workers with their 2013 study of Rh(III)-catalysed oxidative Heck coupling of phenol carbamates with alkenes (see Fig. 9) [49]. Using a M06(THF)//B3LYP protocol, they modelled the reaction of *m*-tolyldimethylcarbamate with ethyl acrylate at $[Rh(OAc)Cp^*]^+$, supported by preliminary work on a Cp model. An initial adduct formed via the carbamate oxygen (**I**, 0.0 kcal/mol) provides access to a one-step C–H activation at the less hindered *ortho*-position with a barrier of 22.2 kcal/mol to give, after loss of HOAc, a 6-membered metallacycle, **II**, at +7.9 kcal/mol. Ethyl acrylate can then undergo 2,1-insertion into the Rh–C bond with an overall barrier of 11.9 kcal/mol resulting in an 8-membered rhodacycle, **III**, at −5.5 kcal/mol. β-H transfer is then uphill ($\Delta G^{\ddagger} = 16.9$ kcal/mol; $\Delta G =$ +16.8 kcal/mol) but it is assumed that the Rh(I)–H intermediate, **IV**, can then release the coupled product in the presence of the $Cu(OAc)_2$ oxidant. C–H activation is therefore computed to be rate limiting, consistent with an intramolecular k_H/k_D KIE of 3.1. Experimentally the modelled reaction requires 24 h at 110 °C [50]. An alternative mechanism based on initial C–H activation of ethyl acrylate was also considered, and this process was found to have a higher barrier of 25.1 kcal/mol.

Scheme 10 Rh-catalysed alkenylation reactions: (**a**) 3-phenylpyrazole with $H_2C=CHR$ (R = Ph, CO_2Me) [51]; (**b**) *N*-methoxybenzamide with dimethyl-2-vinylcyclopropane-1,1-dicarboxylate [52]

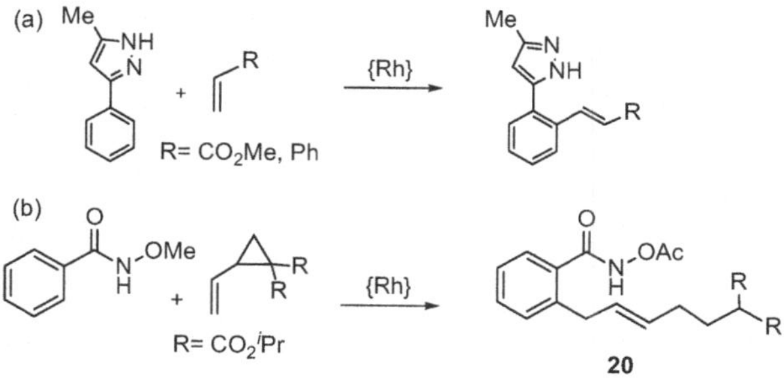

The related coupling of 5-methyl-3-phenylpyrazole with both methyl acrylate and styrene at $Rh(OAc)_2Cp^*$ was modelled by Davies, Macgregor and co-workers using a BP86-D3(DCE) protocol (Scheme 10a) [51]. Starting from alkene adducts of their previously computed cyclometalated intermediate [41], a clear preference for 2,1-insertion was computed for both alkenes, with barriers of ca. 20 kcal/mol for methyl acrylate and ca. 23 kcal/mol for styrene. Insertion produces conformationally flexible rhodacycles from which β-H transfer to both the *cis*- and *trans*-alkene products was characterised, with the formation of the *trans*-isomers being both kinetically and thermodynamically favoured. Overall, for methyl acrylate, similar barriers for insertion and the preceding C–H activation were computed, whereas the higher insertion barrier for styrene makes this process rate limiting in this case. Using $M06(CF_3CH_2OH)$ calculations, Huang, Wang and co-workers showed that the reaction of dimethyl-2-vinylcyclopropane-1,1-dicarboxylate with *N*-methoxybenzamide (Scheme 10b) proceeds via rate limiting 2,1-insertion with an overall barrier of 26.1 kcal/mol. This is preceded by a reversible C–H activation with a barrier of 21.4 kcal/mol [52]. Experimentally, a small kinetic isotope effect value of 1.7 was obtained from a parallel experiment. In this case, subsequent β-H elimination and ring opening of the pendant cyclopropane moiety give access to a highly functionalised (*E*)-but-2-en-1-yl-dimethylmalonate moiety, **20**.

The direct C–H amination of *N-tert*-butylbenzamide with organic azides catalysed by Gp 9 $M(OAc)_2Cp^*$ species has been modelled by Chang, Musaev and co-workers with M06(DCE) calculations (Scheme 11) [53]. Experimentally, $[MCl_2Cp^*]_2$ precatalysts are employed along with an $AgSbF_6$ additive and the latter is assumed to effect the necessary C–H activation (presumably via an electrophilic activation mechanism, i.e. without any specific base assistance) to arrive at a $[M(BA)Cp]^+$ intermediate (where BA is an *O*-bound, C–H activated *N-tert*-butylbenzamide ligand). The reaction proceeds by coordination of PhN_3 followed by loss of N_2 to form a M(V) nitrene that can then insert into the M–C bond. Protodemetallation then proceeds via C–H activation of the benzamide substrate, in which the amido-nitrogen acts as an internal base (i.e., an AMLA-4 step; see **21**, Scheme 11). This releases the amination product and regenerates the $[M(BA)Cp]^+$ active species. When M = Rh, significant barriers are computed for N_2 loss (30.9 kcal/mol) and the final C–H activation step (40.6 kcal/mol). The equivalent barriers for Ir are 29.5 and 33.9 kcal/mol, respectively, with the reduced barriers to C–H activation arising from enhanced charge density on the nitrene nitrogen in this

Scheme 11 Group 9-catalysed amination of *N*-*tert*-butylbenzamide with organic azides, along with the structure of the AMLA-4 C–H activation transition state, **21**, with key distances in Å

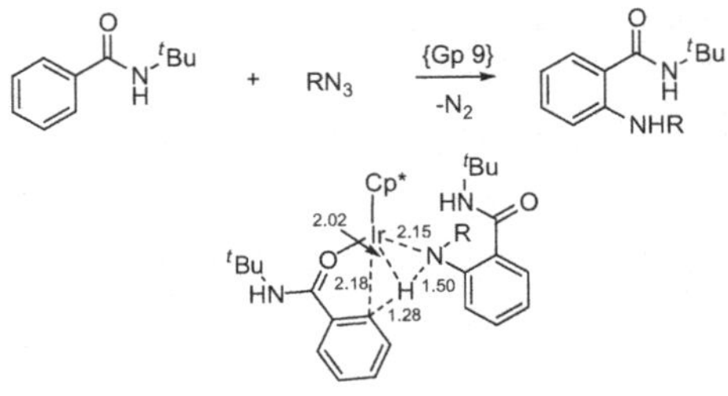

case. Experimentally, the Ir system is more promising with the reaction of ArN_3 ($Ar = 3,5\text{-}(CF_3)_2C_6H_3$) proceeding in 47% yield (90 min, 65 °C); in contrast, Rh gives an 8% yield under these conditions. A significant k_H/k_D KIE is seen observed for the Ir system and $\Delta H^{\ddagger}_{exp}$ is 20.3 kcal/mol, in reasonable agreement with the computed value of 17.9 kcal/mol. In contrast, the small value of $\Delta S^{\ddagger}_{exp}$ (1.4 calmol^{-1} K^{-1}) is at odds with the strongly associative mechanism implied computationally. For the Co(III) analogue, barriers of 31.7 kcal/mol (N_2 loss) and 51.9 kcal/mol (C–H activation) are computed and indeed the Co system is an ineffective catalyst experimentally.

5 Conclusions

Over the period covered by this review, it is clear that computational chemistry has become further embedded as a key tool for the study of C–H bond activation mechanisms. The underlying basis of heteroatom-assisted C–H activation at group 8 and 9 metal centres is now well understood with the combination of an electron-deficient metal centre and a chelating base providing a powerful means to cleave C–H bonds. Intermolecular C–H activation of hydrocarbons, in particular of C(sp^3)–H bonds, remains a challenge and computation is playing an important role in designing promising catalysts for both the key C–H activation step and also the subsequent functionalisation. For intramolecular C–H activation, the computation of full catalytic cycles based on C–H activation and functionalisation is becoming prevalent. Experimentally, such processes are now providing practical methods in organic synthesis and computation is well placed to play an active role in furthering efforts in this area. In doing so, the use of more realistic models that better reflect the systems used experimentally will be important and this is made possible by the availability of increased computational power. However, such models bring with them the dual challenges of a correct treatment of weak interactions and an assessment of conformational flexibility. Close interaction with experiment is still vital in order to benchmark computational protocols against experimental data and the most valuable studies in this chapter have been based on a close alignment of experiment and computation. Improved mechanistic insight therefore demands input from experiment in the form of quantitative experimental data, while computational chemists must be stringent in testing their methods against such data. With this approach, the role of computation in integrating C–H activation into ever more ambitious catalytic schemes has a vibrant future.

Acknowledgements We thank Prof. Dai Davies (University of Leicester) and his group for many fruitful discussions, the EPSRC for financial support through awards EP/J002712/1 (KJTC) and EP/J021911/1 (CLM) and the Université de Lille I, Science et Technologies, for a visiting professorship during which much of this chapter was written.

References

1. Balcells D, Clot E, Eisenstein O (2010) Chem Rev 110:749–823
2. Boutadla Y, Davies DL, Macgregor SA, Poblador-Bahamonde AI (2009) Dalton Trans 14 (30):5820–5831
3. Perutz RN, Sabo-Etienne S (2007) Angew Chem Int Ed 46:2578–2592
4. Vastine BA, Hall MB (2009) Coord Chem Rev 253:1202–1218
5. Lapointe D, Fagnou K (2010) Chem Lett 39:1119–1126
6. Biswas B, Sugimoto M, Sakaki S (2000) Organometallics 19:3895–3908
7. Davies DL, Donald SMA, Macgregor SA (2005) J Am Chem Soc 127:13754–13755
8. Webb JR, Burgess SA, Cundari TR, Gunnoe TB (2013) Dalton Trans 42:16646–16665
9. Guan W, Sayyed FB, Zeng G, Sakaki S (2014) Inorg Chem 53:6444–6457
10. Grimme S, Antony J, Ehrlich S, Krieg H (2010) J Chem Phys 132:154104
11. Grimme S (2006) J Comput Chem 27:1787–1799
12. Chai J-D, Head-Gordon M (2008) Phys Chem Chem Phys 10:6615–6620
13. Zhao Y, Truhlar D (2008) Theor Chem Acc 120:215–241
14. Minenkov Y, Occhipinti G, Jensen VR (2009) J Phys Chem A 113:11833–11844
15. Carr KJT, Davies DL, Macgregor SA, Singh K, Villa-Marcos B (2014) Chem Sci 5:2340–2346
16. Li J, Hu W, Peng Y, Zhang Y, Li J, Zheng W (2014) Organometallics 33:2150–2159
17. Li L, Brennessel WW, Jones WD (2009) Organometallics 28:3492–3500
18. Clot E, Eisenstein O, Jasim N, Macgregor SA, McGrady JE, Perutz RN (2011) Acc Chem Res 44:333–348
19. Ikemoto H, Yoshino T, Sakata K, Matsunaga S, Kanai M (2014) J Am Chem Soc 136:5424–5431
20. Cannon JS, Zou L, Liu P, Lan Y, O'Leary DJ, Houk KN, Grubbs RH (2014) J Am Chem Soc 136:6733–6743
21. Haller LJL, Page MJ, Macgregor SA, Mahon MF, Whittlesey MK (2009) J Am Chem Soc 131:4604–4605
22. Krueger A, Haeller LJL, Mueller-Bunz H, Serada O, Neels A, Macgregor SA, Albrecht M (2011) Dalton Trans 40:9911–9920
23. Meier SK, Young KJH, Ess DH, Tenn WJ III, Oxgaard J, Goddard WA III, Periana RA (2009) Organometallics 28:5293–5304
24. Bischof SM, Ess DH, Meier SK, Oxgaard J, Nielsen RJ, Bhalla G, Goddard WA III, Periana RA (2010) Organometallics 29:742–756
25. Webster-Gardiner MS, Fu R, Fortman GC, Nielsen RJ, Gunnoe TB, Goddard WA III (2015) Catal Sci Technol 5:96–100
26. Cross WB, Razak S, Singh K, Warner AJ (2014) Chem Eur J 20:13203–13209

27. Davies DL, Donald SMA, Al-Duaij O, Fawcett J, Little C, Macgregor SA (2006) Organometallics 25:5976–5978
28. Rhinehart JL, Manbeck KA, Buzak SK, Lippa GM, Brennessel WW, Goldberg KI, Jones WD (2012) Organometallics 31:1943–1952
29. Fu R, Nielsen RJ, Goddard WA III, Fortman GC, Gunnoe TB (2014) ACS Catal 4:4455–4465
30. Fu R, O'Reilly ME, Nielsen RJ, Goddard WA III, Gunnoe TB (2015) Chem Eur J 21:1286–1293
31. O'Reilly ME, Fu R, Nielsen RJ, Sabat M, Goddard WA III, Gunnoe TB (2014) J Am Chem Soc 136:14690–14693
32. Guimond N, Gorelsky SI, Fagnou K (2011) J Am Chem Soc 133:6449–6457
33. Xu L, Zhu Q, Huang G, Cheng B, Xia Y (2012) J Org Chem 77:3017–3024
34. Wodrich MD, Ye B, Gonthier JF, Corminboeuf C, Cramer N (2014) Chem Eur J 20:15409–15418
35. Wu S, Zeng R, Fu C, Yu Y, Zhang X, Ma S (2015) Chem Sci 6:2275–2285
36. Xing Z, Huang F, Sun C, Zhao X, Liu J, Chen D (2015) Inorg Chem 54:3958–3969
37. Yu S, Liu S, Lan Y, Wan B, Li X (2015) J Am Chem Soc 137:1623–1631
38. Chen W-J, Lin Z (2015) Organometallics 34:309–318
39. Zhao D, Shi Z, Glorius F (2013) Angew Chem Int Ed 52:12426–12429
40. Simmons EM, Hartwig JF (2012) Angew Chem Int Ed 51(13):3066–3072
41. Algarra AG, Cross WB, Davies DL, Khamker Q, Macgregor SA, McMullin CL, Singh KJ (2014) Org Chem 79:1954–1970
42. Ruiz S, Villuendas P, Ortuño MA, Lledós A, Urriolabeitia EP (2015) Chem Eur J 21:8626–8636
43. Frasco DA, Lilly CP, Boyle PD, Ison EA (2013) ACS Catal 3:2421–2429
44. Liu L, Wu Y, Wang T, Gao X, Zhu J, Zhao Y (2014) J Org Chem 79:5074–5081
45. Davies DL, Donald SMA, Al-Duaij O, Macgregor SA, Polleth M (2006) J Am Chem Soc 128:4210–4211
46. Kozuch S, Shaik S (2011) Acc Chem Res 44:101–110
47. Park Y, Seo J, Park S, Yoo EJ, Lee PH (2013) Chem Eur J 19:16461–16468
48. Seo J, Park Y, Jeon I, Ryu T, Park S, Lee PH (2013) Org Lett 15:3358–3361
49. Zhang Q, Yu H-Z, Li Y-T, Liu L, Huang Y, Fu Y (2013) Dalton Trans 42:4175–4184
50. Gong T-J, Xiao B, Liu Z-J, Wan J, Xu J, Luo D-F, Fu Y, Liu L (2011) Org Lett 13:3235–3237
51. Algarra AG, Davies DL, Khamker Q, Macgregor SA, McMullin CL, Singh K, Villa-Marcos B (2015) Chem Eur J 21:3087–3096
52. Wu J-Q, Qiu Z-P, Zhang S-S, Liu J-G, Lao Y-X, Gu L-Q, Huang Z-S, Li J, Wang H (2015) Chem Commun 51:77–80
53. Figg TM, Park S, Park J, Chang S, Musaev DG (2014) Organometallics 33:4076–4085

Top Organomet Chem (2016) 55: 77–102
DOI: 10.1007/3418_2015_121

Published online: 10 June 2015

Recent Developments in Pd-Catalyzed Direct Arylations of Heteroarenes with Aryl Halides

Fabio Bellina

Abstract The direct arylation of heteroaromatic compounds with aryl halides using palladium catalysts has significantly been developed as an effective method for making (hetero)aryl-heteroaryl bonds, which are frequently found in biologically active compounds and functional materials. However, issues regarding costly reagents, regioselectivity, and harsh reaction conditions have to be solved to really compete with the classical transition metal-mediated cross-coupling procedures involving stoichiometric organometallic reagents. In this review, the most relevant developments aimed at solving these issues are summarized.

Keywords Arylation · C–H bond activation · Heteroarenes · Palladium catalysts · Regioselectivity

Contents

F. Bellina (✉)
Dipartimento di Chimica e Chimica Industriale, Università Di Pisa, Pisa 56124, Italy
e-mail: fabio.bellina@unipi.it

Abbreviations

Ad	1-Adamanthyl
Bn	Benzyl
Boc	*tert*-Butoxycarbonyl
Cy	Cyclohexyl
dppb	1,4-Bis(diphenylphosphino)butane
dppe	1,2-Bis(diphenylphosphino)ethane
DMA	Dimethylacetamide
DMF	Dimethylformamide
DMSO	Dimethyl sulfoxide
NMP	*N*-Methylpyrrolidone
SEM	2-(Trimethylsilyl)ethoxymethyl

1 Introduction

Arylheteroarenes represent important scaffolds in numerous biologically active compounds [1] and optical and photochromic materials in polymer sciences [2, 3]. Due to their widespread applications, the development of straightforward functional group-tolerant methods for their preparation has attracted organic chemist for over 100 years [4]. Although a variety of routes for the construction of the aryl-heteroaryl Csp^2–Csp^2 bond exist, probably the most general and effective methodology for the synthesis of (hetero)aryl-substituted heteroaromatics involves the use of transition metal-catalyzed cross-coupling reactions between (hetero)aryl halides and (hetero)aryl metal derivatives, including boronic acids, organozinc, organotin, and organosilicon compounds, and Grignard reagents [4]. Although a number of improvements have been developed, and several industrial applications have been reported [5, 6], the preactivation of both the coupling partners involves several manipulations prior to the cross coupling, generating waste from reagents, solvents, and purifications. Furthermore, a stoichiometric amount of metal waste is produced by the coupling process. In some cases, not all the regioisomers of the required organometallic reagents are readily available or are too expensive to buy or synthesize, and in the worst cases they are too instable to participate in the coupling reaction. For all these reasons, there is urgency to continue the search for more efficient, regio- and chemoselective methods for the preparation of arylheteroarenes.

In recent years, the transition metal-catalyzed direct arylation of heteroaryl C–H bonds with aryl halides emerged as an attractive, eco-friendly, and economic alternative to traditional cross-coupling methods [7–9]. These reactions simply replace one of the preactivated coupling partners, the organometallic, with a simple heteroarene, reducing the metal waste generated in the overall process. In particular, electron-rich heteroarenes, such as azoles, constitute the "natural" replacements of heteroarylmetals in the catalytic cycles of these coupling reactions.

An early report of this approach, which contains examples of palladium-catalyzed direct arylation of isoxazole with iodobenzene, was published by Nakamura, Tajima, and Sakai in 1982 [10]. Since then, this chemistry has grown rapidly, in particular in the last decade, and a large variety of new, highly regioselective, and synthetically useful direct inter- and intramolecular reactions involving five- and six-membered heteroarenes have been developed.

In this chapter, it will be specifically examined some of the most relevant results in this field. In particular, the use of cheap aryl chlorides and of heterogeneous palladium catalysts will be discussed. Some significant protocols for lowering reaction temperature and time, and for widening the scope by involving poorly reactive or unreactive heteroaromatic C–H bonds, will be also summarized.

2 Direct Arylations of Heteroarenes with Aryl Chlorides

Significant progress has been made in the use of aryl chlorides as substrates in palladium-mediated cross-coupling reactions, and many reactions with aryl chlorides can now be performed under very mild conditions [11]. In contrast, aryl chlorides as substrates in direct arylation reactions were rarely employed before 2005. In this year, Fagnou and co-workers described in an initial report the use of *N*-heterocyclic carbene (NHC) ligands for intramolecular direct arylation reactions with aryl chlorides [12], but in subsequent studies, they found limitations when sterically hindered substrates were employed [13]. The authors then reinvestigated the intramolecular reaction of a broad range of heteroarenes with aryl iodides, bromides, and chlorides. As regards chlorides, after a screening of bases and ligands, they found that in the presence of 5 mol% $Pd(OAc)_2$, 10 mol% $PCy_3{\cdot}HBF_4$, 2.0 equiv. of K_2CO_3 in DMA at 130°C for 10 h 6*H*-isoindolo[2,1-*a*] indole was obtained in 78% isolate yield starting from 1-(2-chlorobenzyl)-1*H*-indole (Scheme 1) [14].

In 2006 Leclerc and Fagnou reported the successful coupling of a molar excess of diazines *N*-oxides with aryl chlorides (Scheme 2) [15].

In a competition study, the authors found that pyrazine *N*-oxide is approximately twice as reactive as pyridazine *N*-oxide. Pyrazine *N*-oxides may be deoxygenated with ammonium formate and Pd/C in methanol at room temperature, giving the corresponding bases in excellent yields. However, this protocol resulted uncompatible with pyridazine *N*-oxide, and high yields were obtained when catalytic Pd/C in ammonium hydroxide under a hydrogen atmosphere was employed.

During their studies on direct arylations involving azoles, Doucet and co-workers reported that aryl chlorides are able to react with 2-substituted thiazoles in the presence of 1 mol% of the air-stable $PdCl_2(dppb)(C_6H_5)$ complex (Scheme 3) [16].

Among the tested bases, AcONa and AcOK gave the best results, while alkali carbonates or KF resulted ineffective. The authors tried also the use of heterogeneous $Pd(OH)_2/C$ (Pearlman's catalyst) or Pd/C, but with poor results. As regards

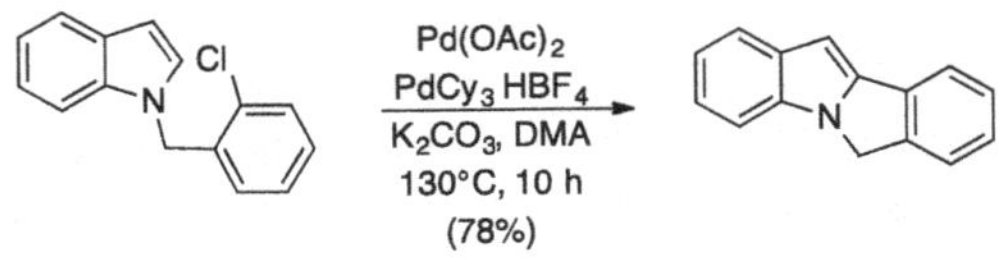

Scheme 1 Synthesis of 6*H*-isoindolo[2,1-*a*] indole by intramolecular direct arylation with an aryl chloride

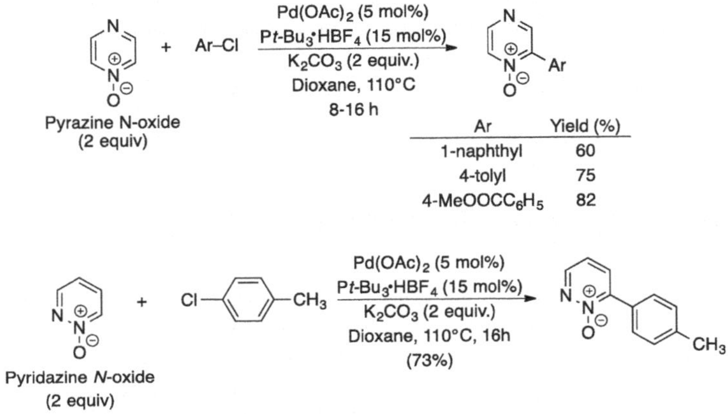

Scheme 2 Synthesis of 2-aryl-substituted diazine *N*-oxides

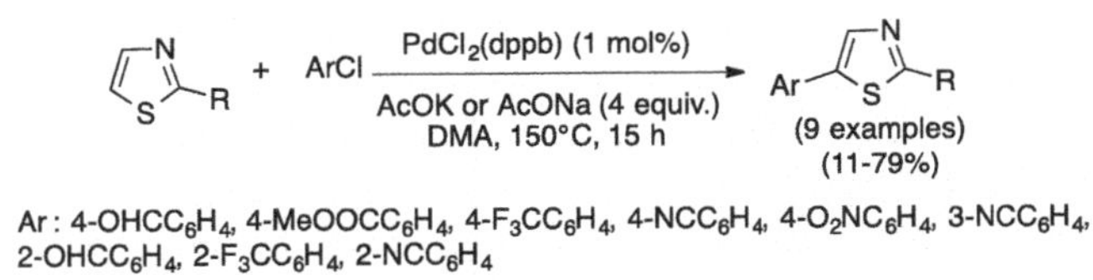

Scheme 3 Direct C5 arylation of 2-substituted thiazoles with aryl chlorides

the precatalyst, "PdCl" resulted better than $Pd(OAc)_2$, and dppb was superior as ligand in respect to dppe, dppf, and the common monodentate ligand PPh_3.

Due to the rapid attainment of high temperatures generally required for direct arylations, microwave irradiation represents an ideal strategy to shorten reaction time and, as a consequence, to increase chemical yields [17]. A microwave-assisted palladium-catalyzed direct arylation of 1,4-disubstituted 1,2,3-triazoles with aryl chlorides was developed by Oshima and co-workers. In details, the treatment of 1,4-disubstituted 1,2,3-triazoles with aryl chlorides in the presence of 1.2 equiv. of K_2CO_3, 5 mol% $Pd(OAc)_2$, 10 mol% Cy_3P, in a mixture of toluene and DMF at 250°C for 15 min in a microwave reactor led to 5-arylated triazoles in good to high yields (Scheme 4) [18].

Ester and hydroxy groups were found to be compatible, among a variety of functional groups. It should be noted, however, that despite the high temperatures easily attained in a microwave apparatus, this approach was not generally applied to

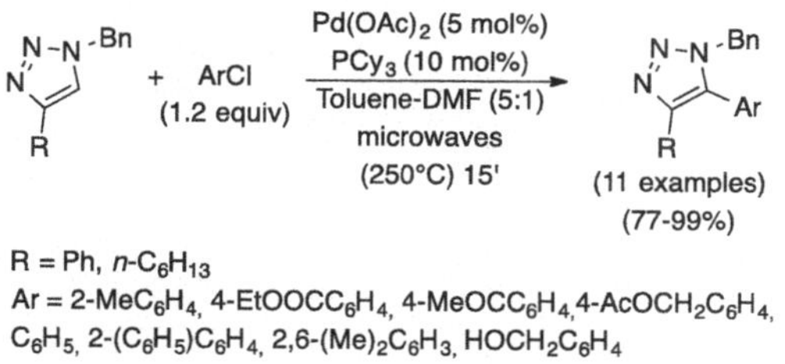

Scheme 4 Microwave-assisted synthesis of 5-arylated 1,2,4-triazoles from aryl chlorides

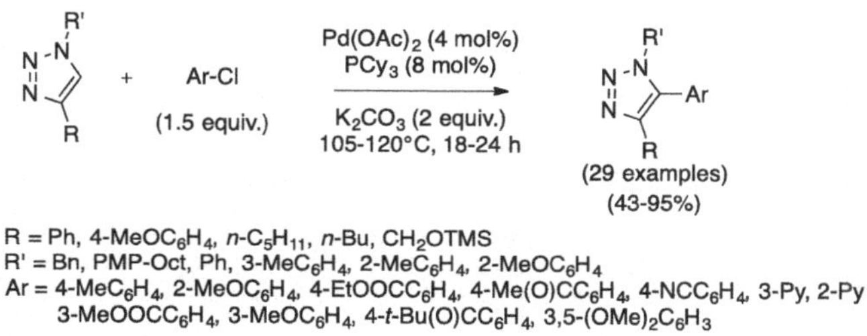

Scheme 5 Synthesis of fully substituted 1,2,3-triazoles through direct arylation under conventional heating

direct arylations of heteroarenes, probably due to scale-up issues and to the elevated costs of microwave reactors, and traditional oil baths are still widely used. As an example of this last statement, one year later, a new generally applicable, palladium-catalyzed direct arylation of 1,2,3-triazoles with aryl chlorides was accomplished through conventional heating at 105–120°C (Scheme 5) [19]. Both *N*-aryl- and *N*-alkyl-substituted 1,4-disubstituted 1,2,3-triazoles were converted to the fully substituted heteroarenes. The catalyst system displayed an elevated tolerance toward important functional groups such as ester and carbonyl groups.

In 2007, Chiong and Daugulis developed what may be considered the first general method for the direct arylation of heteroarenes with aryl chlorides [20]. This procedure made use of a combination of $Pd(OAc)_2$ and $BuAd_2P$ as palladium ligand, K_3PO_4 as the base, in NMP (Scheme 6).

Both electron-rich and electron-poor aryl chlorides can be used but, as expected, electron-poor chlorides are more reactive due to the easier oxidative addition with Pd(0) active catalyst. Some steric hindrance is tolerated in both the coupling partners. As regards the mechanism, the authors suggested that a classical electrophilic aromatic substitution could be the most plausible mechanistic pathway for this specific reaction.

The use of a particular ferrocenyl triphosphine in the direct arylation of heteroarenes with aryl chlorides that allowed low palladium loadings was described by Doucet and co-workers [21]. The couplings were carried out in DMA at 150°C for 16 h, employing 2.0 equiv. of AcOK as the base (Scheme 7). Notably, the authors found that carrying out the reactions under Jeffery's conditions, i.e., in the

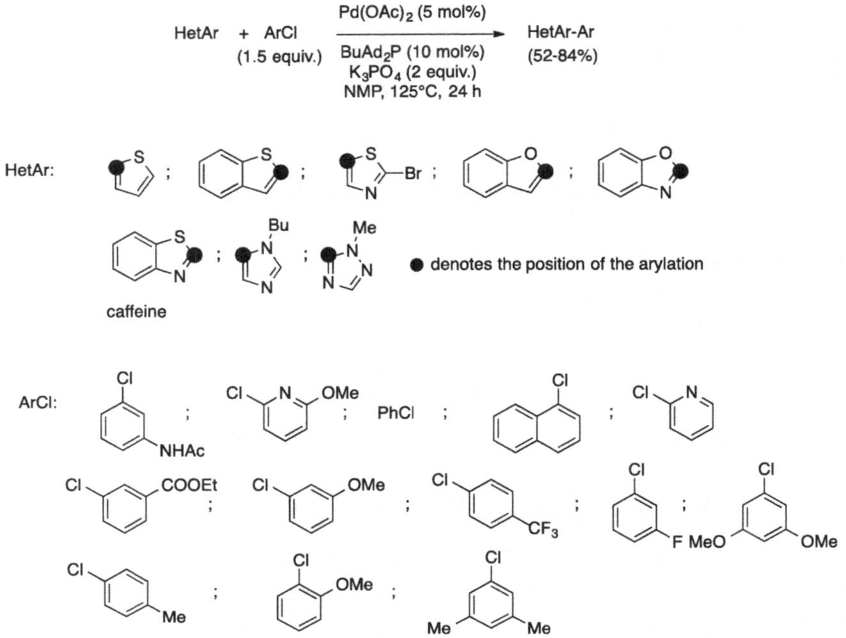

Scheme 6 General method for the Pd-catalyzed direct arylation of heteroarenes with aryl chlorides

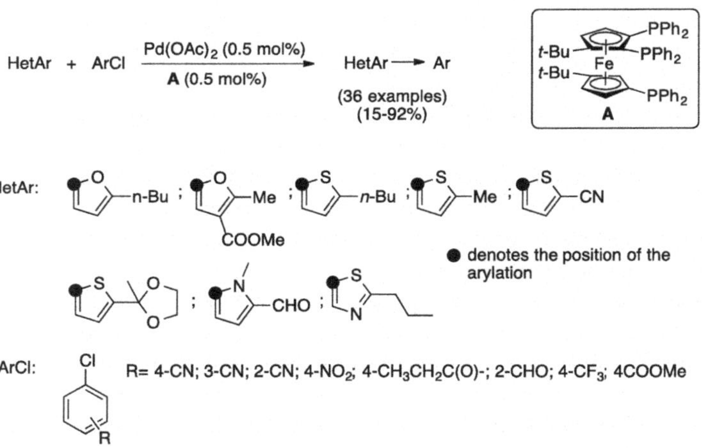

Scheme 7 Low-palladium-loading procedure for the direct arylation of azoles with aryl chlorides

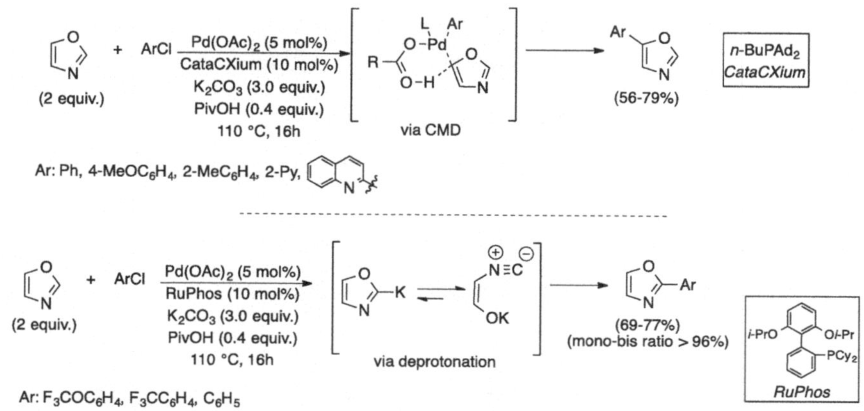

Scheme 8 Regioselective C5 or C2 arylation of oxazoles with aryl chlorides

presence of the ammonium salt $n\mathrm{Bu_4NBr}$, resulted beneficial for the coupling and that ligand-free conditions were ineffective.

Electron-rich, electron-poor, and polysubstituted furans, thiophenes, pyrroles, and thiazoles were arylated by using catalyst loadings of 0.5 mol%. This protocol tolerates important and useful functionalities such as formyl, nitro, cyano, keto, and ester groups, in *ortho*, *meta*, and *para* positions.

In 2010 chemists at Merck developed general conditions for the highly regioselective direct C2 or C5 arylation of oxazoles with aryl chlorides [22]. This method represented the first general method for the C5 arylation of oxazole, which is the relevant structural core in many natural products and pharmaceuticals. The regioselectivity toward C5 or C2 arylation was obtained through a careful selection of palladium ligands and solvents. Two sets of experimental conditions allowed the reaction to follow preferentially two competitive distinct pathways: a CMD transition state which led to C5 arylation and a deprotonative ring-opening ring-closure sequence that gave the C2-arylated oxazoles (Scheme 8).

2-Methylthiophene, 2-(4-methoxyphenyl)thiophene, 3-*n*-hexylthiophene, and benzothiophene were successfully reacted with typical electron-rich and electron-poor aryl chlorides using LiO*t*Bu as the base (Scheme 9) [23]. The combination of a strong base with the basic P*t*$\mathrm{Bu_3}$ ligand allowed the coupling to proceed smoothly, giving rise to the required arylated heteroarenes in 58–92% isolated yields.

The electron-rich NHC-Pd(II) complex **B** was found to be an efficient precatalyst for the regioselective C5 direct arylation of imidazoles with aryl chlorides (Scheme 10) [24].

The authors found that microwave irradiation effectively promoted the coupling, obtaining good yields in only 2 h. When 1-methyl-1*H*-imidazole was used as the coupling partner, low amounts (less than 10%) of the corresponding 2-aryl or 2,5-diaryl-substituted imidazoles were also isolated. The authors, taking into account the higher reactivity of the more nucleophilic 1,2-dimethylimidazole in

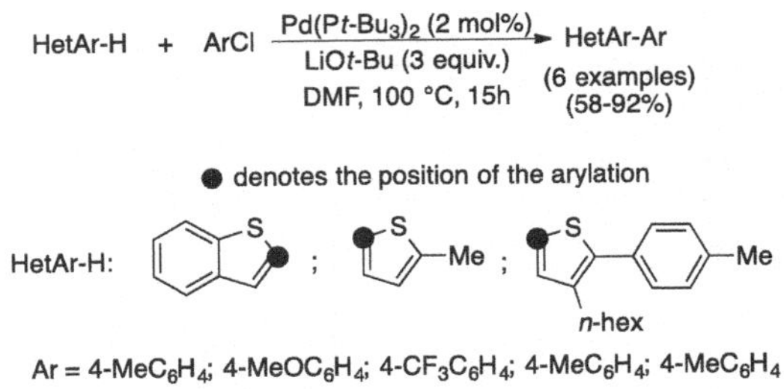

Scheme 9 Direct arylation of thiophene derivatives with Li*Ot*Bu

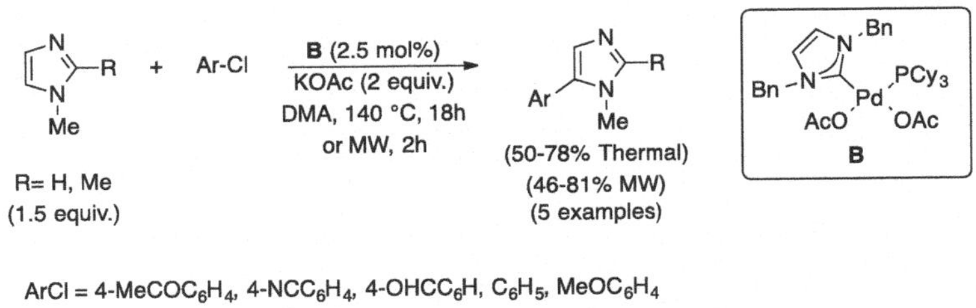

Scheme 10 NHC-Pd(II) promoted direct C5 arylation of imidazoles with aryl chlorides

comparison with 1-methylimidazole, suggested that the probable rate-determining C–H cleavage should occur via an electrophilic aromatic substitution pathway.

In 2011 Daugulis and co-workers found that *N*-substituted indoles, *N*-methylpyrroles, and furans were able to react with a variety of aryl chlorides employing $Pd(OAc)_2$ and 2-(dicyclohexylphosphino)biphenyl as ligand [25]. Arylated heteroaromatics were obtained in moderate to good yields, but the authors noticed that a careful optimization of base, ligand, and solvent was required for each heteroaromatic in order to achieve the best results. As regards indoles, both activated and deactivated aryl chlorides were reactive when used in an elevated molar excess (5.0 equiv.) (Scheme 11), and a variety of 2- or 3-substituted *N*-methylindoles were efficiently arylated at their 3 or 2 free position.

However, the authors evidenced the fact that unprotected indoles afforded mostly *N*-arylated analogues under the developed reaction conditions. Moreover, silicon-containing protective groups were removed, and indoles bearing electron-withdrawing groups on nitrogen were decomposed or gave low yields.

When *N*-methylpyrroles were reacted with aryl chlorides, it was found that a severe molar excess of azole (5.0 equiv.) was required to avoid diarylation, but electron-poor and electron-rich chlorides were found to be equally reactive. The optimized conditions for *N*-methylpyrroles required K_3PO_4 as the base and NMP as the solvent (Scheme 12).

Several equivalents of furan were necessary to avoid diarylation, while for monosubstituted furans an excess of aryl chloride was required (Scheme 13).

In an effort to prepare new organic fluorophores containing the indolizine core, in 2012 Yuo, Lau, and co-workers developed a straightforward method for the direct

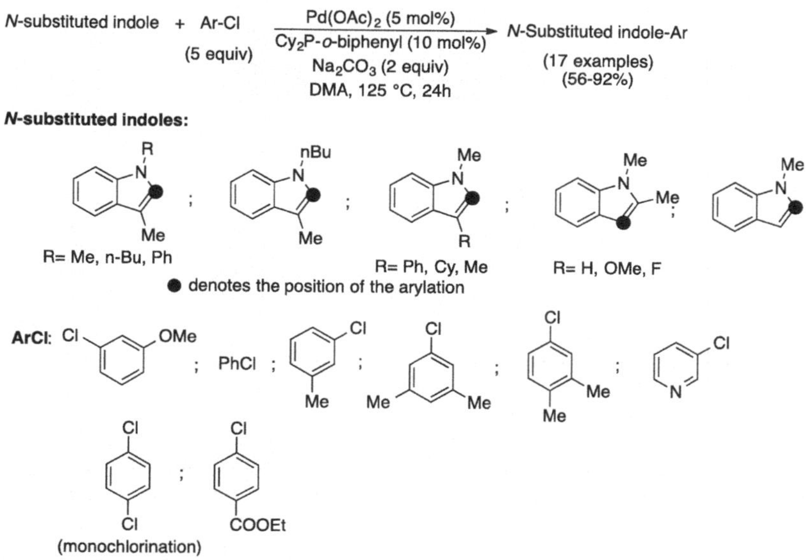

Scheme 11 Direct arylation of *N*-substituted indoles with aryl chlorides

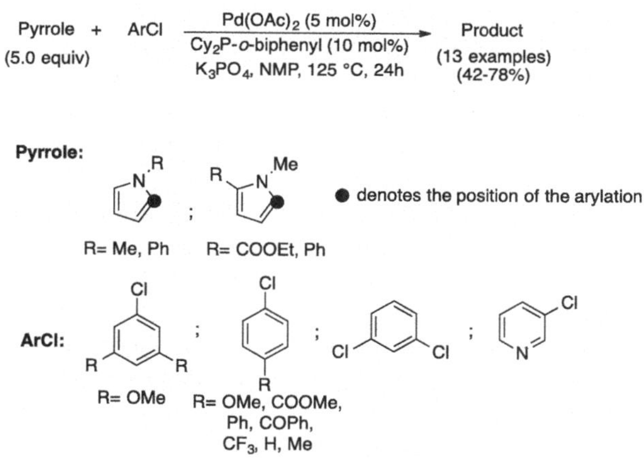

Scheme 12 Direct arylation of *N*-methylpyrroles with aryl chlorides

arylation not only of a variety of substituted indolizines but also amenable to the arylation of several electron-rich as well as electron-deficient heteroarenes. In fact, xanthines such as caffeine, theophylline and theobromine, purines, imidazoles, thiazoles, oxazoles, 1,2,3-triazoles, and *N*-heteroarene *N*-oxides were efficiently arylated using *p*-chlorotoluene as a model electrophilic partner (Scheme 14) [26].

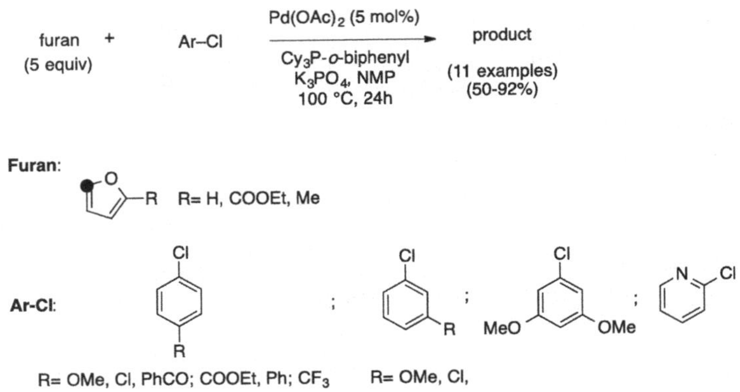

Scheme 13 Direct arylation of furans

Once again, the key of the success of this arylation procedure is represented by a careful optimization of the reaction system, including base, palladium ligand, and solvent.

In the same year, Gosh and Lee observed that all the previously published methods involving aryl chlorides suffered from low yields when hindered chlorides were used. Hence, they developed an interesting protocol for the direct arylation of heteroarenes with sterically hindered aryl chlorides where the Pd(II) complex **B** bearing a functionalized NHC and a PCy_3 had a key role [27]. In the presence of 2.0 mol% **B**, 30 mol% PivOH, 1.5 equiv. K_2CO_3 in DMA for 12 h at 110°C, 2-substituted furans, thiophenes, and *N*-methylpyrroles were successfully coupled with a variety of aryl chlorides, including several sterically encumbered ones (Scheme 15).

Interestingly, free aliphatic hydroxy, formyl, and keto functional groups were well tolerated by these catalytic systems. Under this protocol, a twofold C–H bond arylation was also successfully accomplished (Scheme 16).

Through competitive and kinetic isotopic experiments, the authors concluded that a concerted metalation-deprotonation (CMD) pathway rather than an electrophilic aromatic substitution or radical processes should be involved in the rate-determining C–H cleavage step.

Finally, in 2014, Sheo and co-workers were able to achieve a clean direct arylation of oxazoles and benzoxazoles with aryl chlorides using a well-defined palladium(II) complex **C** bearing an NHC and an unusual *N*-methylimidazole as ancillary ligand [28]. This procedure allowed the preparation of the corresponding 2-arylated (benzo)oxazoles in modest to good yields (Scheme 17).

The optimal reaction conditions were selected through the usual careful screening of bases and solvents, using benzoxazole and chlorobenzene as model coupling partners. It is intriguing to observe the fact that the precatalyst bears an imidazole as ligand, which at least in principle may be reactive under the proposed reaction conditions at one of its C–H bonds. It may then be argued, from the reported results, that coordination of azoles to palladium may significantly influence the outcome of

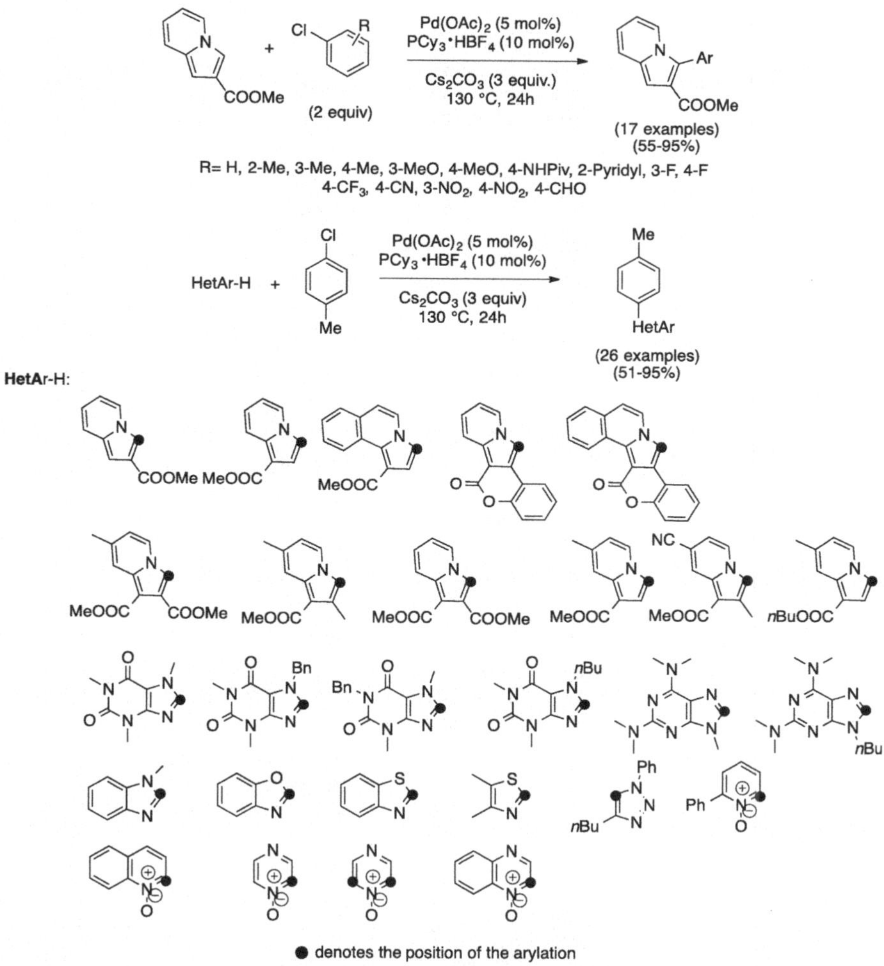

Scheme 14 Direct arylation of indolizines and other heteroarenes with aryl chlorides

direct arylations. The beneficial effect of using an excess of azoles may hence be related not only to an enhancement of the selectivity toward monoarylated products (as it is generally stated) but also to the formation of more active catalysts. On the other hand, the coordination of azoles to palladium catalysts may also lead to a deleterious effect, as it may be hypothesized in the case of direct arylations involving free-NH imidazole, which is extraordinary unreactive with aryl halide in the presence of palladium catalysts and bases (while it may be arylated at C2 under ligandless and base-free conditions using a mixed Pd-Cu co-catalysis [29]).

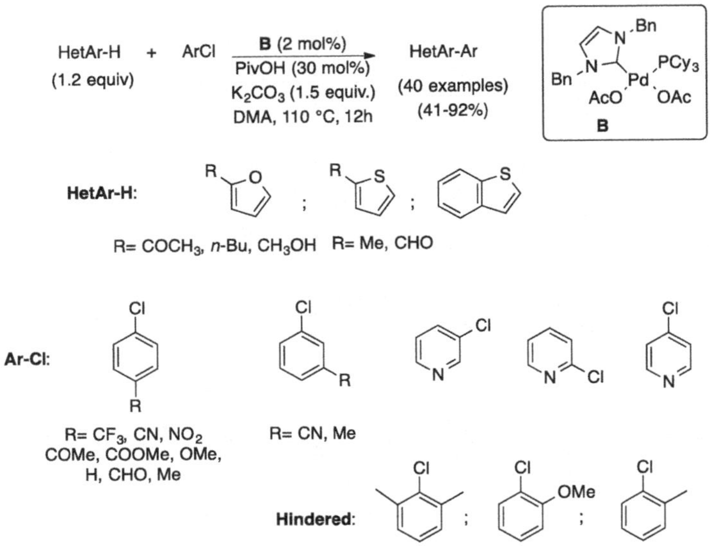

Scheme 15 Direct arylation of heteroarenes with sterically demanding aryl chlorides

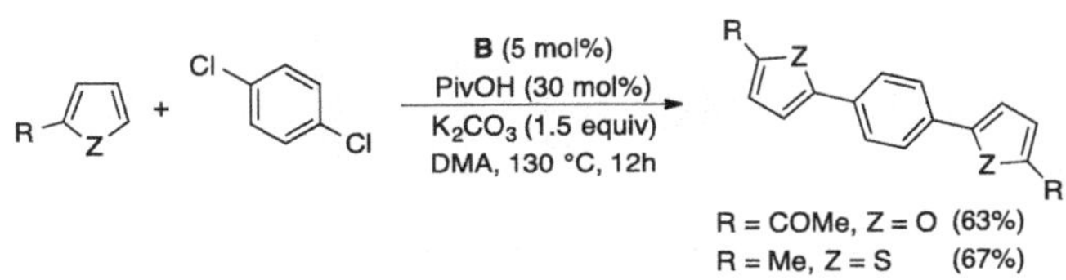

Scheme 16 Double direct arylation with aryl chlorides

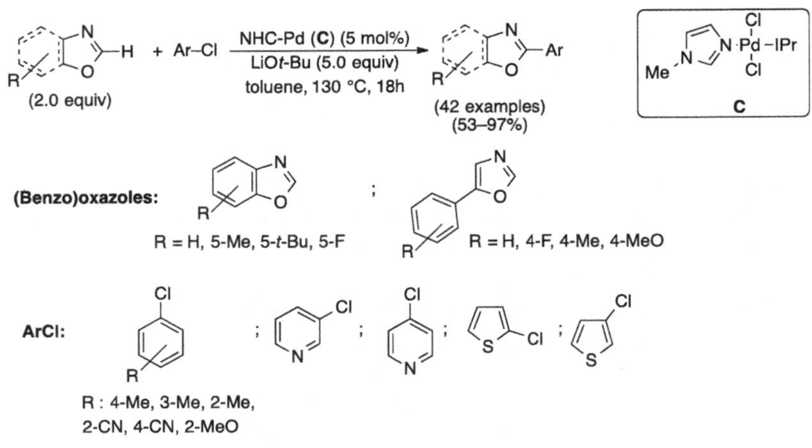

Scheme 17 Direct arylation of heteroarenes promoted by a mixed NHC-*N*-methylimidazole Pd (II) complex

3 Blocking/Activating Removable Functional Groups

3.1 Functional Group on Nitrogen

While *N*-substituted azoles are generally reactive in direct arylations, the use of *N*-unprotected derivatives has been found to be difficult when not impossible, probably due to palladium poisoning. Till today, for example, as stated in the previous chapter, free NH-imidazole is unable to react with aryl halides in the presence of a palladium catalyst alone. Hence, various protective groups have been used to allow the recovery, after deprotection, of arylated azoles bearing free nitrogen-hydrogen bonds. The role of these groups, as it will be discussed in the next paragraphs, frequently is not limited to a simple protection of N–H bonds of heteroarenes, but they may have a significant effect on the reactivity of the heteroarene and on the regioselectivity of the coupling.

The use of deprotonated azoles in direct arylations may represent the simplest "protection" of the NH moiety of azoles. The first use of an azole anion in direct arylation was described by Sadighi and co-workers in 2004 [30]. They reported that using a palladium precatalyst and sterically demanding phosphinobiphenyl ligands, pyrrolylzinc chlorides may be successfully coupled with a wide range of aryl bromides and chlorides, including highly sterically encumbered ones. The pyrrolylzinc anions were generated in situ from pyrrolylsodium (from pyrrole and NaH) and zinc dichloride (Scheme 18).

A combination of $Pd(OAc)_2$/JohnPhos and 1.6 equiv. of anion was generally used, while in cases of more sterically demanding substrates, the use of $Pd_2(dba)_3$ in combination with Cyclohexyl JohnPhos resulted more effective.

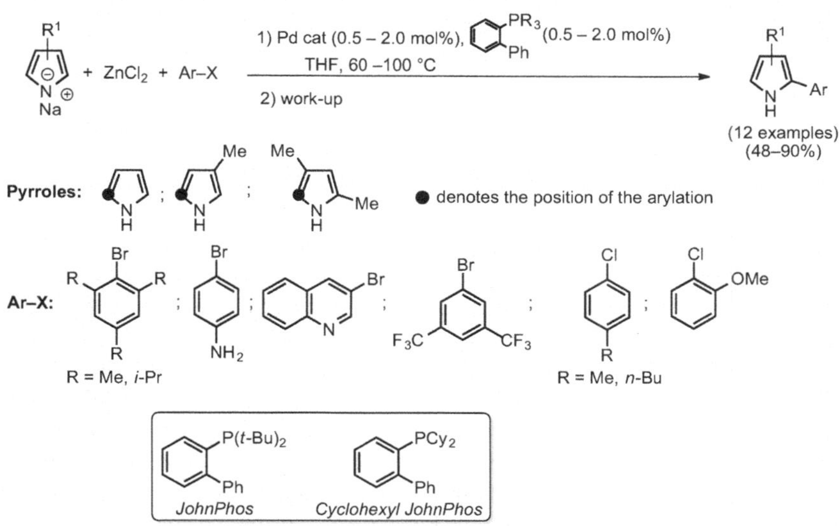

Scheme 18 Direct arylation of pyrrolylzinc chlorides with aryl halides

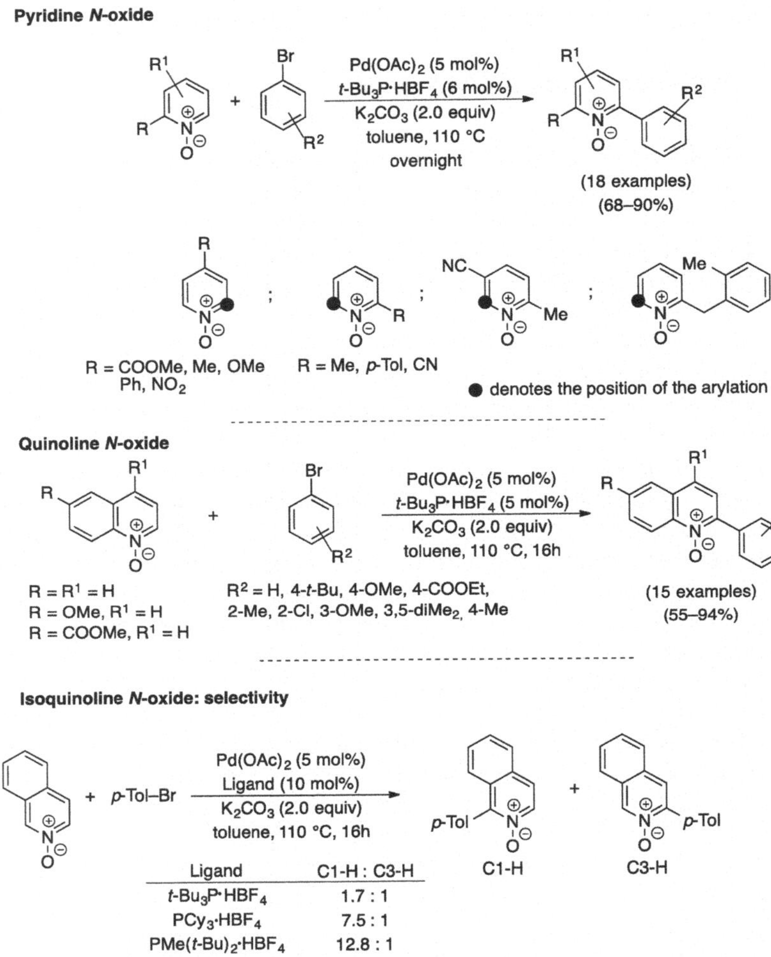

Scheme 19 Direct arylation of azine *N*-oxides with aryl halides

Fagnou and co-workers disclosed, in a series of papers dated back to 2005, the synthetic potential of azine and azole *N*-oxides in *ortho* direct arylations [31–33]. The results of these studies showed that for azine *N*-oxides, the regioisomeric distribution is governed by the nature of the azine and may be influenced by the choice of the palladium ligand (Scheme 19).

For what concerns azole *N*-oxides, preferential reaction occurred at C2 under very mild reaction conditions. Subsequent arylation involved C5-H, followed by C4-H. As an example of the importance of the optimized protocols for the sequential C2 → C5 → C4 arylations for thiazole *N*-oxides and C2 → C4 for imidazole *N*-oxides, the synthesis of a Tie2 tyrosine kinase inhibitor was successfully achieved (Scheme 20) [32].

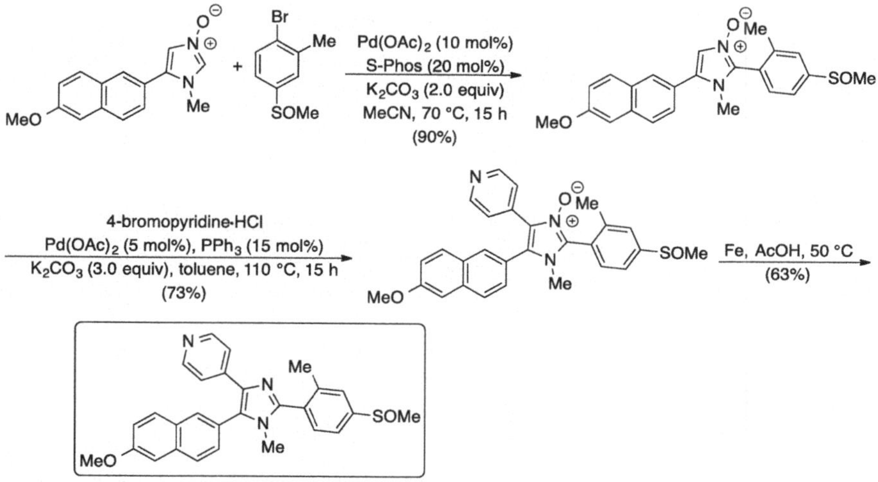

Scheme 20 Synthesis of Tie2 tyrosine kinase inhibitor by sequential direct arylations of 1,3-azole *N*-oxides with aryl halides

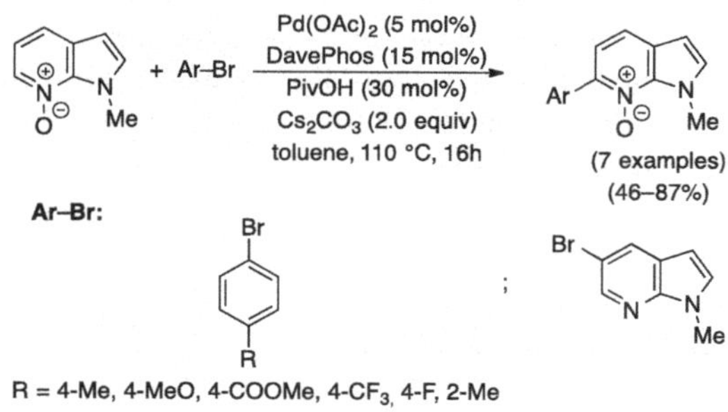

Scheme 21 Direct arylation of 6- and 7-azaindole *N*-oxides

Note that at the end of the synthetic sequence, the *N*-oxide function was easily removed by reduction with Fe in AcOH.

The regioselective direct arylation of 6- and 7-azaindole cores could be achieved using *N*-oxide activation strategy, which led to the functionalization of C–H bonds usually unreactive on the parent heteroarenes (Scheme 21) [33].

The synthetic importance of hetarene *N*-oxides, disclosed by Fagnou's group, has received increasing attention due to the possibility of activating C–H bonds usually unreactive and to the different methods available for the subsequent deoxygenation of the arylated *N*-oxides. Their reactivity has been widely explored, and at present not only the arylation is feasible on a wide range of azine and azole *N*-oxides, but also alkenylation, alkylation, acyloxylation, amination, and sulfonylation involving these interesting synthetic intermediates may be easily carried out [34].

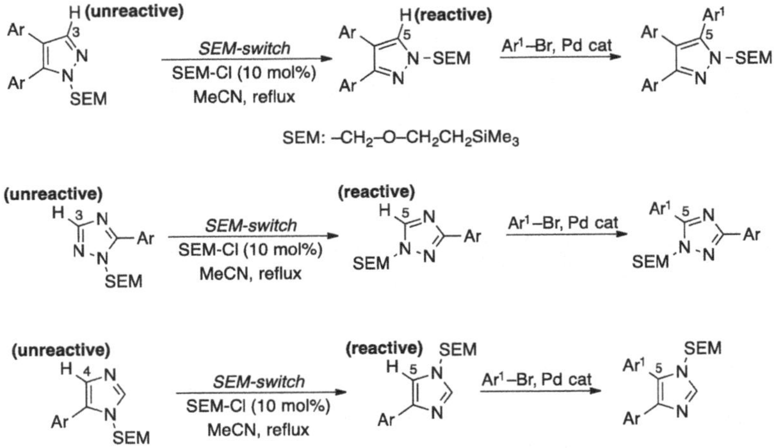

Scheme 22 Direct arylation of *N*-SEM-protected azoles through switch of the protective group

In a series of papers, Sames and co-workers described effective protocols for performing multiple arylations of pyrazoles [35], imidazoles [36], and 1,2,4-triazoles [37]. The key feature of these methods is the use of the transposition of the SEM protective group of nitrogen atoms of these azoles, allowing the functionalization of the otherwise unreactive C3–H bond of pyrazoles and 1,2,4-triazoles and of C4–H bond of imidazoles (Scheme 22).

3.2 Functional Group on Carbon

In contrast with direct arylations involving arenes, the presence of heteroatoms in heteroarenes often introduces significant differences in the reactivity of the various C–H bonds, allowing at least in principle the attainment of good regioselectivities. However, classical direct arylation protocols carried out on simple heteroarenes generally involve only one of the different C–H bonds present on the heteroaromatic core. Depending on the active mechanistic pathway, usually only the most nucleophilic carbon atom or the most acidic C–H bond is involved in the coupling with aryl halides, while the other positions remain unreactive. Hence, when these last positions have to be arylated due to synthetic needs, a common strategy to overcome this regioselectivity issue is represented by the employment of specific functional groups that are able to block the most reactive position, allowing the arylation to involve otherwise unreactive C–H bonds. Silylated and carboxyalkyl functionalities, or a chlorine atom, are among the most common activating/blocking groups due to their stability under the arylation reaction conditions and to the methods available for their removal.

Scheme 23 Direct arylation of a 2-TMS-protected thiophene

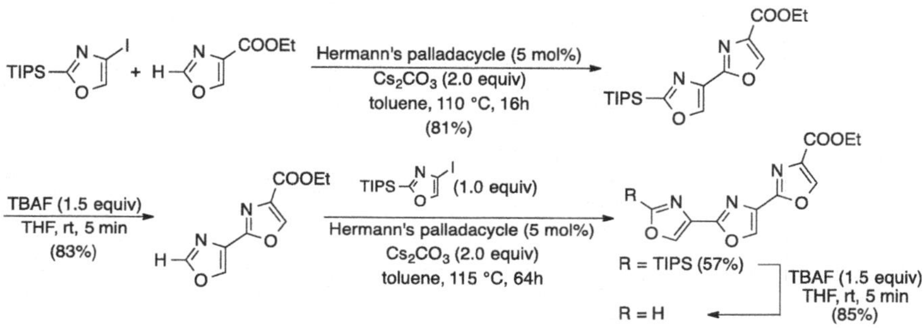

Scheme 24 Direct arylation involving a C2 TIPS-protected oxazole toward the synthesis of the trisoxazole core of ulapualide A

As the desilylation of heteroaromatics is relatively easy, the presence of silyl substituents might be useful to block reactive positions, enabling, at the same time, the functionalization of unreactive C–H bonds. This approach was demonstrated by the arylation of 2-(trimethylsilyl)-5-methylthiophene at its C4 position (Scheme 23) [38].

Searching for direct arylation routes to bioactive polyazoles, Greaney and co-workers synthesized C2 TIPS-protected 4-iodooxazole, which was then employed in a short synthesis of the trisoxazole core of the antifungal ulapualide A (Scheme 24) [39].

Alternatively to the two-step sequential direct arylation-desilylation depicted in Scheme 24, the deprotection of the bisoxazole may be performed directly on the crude reaction mixture obtained by direct C2 arylation of 4-carboxyethyloxazole, by diluting it with THF and reacting with TBAF (1.0 equiv.) for 5 min at room temperature. In this case, the combined yield over the two steps was 71%.

Chloride atom may be used to improve and/or to induce regioselectivity and may be easily removed at the end of the synthetic sequence. For example, while the direct arylation of 3-hexylthiophene with 4-bromonitrobenzene gave two regio-isomeric derivatives in a 1.3:1 molar ratio, blocking one of the two α-position on thiophene allowed a clean arylation at the remaining free C–H bond (Scheme 25) [40].

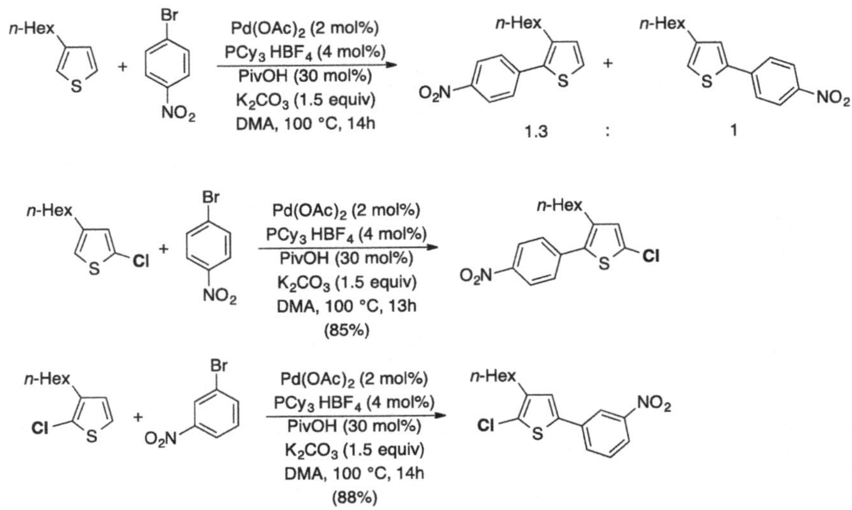

Scheme 25 Regioselective arylation of chloro-substituted thiophenes

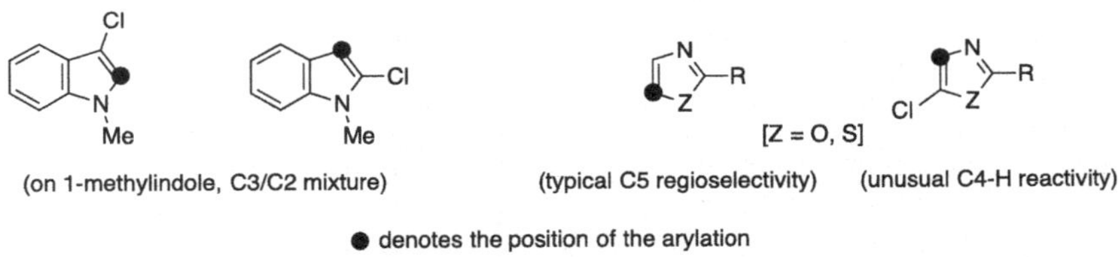

Fig. 1 Regioselectivity of direct arylations involving chlorine-substituted azoles

The deprotective reduction was easily carried out with 10 mol% Pd/C, H_2 (1 atm), Et_3N (1.2 equiv.) in MeOH at room temperature for 5 h, giving rise to the required dehalogenated products in yields higher than 85%.

In a similar way, other chlorinated azoles were efficiently arylated also at positions generally not reactive when the same reaction conditions were applied to the direct arylation of chlorine-free analogues (Fig. 1) [40].

A carbomethoxy group, which could be easily removed by base-promoted decarboxylation, was used by Doucet and co-workers to block position 2 on 3-aminothiophene, allowing for the first time the direct arylation of a heteroarene bearing a free NH_2 group (Scheme 26) [41].

Employing KOH along with KOAc, only the 2-arylated derivatives were observed through decarboxylation-arylation. The authors argued that in these cases the amino group may play a role in coordinating palladium species and directing the electrophilic palladation at C2. For example, the reaction of 3-amino-2-carboxymethyl-4-methylthiophene with 4-bromotoluene gave the corresponding 2-arylated thiophene in 64% isolated yield (Eq. 1).

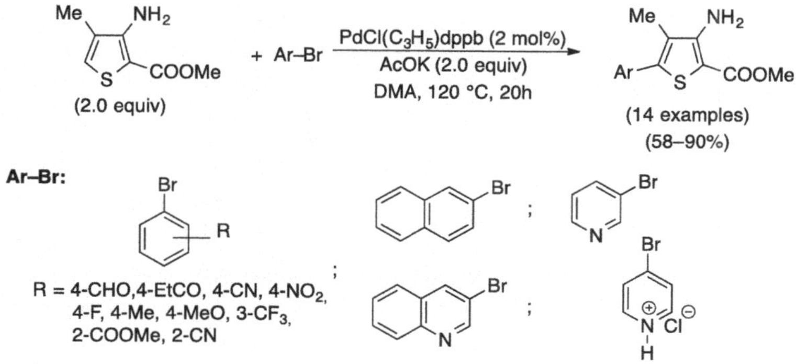

Scheme 26 Regioselective C5 arylation of 3-amino-2-carbomethoxythiophene

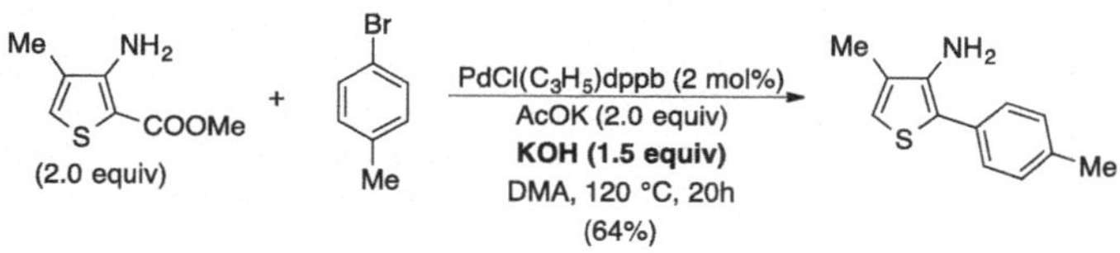

(1)

4 Heterogeneous Palladium Catalyst Systems

With the aim to develop more sustainable direct C–H arylations, the use of heterogeneous catalysts is particularly attractive, although the inertness of the C–H bond has rendered this promising implementation rather difficult [42]. However, the use of heterogeneous catalysts gives a number of advantages: (1) easy catalyst recovery, (2) metal recycling or reprocessing, and (3) reduced metal contamination of solvents and products. Some significant examples of the use of heterogeneous palladium systems are summarized below.

A simple heterogeneously palladium-catalyzed procedure for the selective C3 arylation of indoles was reported by Cusati and Djakovitch in 2008 [43]. Using only 1 mol% of $[Pd(NH_3)_4]^{2+}/NaY$ as catalyst and K_2CO_3 as the base, in dioxane at reflux, indoles substituted or not at their position 2 gave up to 92% conversion and up to 85% isolated yields of the expected 3-arylated indoles. Notably, no conversion was observed for 2-methyl and 2-phenyl indoles when 4-bromotoluene and 4-bromoanisole were used as coupling partners (Scheme 27).

The authors supposed the coexistence of two concurrent mechanisms: an electrophilic substitution (ES) vs. an S_NAr mechanism. The ES mechanism, which occurs through the coordination of an Ar-Pd-X complex to indole nucleus, is sensible to steric hindrance at position 2. The second mechanism should occur when activated (i.e., electron-poor) aryl bromides are employed, and mainly with 2-methylindole, which has been demonstrated to be more reactive than indole or 2-phenylindole in such a mechanism. The catalyst was obtained by ion exchange of a NaY zeolite using an aqueous solution of $[Pd(NH_3)_4]^{2+}/2Cl^-$.

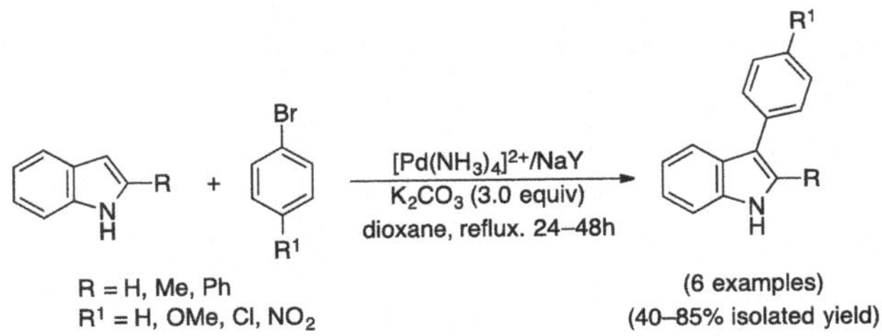

Scheme 27 Direct C3 arylation of indoles promoted by heterogeneous $[Pd(NH_3)_4]^{2+}$/NaY

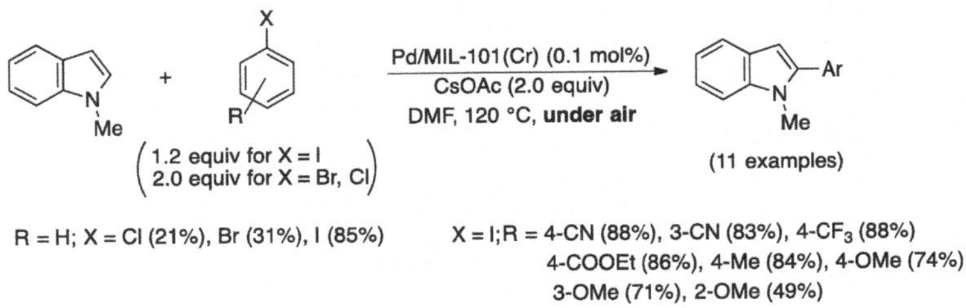

Scheme 28 Direct C2 arylation of indoles promoted by heterogeneous Pd NPs

An efficient heterogeneous catalyst system for the C2 direct arylation of indoles is represented by palladium nanoparticles (Pd NPs) encapsulated in the mesoporous MOF MIL-101(Cr) [44]. The direct arylation required only 0.1 mol% Pd, affording the required C2-arylated indoles in good yields (Scheme 28).

The heterogeneous catalyst was prepared by simple impregnation with an aqueous solution of $Pd(NO_3)_2$ of ML-101(Cr) $\{Cr_3(F,OH)(H_2O)_2O[(O_2C)–C_6H_4–(CO_2)]_3 \cdot nH_2O$ ($n = 25$), MIL: Matérial Institut Lavoisier}. The Pd leaching resulted very low (<0.4 ppm), and no Cr was released by the MOF. Two clear advantages of using this catalyst are the fact that the reactions may be carried in the open air and that the catalyst may be recovered by centrifugation. Moreover, the catalyst still exhibits high reactivity after five runs (a decrease of only 4% was observed).

Pearlman's catalyst ($Pd(OH)_2/C$) was found to catalyze direct arylation reactions of several azoles and 2-formylfuran with aryl bromides (Scheme 29) [45].

Authors proved that, in this case, a homogeneous catalyst is generated under the reaction conditions and that this species is responsible for the observed catalysis. This may account also for the high catalyst loading required by this method, in contrast with homogeneous catalyst systems ($\leq$5 mol%).

The effectiveness of the use of heterogeneous Pd/C in direct arylation of heteroarenes has been recently demonstrated by the regioselective polycondensation of thiophenes [46]. Linear π-conjugated alternating copolymers were obtained with high molecular weight in high yields (Scheme 30).

Palladium nanoparticles embedded in spherical polypyrrole globules have been successfully used by Hierso and co-workers for the direct arylation of 2-butylfuran

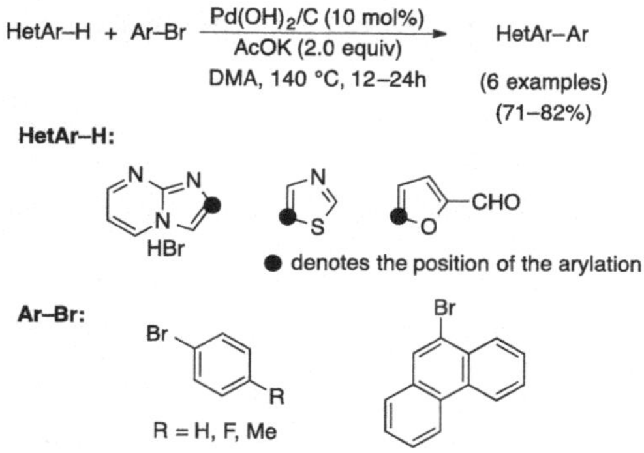

Scheme 29 Direct arylation of heteroarenes promoted by Pearlman's catalyst

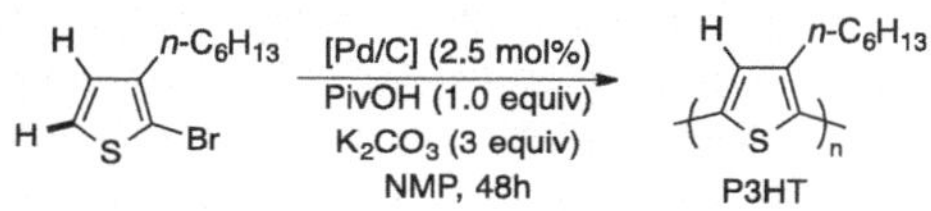

[Pd/C] cat	T (°C)	M_n (kDa)	Isolated yield (%)
10% Pd/C (wetted with 55% water)	120	16.3	91
2.5% Pd(OH)$_2$/C (wetted with 55% water)	100	18.9	99

Scheme 30 Polycondensation of thiophenes in the presence of heterogeneous Pd/C

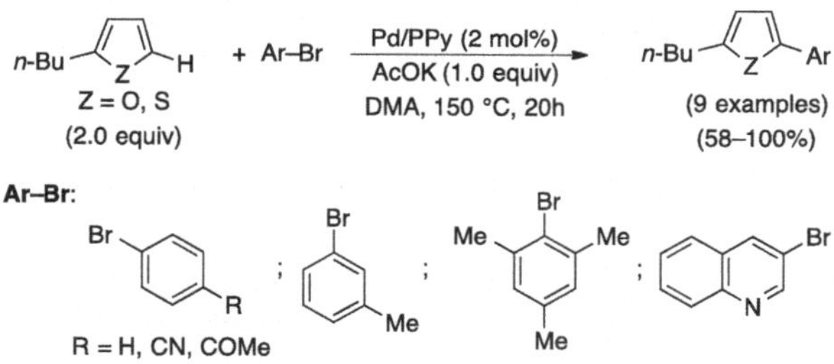

Scheme 31 Direct C5 arylation of 2-butylfuran and 2-butylthiophene promoted by Pd/PPy

or 2-butylthiophene with activated or deactivated electron-poor and electron-rich functionalized bromoarenes (Scheme 31) [47].

The catalyst was recovered by simple filtration, but recycling tests were successful only for few substrates. This was due to a growth of the particles size after the first cycle, which dramatically reduces the catalyst activity.

5 Mild Protocols for the Direct Arylation of Heteroarenes

One issue of direct arylation procedures is represented by the need for drastic conditions, elevated temperatures (from 110°C up to 160°C and higher when microwave irradiation is employed), and long reaction times (24 to 48 h and more are common).

To solve this problem, Larrosa and Lebrasseur proposed a new approach for the direct C2 arylation of indoles with aryl iodides at room temperature without phosphines or other ligands [48]. The key features of this protocol were the combined use of Ag_2O as the base, which enabled the formation of an electrophilic cationic palladium complex, and *ortho*-nitrobenzoic acid, which generates in situ a silver carboxylate particularly effective in promoting a CMD mechanistic pathway (Scheme 32).

All the reactions were performed at 25°C, except for the direct arylation involving free NH-indole, which required 38 h at 50°C to go to completion (68% isolated yield). The authors said that both a P(0)/Pd(II) and a Pd(II)/Pd(IV) catalytic cycle are conceivable even if they propended for the first mechanistic pathway. An important specificity for solvent was also found: DMF, a highly coordinating solvent, resulted superior (95% yield) to CH_2Cl_2 (27%), MeCN (14%), AcOH (50%), and DMA (50%).

A similar approach, i.e., the use of a Ag(I) base to promote the formation of an active cationic Pd(II) complex, allowed the direct arylation of some halothiophenes at only 60°C using an unusual biphasic AcOEt/H_2O (2.5:1) solvent system (Scheme 33) [49].

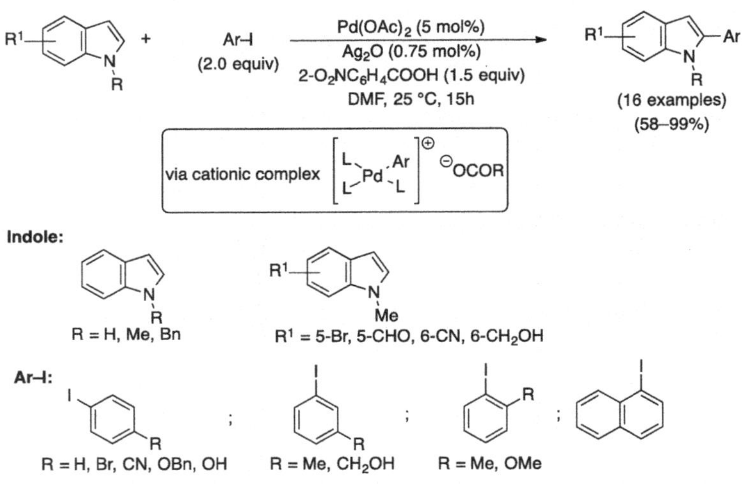

Scheme 32 Room temperature direct C2 arylation of indoles

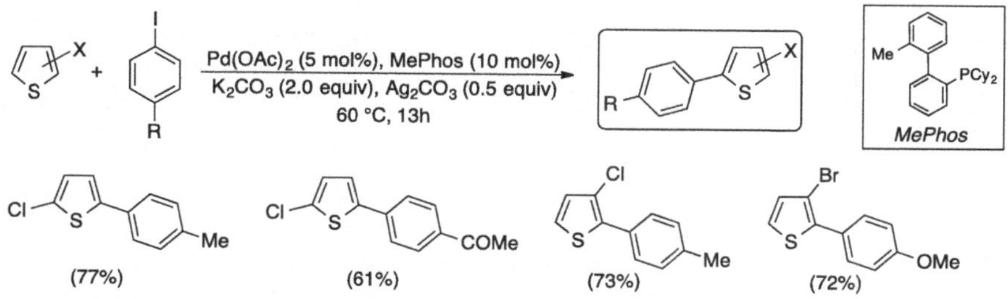

Scheme 33 Mild direct arylation of halothiophenes

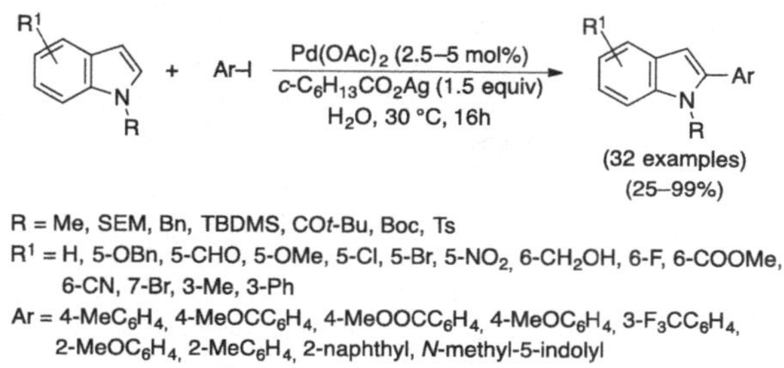

Scheme 34 Mild direct arylation of indoles “on water”

It is interesting to note that 3-halothiophenes led regioselectively to the 2-arylated derivatives and that C–Br bonds of 3-bromothiophene resulted unreactive under this mild reaction conditions.

The use of Ag(I) salts to promote direct arylations was proved to be successful also employing an “on water” approach. Several 2-arylindoles were obtained in high yields by reacting the corresponding indoles with aryl iodides in neat water at 30°C for 16 h in the presence of *c*-$C_6H_{11}CO_2Ag$ (1.5 equiv.) (Scheme 34) [50]. Notably, the reaction did not work when free NH-indole was employed.

In 2013, a mild, general, and convenient phosphine-free direct arylation of 1-methylpyrazole, oxazole, and thiazole promoted by tetrabutylammonium acetate (Bu_4NOAc) was developed by Bellina and co-workers [51]. Electron-poor and electron-rich aryl bromides were found to react at only 70°C in DMA, allowing the isolation of the corresponding 5-substituted azoles in good yields (Scheme 35).

This protocol, paired with the regioselective direct C2 arylation of azoles mediated by palladium and copper, was applied to the synthesis of the natural products balsoxin and texalin through a one-pot sequential C5–C2 double arylation [51] and, more recently, to the preparation of several imidazole analogues of resveratrol able to reduce the growth of human tumor cell lines [52].

Scheme 35 Mild direct C5 arylation of azoles promoted by Bu_4NOAc

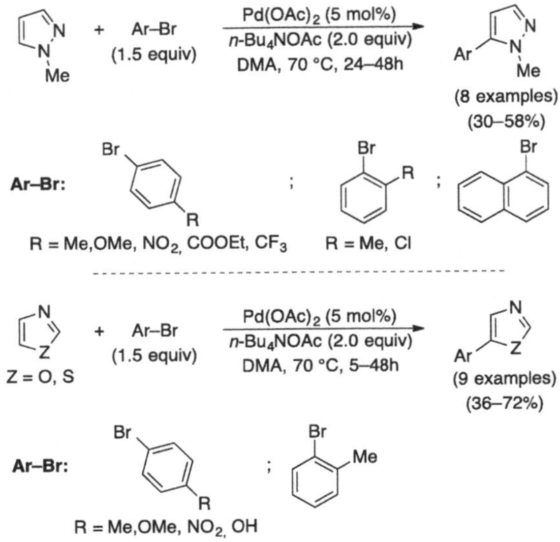

6 Conclusions

The methods of palladium-catalyzed direct arylation of heteroarenes with (hetero) aryl halides displayed an exponential growth in less than 20 years. Several significant progresses have been done since 1998, when Miura and co-workers published their seminal paper in which they describe for the first time the role of the base and of the catalyst in the activation of heteroaryl C–H bonds [53]. As summarized in this account, methods have been developed to react cheap aryl chlorides instead of the more expensive bromides or iodides, heterogeneous palladium catalysts have been employed in order to reduce metal waste and to allow recycling, procedures for lowering the reaction temperature have been set up, and removable functional group have been selected to increase the efficiency and the regioselectivity of the direct arylations. Many efforts should be done, however, in further improving the regioselectivity of the process in order to reduce or eliminate the need for chromatographic purifications in view of processes scale-up and industrial applications and also to increase the catalyst efficiency with the aim of widening the substrate scope.

References

1. Horton DA, Bourne GT, Smythe ML (2003) Chem Rev 103:893. doi:10.1021/cr020033s
2. Jiang W, Li Y, Wang Z (2013) Chem Soc Rev 42:6113. doi:10.1039/c3cs60108k
3. Mercier LG, Leclerc M (2013) Acc Chem Res 46:1597. doi:10.1021/ar3003305
4. Hassan J, Sévignon M, Gozzi C, Schulz E, Lemaire M (2002) Chem Rev 102:1359. doi:10.1021/cr000664r
5. Torborg C, Beller M (2009) Adv Synth Catal 351:3027. doi:10.1002/adsc.200900587

6. Corbet JP, Mignani G (2006) Chem Rev 106:2651. doi:10.1021/cr0505268
7. Hussain I, Singh T (2014) Adv Synth Catal 356:1661. doi:10.1002/adsc.201400178
8. Rossi R, Bellina F, Lessi M, Manzini C (2014) Adv Synth Catal 356:17. doi:10.1002/adsc.201300922
9. Yamaguchi J, Yamaguchi AD, Itami K (2012) Angew Chem Int Ed 51:8960. doi:10.1002/anie.201201666
10. Nakamura N, Tajima Y, Sakai K (1982) Heterocycles 17:235
11. Littke AF, Fu GC (2002) Angew Chem Int Ed 41:4176. doi:10.1002/1521-3773(20021115)41:22<4176::AID-ANIE4176>3.0.CO;2-U
12. Campeau LC, Thansandote P, Fagnou K (2005) Org Lett 7:1857. doi:10.1021/ol050501v
13. Leblanc M, Fagnou K (2005) Org Lett 7:2849. doi:10.1021/ol0505959
14. Campeau LC, Parisien M, Jean A, Fagnou K (2006) J Am Chem Soc 128:581. doi:10.1021/ja055819x
15. Leclerc JP, Fagnou K (2006) Angew Chem Int Ed 45:7781. doi:10.1002/anie.200602773
16. Gottumukkala AL, Doucet H (2007) Eur J Inorg Chem 2007:3629. doi:10.1002/ejic.200700420
17. Sharma A, Vacchani D, Van der Eycken E (2013) Chem Eur J 19:1158. doi:10.1002/chem.201201868
18. Iwasaki M, Yorimitsu H, Oshima K (2007) Chem Asian J 2:1430. doi:10.1002/asia.200700206
19. Ackermann L, Vicente R, Born R (2008) Adv Synth Catal 350:741. doi:10.1002/adsc.200800016
20. Chiong HA, Daugulis O (2007) Org Lett 9:1449. doi:10.1021/ol0702324
21. Roy D, Mom S, Beauperin M, Doucet H, Hierso JC (2010) Angew Chem Int Ed 49(37):6650–6654. doi:10.1002/anie.201002987
22. Strotman NA, Chobanian HR, Guo Y, He J, Wilson JE (2010) Org Lett 12:3578. doi:10.1021/ol1011778
23. Tamba S, Okubo Y, Tanaka S, Monguchi D, Mori A (2010) J Org Chem 75:6998. doi:10.1021/jo101433g
24. Kumar PV, Lin W-S, Shen J-S, Nandi D, Lee HM (2011) Organometallics 30:5160. doi:10.1021/om200490k
25. Nadres ET, Lazareva A, Daugulis O (2011) J Org Chem 76:471. doi:10.1021/jo1018969
26. Liu B, Wang Z, Wu N, Li M, You J, Lan J (2012) Chem Eur J 18:1599. doi:10.1002/chem.201103329
27. Ghosh D, Lee HM (2012) Org Lett 14:5534. doi:10.1021/ol302635e
28. Shen XB, Zhang Y, Chen WX, Xiao ZK, Hu TT, Shao LX (2014) Org Lett 16:1984. doi:10.1021/ol500531m
29. Bellina F, Cauteruccio S, Rossi R (2006) Eur J Org Chem 2006:1379–1382. doi:10.1002/ejoc.200500957
30. Rieth RD, Mankad NP, Calimano E, Sadighi JP (2004) Org Lett 6:3981. doi:10.1021/ol048367m
31. Campeau LC, Bertrand-Laperle M, Leclerc JP, Villemure E, Gorelsky S, Fagnou K (2008) J Am Chem Soc 130:3276. doi:10.1021/ja7107068
32. Campeau LC, Stuart DR, Leclerc JP, Bertrand-Laperle M, Villemure E, Sun HY, Lasserre S, Guimond N, Lecavallier M, Fagnou K (2009) J Am Chem Soc 131:3291. doi:10.1021/ja808332k
33. Huestis MP, Fagnou K (2009) Org Lett 11:1357. doi:10.1021/ol900150u
34. Yan G, Borah AJ, Yang M (2014) Adv Synth Catal 356:2375. doi:10.1002/adsc.201400203
35. Goikhman R, Jacques TL, Sames D (2009) J Am Chem Soc 131:3042. doi:10.1021/ja8096114
36. Joo JM, Toure BB, Sames D (2010) J Org Chem 75:4911. doi:10.1021/jo100727j
37. Joo JM, Guo P, Sames D (2013) J Org Chem 78:738. doi:10.1021/jo3021677
38. Chen L, Roger J, Bruneau C, Dixneuf PH, Doucet H (2011) Chem Commun (Camb) 47:1872. doi:10.1039/c0cc04302h
39. Flegeau EF, Popkin ME, Greaney MF (2008) Org Lett 10:2717. doi:10.1021/ol800869g

40. Liegault B, Petrov I, Gorelsky SI, Fagnou K (2010) J Org Chem 75:1047. doi:10.1021/jo902515z
41. Derridj F, Roger J, Djebbar S, Doucet H (2010) Org Lett 12:4320. doi:10.1021/ol101758w
42. Djakovitch L, Felpin F-X (2014) ChemCatChem 6:2175. doi:10.1002/cctc.201402288
43. Cusati G, Djakovitch L (2008) Tetrahedron Lett 49:2499. doi:10.1016/j.tetlet.2008.02.130
44. Huang Y, Lin Z, Cao R (2011) Chem Eur J 17:12706. doi:10.1002/chem.201101705
45. Parisien M, Valette D, Fagnou K (2005) J Org Chem 70:7578. doi:10.1021/jo051039v
46. Hayashi S, Kojima Y, Koizumi T (2015) Polym Chem 6:881. doi:10.1039/c4py01426j
47. Zinovyeva VA, Vorotyntsev MA, Bezverkhyy I, Chaumont D, Hierso J-C (2011) Adv Funct Mater 21:1064. doi:10.1002/adfm.201001912
48. Lebrasseur N, Larrosa I (2008) J Am Chem Soc 130:2926. doi:10.1021/ja710731a
49. Rene O, Fagnou K (2010) Org Lett 12:2116. doi:10.1021/ol1006136
50. Islam S, Larrosa I (2013) Chem Eur J 19:15093. doi:10.1002/chem.201302838
51. Bellina F, Lessi M, Manzini C (2013) Eur J Org Chem 2013:5621. doi:10.1002/ejoc.201300704
52. Bellina F, Guazzelli N, Lessi M, Manzini C (2015) Tetrahedron 71:2298. doi:10.1016/j.tet.2015.02.024
53. Pivsa-Art S, Satoh T, Kawamura Y, Miura M, Nomura M (1998) Bull Chem Soc Jpn 71:467. doi:10.1246/bcsj.71.467

Top Organomet Chem (2016) 55: 103–118
DOI: 10.1007/3418_2015_137

Published online: 23 July 2015

New Arylating Agents in Pd-Catalyzed C–H Bond Functionalization of 5-Membered Ring Heteroarenes

Jean-François Soulé and Henri Doucet

Abstract In recent years, the palladium-catalyzed direct arylation of heteroaromatics via C–H bond activation has become a popular method for generating carbon–carbon bonds. In most cases, aryl halides were employed as the coupling partners. However, since a few years, several new coupling partners have emerged as useful alternatives for such couplings. This chapter focuses on the recent advances on the discovery of new arylating agents in the palladium-catalyzed arylation of 5-membered ring heteroaromatics.

Keywords Aryl silanes · Arylboronic acids · Arylsulfonyls · Benzoic acids · C–C bond formation · Heteroaromatic compounds · Palladium

Contents

J.-F. Soulé (✉) and H. Doucet (✉)
Institut des Sciences Chimiques de Rennes, UMR 6226, CNRS-Université de Rennes 1, “Organométalliques, matériaux et Catalyse”, Campus de Beaulieu, 35042 Rennes, France
e-mail: jean-francois.soule@univ-rennes1.fr; henri.doucet@univ-rennes1.fr

1 Introduction

Since the pioneering work on the palladium-catalyzed arylation of isoxazoles reported in 1982 by Nakamura, Tajima, and Sakai [1] and on the arylation of several other heterocycles described by Ohta et al. in 1985–1992 [2, 3], the direct arylation of heteroaromatics via a C–H bond activation has been demonstrated to be a powerful method for a simpler access to the corresponding aryl–heteroaryl derivatives. In most cases, aryl halides have been employed as the coupling partners [4–8]. Hypervalent iodine derivatives [9], aryltriflates [10–12], or aryltosylates [13] have also been employed successfully. Arenediazonium salts have also been used as aryl sources through palladium-catalyzed non-radical direct C–H arylations of heterocycles [14, 15]. However, the couplings with such arylating agents still suffer of limitations in terms of regioselectivity for some of these couplings and also to the limited functional group tolerance. For example, C4-arylated thiazoles cannot be easily obtained using aryl halides as the coupling partners [16]. In addition, the reactions with benzofurans afford mixtures of C2 and C3 arylated products [17]. Moreover, the presence of several halide functional groups on arenes is generally not well tolerated. Since a few years, several new alternative coupling partners have been successfully employed for such couplings such as $ArB(OH)_2$, $ArBF_3K$, $ArCO_2H$, $ArSiR_3$, $ArSO_2Cl$, $ArSO_2Na$, or even $ArSO_2NHNH_2$ using palladium catalysts to afford a variety of arylated heteroarenes which in several cases cannot be obtained using aryl halides as the coupling partners. In this chapter, we will outline the recent advances in the Pd-catalyzed direct arylation of 5-membered ring heteroarenes using these new alternative arylating agents.

2 Arylboronic Acids or Aryl Trifluoroborates

Shi and co-workers demonstrated in 2008 that arylboronic acids allow the direct construction of biaryl C–C bonds by a palladium-catalyzed cross coupling via C–H bond activation of heteroarenes (Scheme 1) [18]. The reaction was performed in the presence of 1 equiv. of $Cu(OAc)_2$ and O_2 was used in the transformation as the final oxidant. A wide range of heteroarenes was tolerated such as benzothiophenes, benzofurans, indoles, or pyrroles. Both electron-withdrawing and electron-donating substituents on the arylboronic acid were tolerated. Palladium nanoparticles

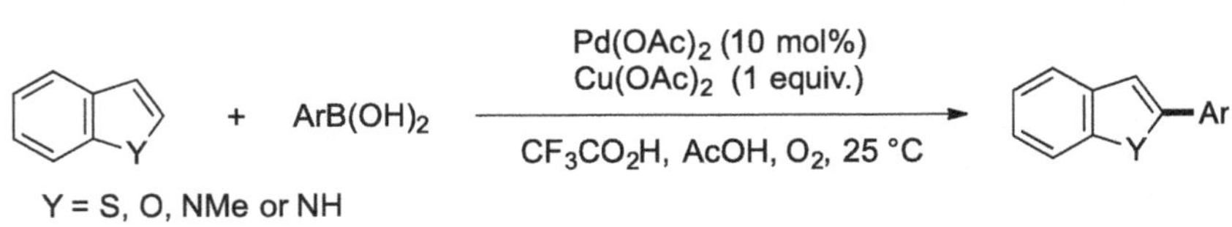

Scheme 1 C2 arylation of heteroarenes using arylboronic acids

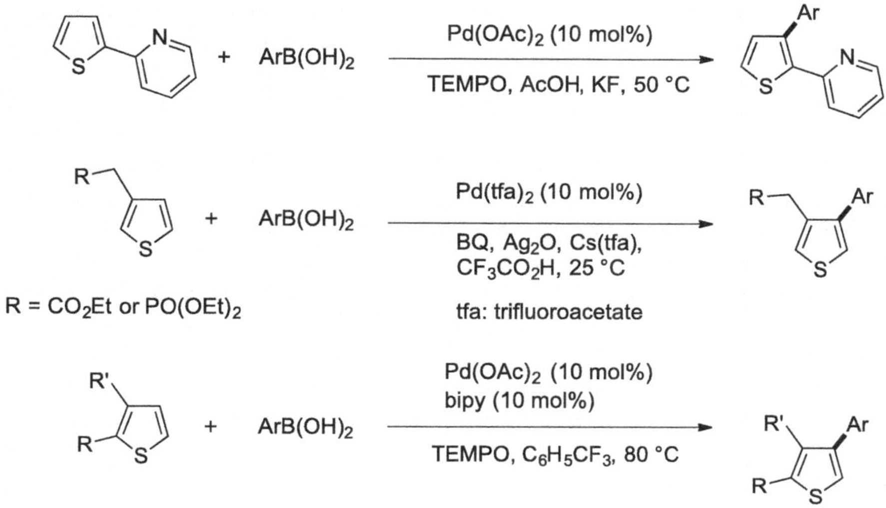

Scheme 2 C3 arylation of heteroarenes using arylboronic acids

encapsulated with O_2 or TEMPO as an external oxidant and without addition of copper were also found to promote the C2 arylation of indoles [19].

The same year, Zhang et al. reported the palladium-catalyzed direct arylation of indoles with potassium aryl trifluoroborates using copper as a cocatalyst and air as the oxidant [20]. Indolizines were also arylated using potassium aryl trifluoroborates under slightly different conditions as AgOAc was used as the oxidant [21].

The C2 arylation of indoles with arylboronic acids also proceed using $Pd(OAc)_2$ as catalyst with 2,2,6,6-tetramethylpiperidine-*N*-oxyl radical (TEMPO) and KF in propionic acid at room temperature [22]. In 2010, a wide range of azoles such as benzoxazoles, benzothiazoles, benzimidazoles, or caffeine were arylated using arylboronic acids in the presence of $Pd(OAc)_2$ as catalyst associated to $Cu(OAc)_2$, CuCl, and BQ [23]. A slightly simpler procedure was reported in 2011 for the arylation of benzoxazoles and benzothiazoles [24]. In 2013, Kuang et al. reported the $Pd(OAc)_2$-catalyzed synthesis of 3,4-diarylsydnones and of 5-aryltriazoles by coupling with arylboronic acids [25, 26].

One of the major advantages of the use of arylboronic acids for the direct arylation of heteroarenes appears to be the ability to give access to regioisomers different from that obtainable when aryl halides are employed. For example, Studer et al. reported the oxidative coupling of 2-pyridylthiophene with arylboronic acids using $Pd(OAc)_2$ as catalyst. TEMPO was used as a stoichiometric oxidant (Scheme 2, top) [27]. The 2-pyridyl group served as *ortho*-directing groups to promote an unusual β-arylation of the thienyl moiety. Bach et al. reported in 2011 that thiophenes substituted at C3 by $CH_2PO(OEt)_2$ or CH_2COOEt undergo a regioselective oxidative coupling reaction at C4 with aryl boronic acids in the presence of Ag_2O, cesium trifluoroacetate [Cs(tfa)], benzoquinone (BQ), and Pd $(tfa)_2$ catalyst in trifluoroacetic acid as the solvent (Scheme 2, middle) [28]. They recently extended this procedure to a broad range of substrates [29, 30]. Studer and Itami also described a procedure for the β-arylations of thiophenes with arylboronic acids under Pd/TEMPO catalysis (Scheme 2, bottom) [31]. Even sterically very hindered arylboronic acids were successfully coupled [32]. This procedure also promotes the 4-arylation of thiazoles [33–36]. The use of Pd(II)–sulfoxide–oxazoline associated to iron–phthalocyanine as catalytic system, with air as terminal oxidant, also let to β-arylated thiophenes [37].

In summary, under oxidative conditions, arylboronic acids and also aryltrifluoroborates can be employed as coupling partners for arylation of heteroarenes. The major advantage appears to be their ability to promote arylations at unusual positions such as 3-arylation of thiophenes or 4-arylation of thiazoles even in the absence of directing group.

3 Benzoic Acids

During the last decade, benzoic acids have been widely used as starting materials in cross coupling reactions [38, 39]. The use of such substrates in organic synthesis is very attractive due to their low cost and wide availability. In the last few years, C–H bond functionalization via decarboxylative coupling has been explored by a few groups.

The first example was reported in 2010 by Tan et al. (Scheme 3) [40]. A palladium-catalyzed decarboxylative coupling of benzoxazoles and thiazoles with various benzoic acid derivatives using $PdCl_2$ associated to PPh_3 as the catalyst and Ag_2CO_3 as the base and the oxidant affords the 2-arylbenzoxazoles or the 2-arylbenzothiazoles. The scope of the reaction is limited to the use of benzoic acids bearing a C2 substituent such as 2-nitrobenzoic acid.

The same year, Su et al. reported the arylation of indoles using also benzoic acids as arylating agents (Scheme 4) [41]. The reaction of *N*-pivaloylindole with 5 equiv. of benzoic acids conducted in dioxane at 80°C using $Pd(tfa)_2$ as catalyst and 2 equiv. of Ag_2CO_3, using tetramethylene sulfoxide (TMSO) (1.5 equiv.), and propionic acid (0.25 equiv.) as additives led to the coupling products in good yields. From electron-rich benzoic acids, 2-arylindoles were formed regioselectively, while 3-arylindoles were exclusively generated from electron-deficient benzoic acids.

In 2012, Su et al. enlarged the substrate scope to thiophene derivatives. They found that the $PCy_3/Pd(tfa)_2$ catalytic system, in combination with Ag_2CO_3,

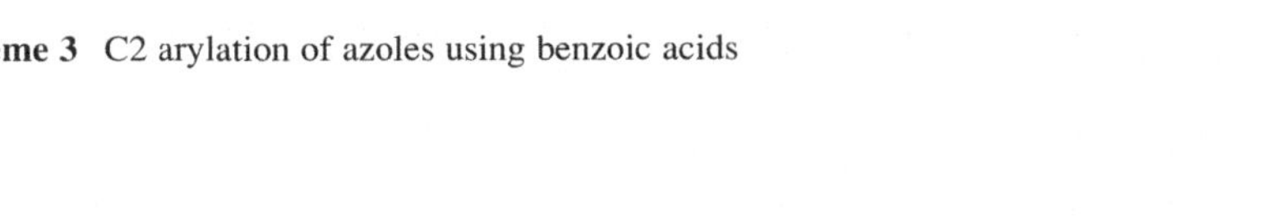

Scheme 3 C2 arylation of azoles using benzoic acids

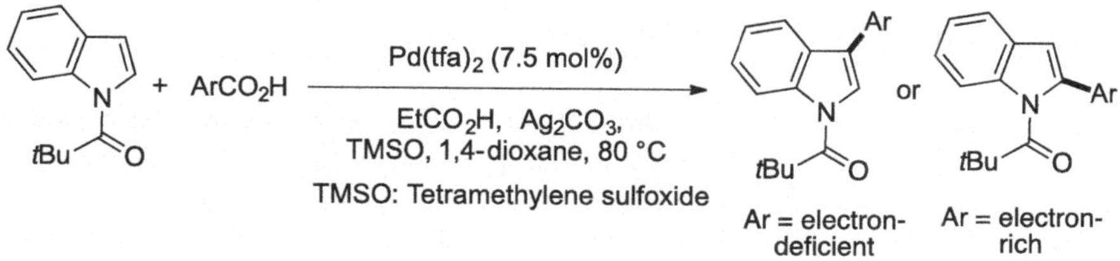

Scheme 4 Arylation of indoles using benzoic acids

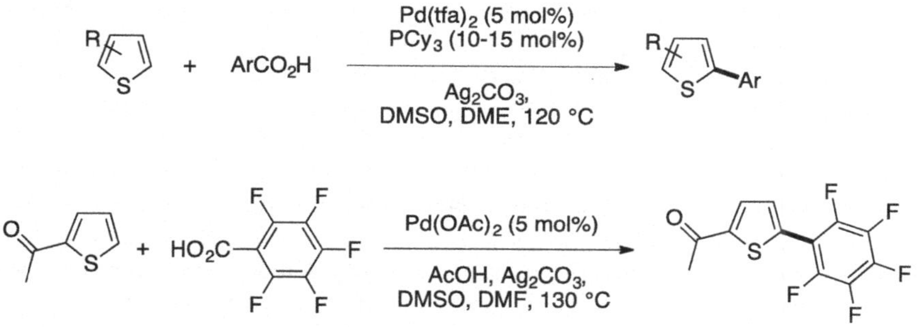

Scheme 5 Arylation of thiophenes using benzoic acids

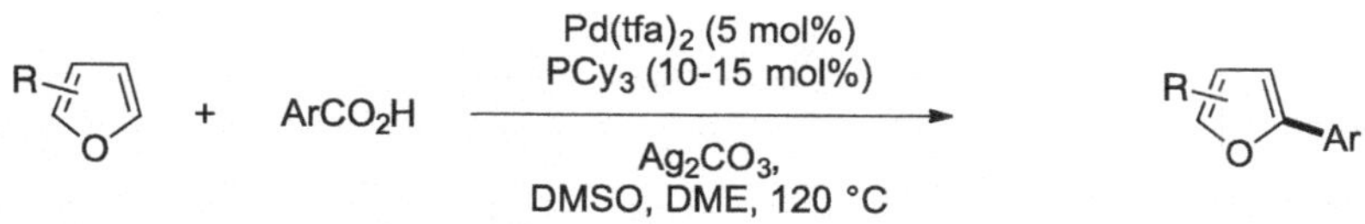

Scheme 6 Arylation of furans using benzoic acids

efficiently promotes the decarboxylative C–H bond arylation of thiophenes with benzoic acids as arylating agents (Scheme 5, top) [42]. This protocol exhibits a quite broad substrate scope for the thienyl moiety as it is compatible with keto, formyl, ester, trifluoromethyl, alkenyl, nitro, fluoro, and chloro substituents. Again, the reaction is limited to the use of 2-substituted benzoic acid derivatives such as 2-nitro-, 2-methoxy-, or 2-halo-benzoic acids. The non-*ortho*-substituted benzoic acids did not afford the coupling products due to their poor reactivity in the decarboxylation compared to *ortho*-substituted benzoic acids. A quite similar procedure was employed by Loh et al. for the coupling of pentafluorobenzoic acid with 2-acetylthiophene (Scheme 5, bottom) [43].

Then, in 2014, Su et al. applied their reaction conditions to the coupling of benzoic acids with furan, benzofuran, pyrrole, and thiazole derivatives (Scheme 6) [44]. Again the reaction was limited to the use of 2-substituted benzoic acids.

In summary, several 5-membered heteroaromatics can be arylated using benzoic acids as arylating agents, but only 2-substituted benzoic acids were reactive for such couplings.

4 Aryl Silanes and Aryl Siloxanes

Following the pioneering work reported by Shi in 2006, on the use of trialkoxyaryl-silanes for palladium-catalyzed direct arylation of acetanilides [45], in 2010 Zhang et al. described conditions to achieve a $Pd(OAc)_2$-catalyzed regioselective C2 arylation of indoles using aryl siloxanes as arylating agents, in the presence of $n\mathrm{Bu_4NF}$ and Ag_2O in acidic medium (Scheme 7) [46]. Both electron-donating and electron-withdrawing substituents on the reaction partners were tolerated.

One year later, Ofial et al. extended the scope of the arylation with aryl siloxanes to a broad range of azole derivatives using $Cu(OAc)_2$ and AgF as additives (Scheme 8) [47].

In 2012, Oi et al. described the direct arylation of (benzo)thiophenes with aryltrimethylsilanes (Scheme 9). The use of $PdCl_2(MeCN)_2$ as pre-catalyst in the presence of $CuCl_2$ as oxidant gives the β-arylated (benzo)thiophenes [48].

In summary, so far the use of $ArSiR_3$ derivatives as arylating agents in palladium-catalyzed direct arylation of heteroaromatics requires the presence of stoichiometric amount of copper and/or silver salts as additives making these processes less attractive than the reactions with aryl halides. However, as it was demonstrated by Oi, such coupling partners promote challenging β-arylations of (benzo)thiophenes instead of the α-arylations obtained with most procedures involving aryl halides, but their reactivity deserves to be investigated in more details.

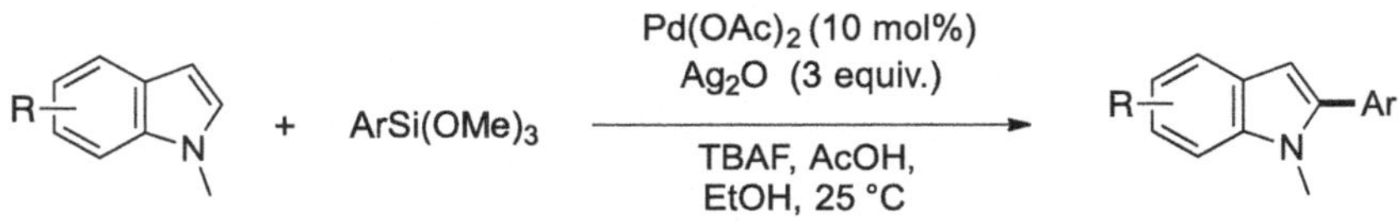

Scheme 7 C2 arylation of indoles using arylsiloxanes

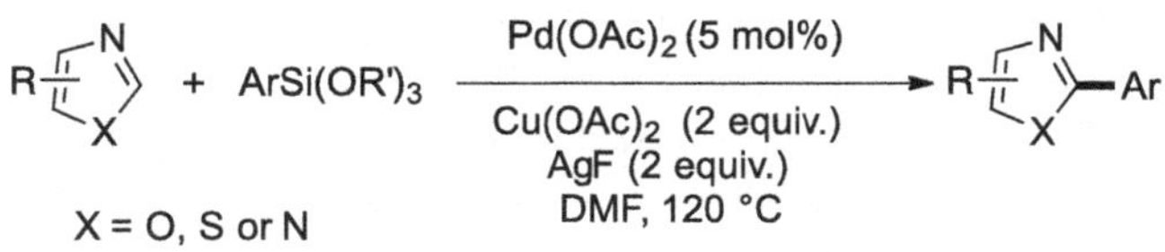

Scheme 8 C2 arylation of azoles using arylsiloxanes

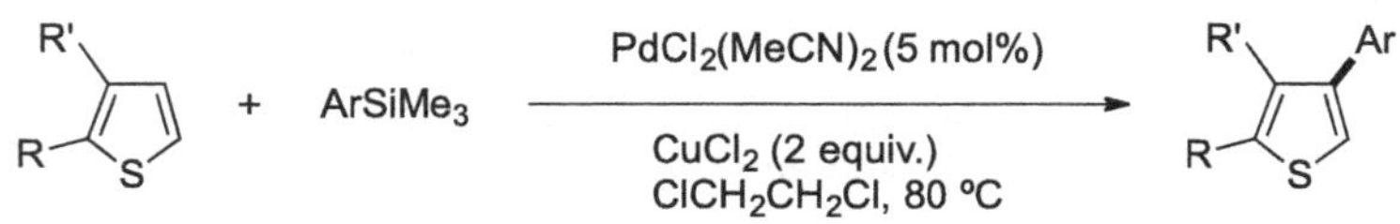

Scheme 9 C4 arylation of thiophenes using arylsilanes

5 Arylsulfonyl Chlorides

In the late 1980s, Kasahara [49] and Miura [50, 51] have independently introduced arylsulfonyl chlorides in palladium-catalyzed C–C bond formation through a desulfitative cross coupling. Then, these coupling partners were popularized by Vogel for Stille [52, 53], Suzuki–Miyaura [54], Sonogashira–Hagihara [55], Mizoroki–Heck [56], Negishi [57], and Corriu–Kumada reactions [58]. Advantages of these substrates are that many of them are commercially available at an affordable cost or they can also be easily prepared from sulfonic acids or sulfur substrates by chlorination [59–63]. Their uses as coupling partners in C–H bond functionalization are more recent. Based on Dong's pioneering work on Pd-catalyzed C–H bond sulfonylation of 2-phenylpyridines using benzenesulfonyl chlorides as coupling partners, in which they observed a desulfitative coupling in one case [64], Cheng and co-workers reported the use of benzenesulfonyl chlorides as aryl sources for the direct arylation of benzoxazoles (Scheme 10) [65]. The reaction displays a broad scope including substrates bearing C–Br bonds. However, lower yields were observed for the direct arylation of benzothiazole under the same reaction conditions.

In 2013, Jafarpour and co-workers reported two examples of direct desulfitative arylation of *N*-methylpyrrole in the course of their investigation on regioselective arylation of coumarins using benzenesulfonyl chlorides as coupling partners (Scheme 11) [66]. Their optimized reaction conditions were $Pd(OAc)_2$ as catalyst associated to 2 equiv. of $Cu(OAc)_2$ in DMA in the presence of KOAc as base and afforded the α-arylated pyrroles in moderate yields.

In 2014, Doucet and co-workers reported the Pd-catalyzed direct desulfitative arylation of thiophenes with benzenesulfonyl chlorides as coupling partners [67]. In contrast to other coupling partners, which allowed direct C2 arylation [28, 29, 31, 32, 37, 48, 68–70], the reaction exclusively occurred at the C-3 position (Scheme 12, top). The best reaction conditions were found using 5 mol%

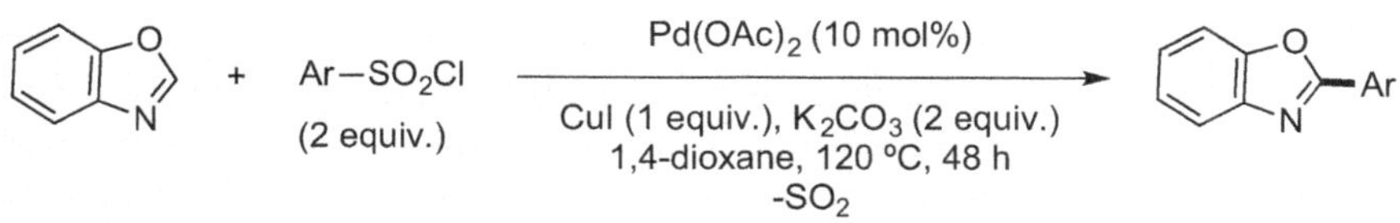

Scheme 10 C2 arylation of benzoxazole using benzenesulfonyl chlorides

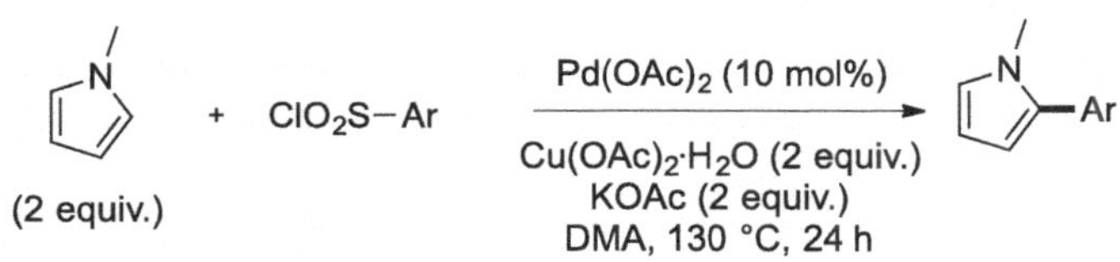

Scheme 11 C2 arylation of *N*-methylpyrrole using benzenesulfonyl chlorides

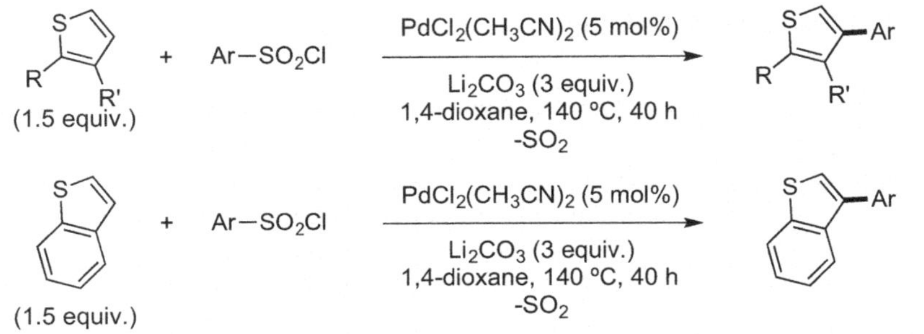

Scheme 12 C3 or C4 arylation of thiophene and benzothiophene using benzenesulfonyl chlorides

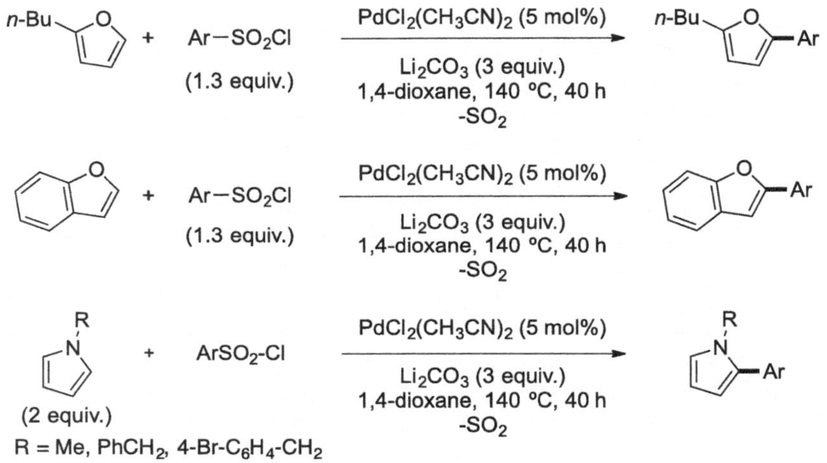

Scheme 13 C2 or C5 arylation of furans, benzofurans, and N-protected pyrroles using benzenesulfonyl chlorides

$PdCl_2(CH_3CN)_2$ in the presence of 3 equiv. of lithium carbonate (Li_2CO_3) as the base at 140°C in dioxane. Sensitive functional groups such as cyano, bromo, nitro, and ester on both coupling partners were tolerated, offering a broad substrate scope. Thiophene bearing a bromine atom also chemoselectively reacted without cleavage of the C–Br bond. Following the same procedure, benzothiophene was arylated at the β-position in high yields (Scheme 12, bottom). The authors explained the β-regioselectivity through a mechanism involving a Pd(II)/Pd(IV) catalytic cycle and an electrophilic palladation.

Continuing to investigate this area, Doucet and co-workers applied their reaction conditions to the direct desulfitative arylation of furans [71], benzofurans [17], and *N*-protected pyrroles [72]. In all cases, both good yields and high α-regio-selectivities were obtained (Scheme 13). It is important to note that copper-free conditions have been used for such direct desulfitative cross coupling reactions.

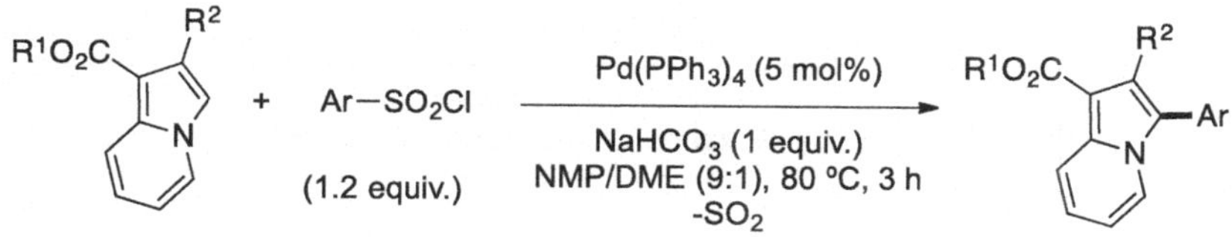

Scheme 14 C3 arylation of indolizines using benzenesulfonyl chlorides

In 2015, Doucet and co-workers applied their methodology to the synthesis of (poly)halo-substituted bi(hetero)aryls using (poly)halobenzenesulfonyl chlorides as coupling partners [73]. In all cases, the reaction was very chemoselective as C–Br and C–I bonds were not involved in the direct arylation process. In the same years, they disclosed that such desulfitative direct arylations of thiophenes, furans, and pyrroles could be performed in green solvents such as CPME or diethyl carbonate (DEC) and even in some case under free-solvent conditions in high yields using the same catalytic system, namely, 5 mol% $PdCl_2(CH_3CN)_2$ in the presence of 3 equiv. of lithium carbonate (Li_2CO_3) as base at 140°C [74].

In 2015, Zhao and co-workers reported the direct C-3 arylation of indolizine derivatives with benzenesulfonyl chlorides as aryl sources (Scheme 14) [75]. This transformation was performed using 5 mol% of $Pd(PPh_3)_4$ in the presence of $NaHCO_3$ as base in a mixture of 1-methyl-2-pyrrolidone (NMP) and dimethoxyethane (DME) as solvents and afford a new orthogonal approach to C-3-arylated indolizines even using 4-bromobenzenesulfonyl chloride.

In summary, benzenesulfonyl chlorides have been successfully used as alternative coupling partners and allow orthogonal reactions owing to their C–halo bond tolerances. Moreover, the benzenesulfonyl chlorides offered sometimes unexpected regioselectivity such as β-arylation of thiophene derivatives. Copper-free conditions or the use of DEC provides green synthetic approaches for the synthesis of aryl–heteroaryl derivatives, albeit a high palladium catalyst loading is generally required.

6 Sodium Arylsulfinates

Sodium arylsulfinates have also been largely used as coupling partners as they show high functional group tolerances and are easy to handle. However, only a few of them are commercially available at affordable cost, but they can be prepared by the reduction of the corresponding RSO_2Cl [76]. Deng and co-workers were the first to report a Pd-catalyzed direct desulfitative arylation of azoles using sodium sulfinates as aryl source (Scheme 15, top) [77]. Several azoles such as benzothiazoles, benzoxazoles, thiazoles, or caffeine were arylated in moderate to high yields using 2.5 mol% $Pd(OAc)_2$ as catalyst in the presence of 2 equiv. of $Cu(OAc)_2$ in mixed 1,4-dioxane diglyme at 120°C. Diverse substituents on the sulfinate benzene ring were tolerated, albeit the presence of electron-withdrawing groups decreased

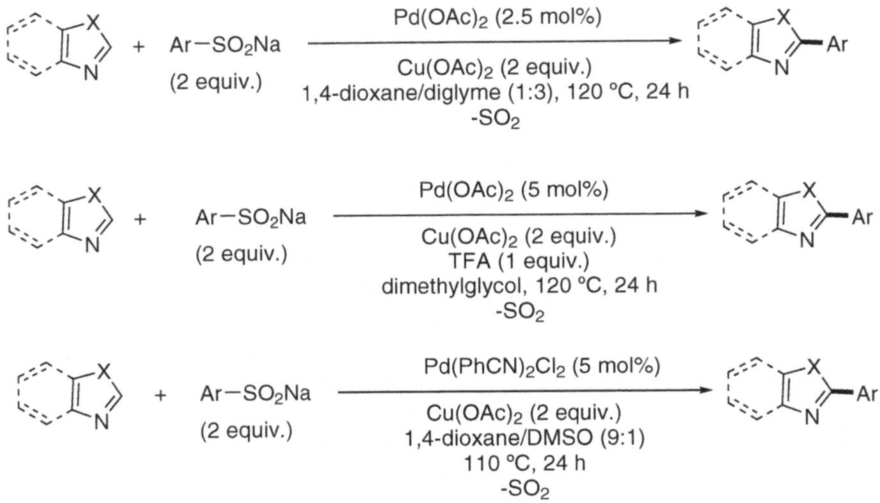

Scheme 15 C2 arylation of azoles using sodium sulfinates

the reaction yields. In this reaction, copper salts have a dual role, namely, act as base and oxydant (i.e., reoxidize Pd(0) into Pd(II)). Inspired by this work, other research groups developed similar reaction conditions for the direct arylation of azoles using sodium sulfinates. In 2012, Wang and co-workers reported similar reaction conditions, albeit they used acidic conditions (Scheme 14, middle) [78]. However, the reaction again needs an oxidant by the presence of copper salts. Later, You and co-workers also employed sodium sulfinates for the direct desulfitative arylation of five-membered ring heteroarenes containing more than one heteroatom (Scheme 14, bottom) [79]. The reaction conditions were similar to those reported by Deng, except the use of 5 mol% $PdCl_2(PhCN)_2$ in DMSO. Caffeine derivatives, purine derivatives, imidazoles, and quinoxaline N-oxide were regioselectively arylated at C2 position in high yields.

In 2014, Liu and co-workers disclosed the Pd-catalyzed direct arylation of thiazolo[3,2-*b*]-1,2,4-triazoles using sodium sulfinates (Scheme 16) [80]. By using 5 mol% $Pd(OAc)_2$ as catalyst in the presence of 2 equiv. of $Cu(OAc)_2$ as oxidant/base in dioxane/diglyme at 120°C, they obtained the desired desulfitative arylation products in moderate to high yields.

The scope of the Pd-catalyzed direct desulfitative arylation using sodium sulfinates is not limited to five-membered ring heteroarenes containing more than one heteroatom. In 2012, Deng, Luo, and co-workers reported the first arylation of indoles with arylsulfinates (Scheme 17, top) [81]. The use of 5 mol% $Pd(OAc)_2$ catalyst in the presence of 2 equiv. of copper(II) salts as oxidant in toluene/dioxane as mix-solvent at 100°C allowed to obtain the desired C2-arylated indoles in high yields. The reaction showed a broad functional group tolerance such as methyl, methoxy, bromo, fluoro, nitro, and even ester moieties on the *N*-methylindole part.

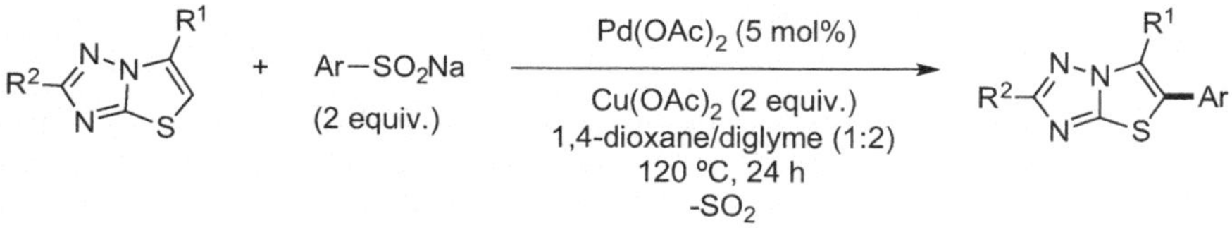

Scheme 16 C2 arylation of thiazolo[3,2-*b*]-1,2,4-triazoles using sodium sulfinates

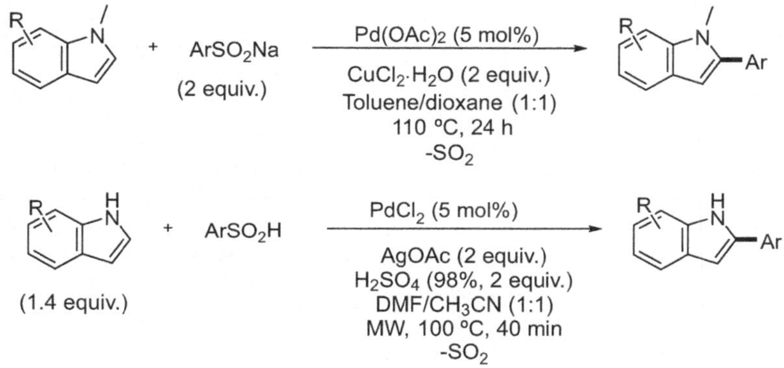

Scheme 17 C2 arylation of indoles using sodium sulfinates

Moreover, other *N*-protected indoles such as *N*-ethylindole or *N*-acetylindole reacted also in good yields, while no reaction occurred using free *NH*-indole. Again, the reaction conditions tolerated the presence of C–Br bonds during the reaction allowing further metal-catalyzed transformations. The reaction of free (*NH*)-indoles was solved by Wang and co-workers using microwave heating and slightly modified reaction conditions, namely, silver acetate (AgOAc, 2 equiv.) as oxidant and sulfuric acid (2 equiv.) as additive, which might prevent the decomposition of indoles (Scheme 17, bottom) [82]. It is important to note that the direct desulfitative arylation of indoles using benzenesulfonyl chlorides gives mixtures of C2- and C3-arylated indoles [72, 74].

As seen in this section, sodium arylsulfinates were successfully used in C–H bond reactions. Compared with benzenesulfonyl chlorides, these substrates prevent sulfonation reactions allowing a high functional group tolerance such as *N*H-free indoles. However, such coupling partners require the use of a stoichiometric amount of copper salts as additive in order to reoxidize Pd(0) into Pd(II).

7 Arylsulfonyl Hydrazides

In 2012, Wan and co-workers introduced the use of arylsulfonyl hydrazides as aryl sources in Pd-catalyzed desulfitative arylation of azoles (Scheme 18, top) [83]. The optimized conditions were found using $PdCl_2(CH_3CN)_2$ associated to 1,10-phenanthroline as ligand in the presence of a large excess of copper and 1.5 equiv. of Na_2CO_3 as base in a mixture of dioxane/DMSO in 9:1 ratio. Tetra-*n*-butylammonium bromide was also used as additive in order to stabilize the active palladium species. Benzoxazoles, oxazoles, benzothiazoles, thiazoles, and caffeine were regioselectively arylated between the two heteroelements, and no effect of the substitution on the arylsulfonyl hydrazide partners was observed. The authors explained that copper probably mediates the denitrogenative and desulfitative oxidation of $TsNHNH_2$ affording copper(II) aryl sulfonyl species, which undergoes transmetalation to afford a palladium *p*-tolyl intermediate. During the same period, Kwong and co-workers reported a simpler catalytic system for this reaction, albeit the desired C2-arylated product was isolated in lower yields (Scheme 18, middle) [84]. The optimized reaction conditions were 5 mol% $Pd(OAc)_2$ associated with triphenylphosphine, in the presence of only $Cu(OAc)_2$ (2 equiv.) as additive, in dioxane under air atmosphere. Using the same reaction conditions, You and co-workers regioselectively arylated caffeine derivatives, purine derivatives, imidazoles, and quinoxaline *N*-oxide (Scheme 18, bottom).

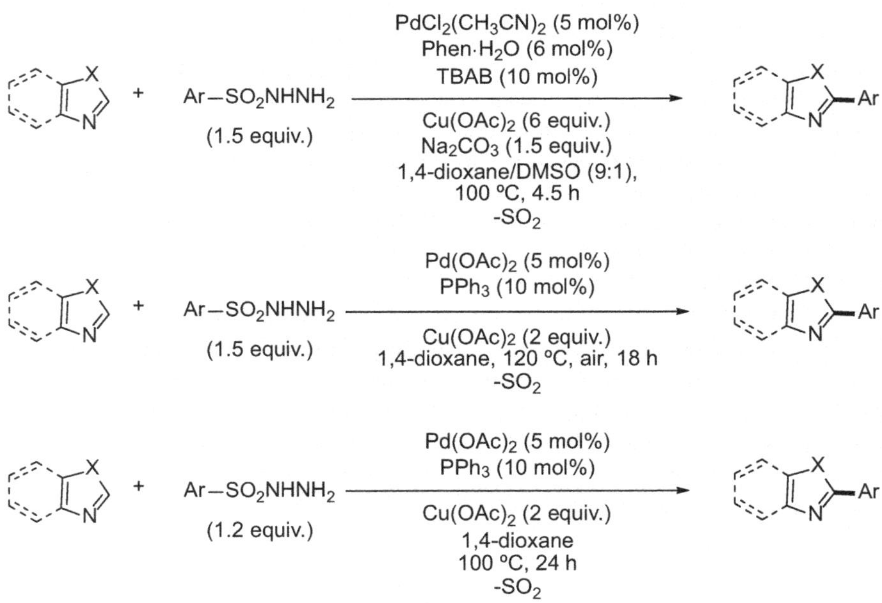

Scheme 18 C2 arylation of azoles using arylsulfonyl hydrazides

In conclusion, arylsulfonyl hydrazides have been introduced for desulfitative direct arylation of azoles and have a similar substrate scope than arenesulfinate salts. From a green chemistry point of view, the use of a stoichiometric amount of copper(II) as additive remains a major drawback.

8 Other Coupling Partners

In 2010, Ackermann and co-workers introduced the use of aryl imidazolylsulfonates as aryl sources in the palladium-catalyzed C2-direct arylation of azoles (Scheme 19) [85]. These derivatives were not moisture sensitive and the generated side product (e.g., imidazolesulfonic acid) was nongenotoxic. A wide range of azoles were arylated in high yields at the C2-position using a simple catalytic system, namely, 5 mol% $Pd(OAc)_2$ associated to dppe as phosphine ligand in the presence of Cs_2CO_3 as base in NMP at 100°C. In addition, these conditions were applied to the direct benzylation and alkenylation of azoles using easily accessible benzyl and alkenyl phosphates as electrophilic coupling partners.

On the other hand, Van der Eycken and co-workers reported an interesting aryl source—based on tautomerization of amides and C–OH activation—in palladium-catalyzed direct C2 arylation of azoles (Scheme 20) [86]. The procedure operates in

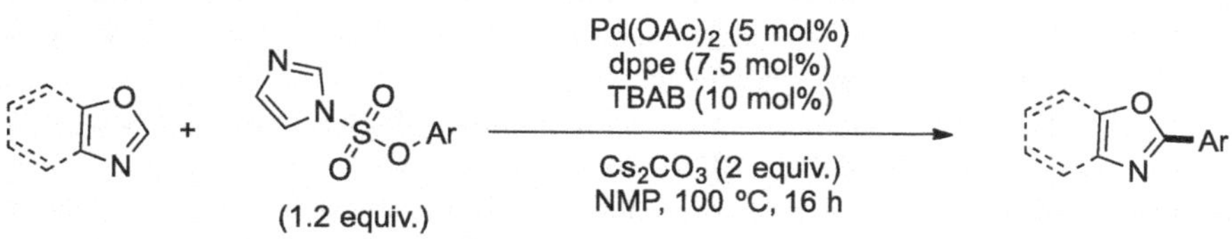

Scheme 19 C2 arylation of azoles using aryl imidazolylsulfonate

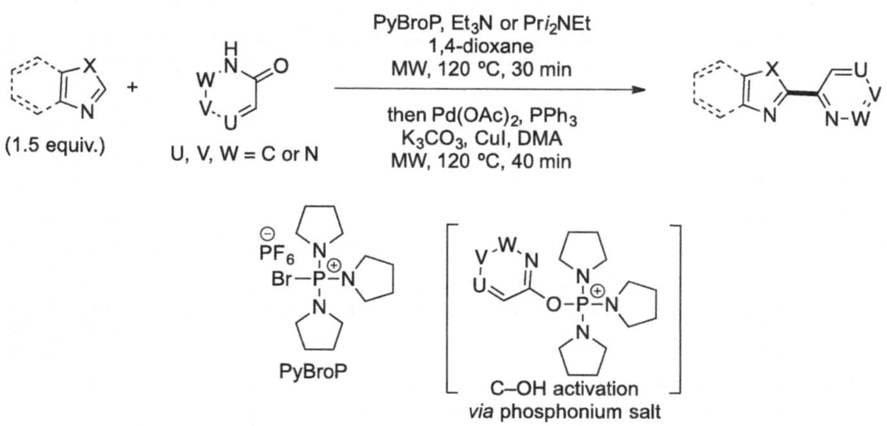

Scheme 20 C2 arylation of azoles using tautomerizable heterocycles

a sequential manner. In a first operation, a phosphonium salt was generated from a tautomerizable heterocycle by the action of bromotri(pyrrolidin-1-yl)phosphonium hexafluorophosphate(V) and a base (e.g., Et_3N or Pri_2NEt) in 1,4-dioxane under microwave irradiation. Then, azole derivative was introduced to the reaction vessel with the catalytic system (i.e., $Pd(OAc)_2$ associated to PPh_3) in the presence of K_2CO_3 and CuI in DMA and was heated at 120°C under microwave irradiation. The reaction conditions tolerated a wide range of azoles including 1,3,4-thiadiazole.

9 Conclusion

In summary, during the last decade, the scope of the arylation coupling partners for palladium-catalyzed direct functionalization of 5-membered ring heteroarenes has been largely extended. The recent modifications of these coupling partners, such as the use of $ArSO_2Cl$ or $ArB(OH)_2$, present several advantages compared to the reactions with aryl halides. In some cases, such as for the Pd-catalyzed arylation of thiophenes or thiazoles with $ArB(OH)_2$, unusual C3 or C4 arylation products can be obtained. A change in the regioselectivity was also observed in the course of the arylation of benzofurans, as the use of $ArSO_2Cl$ selectively affords the C2-arylated benzofurans instead of the mixtures of C2- and C3-arylated and also C2 and C3-di-arylated benzofurans obtained with aryl halides. Moreover, the coupling with most $ArSO_2R$ derivatives tolerates several functional groups including bromo- and iodo-substituents allowing further transformations.

Despite the numerous recent reports, a number of challenges remain. The influence of the nature of these alternative coupling partners on the reactivities and selectivities need to be explored in more details. Concerning catalytic cycles, a large number of interrogations are not addressed. If Pd(0)/Pd(II) catalytic cycles are certainly involved in most couplings with aryl halides, with some of these new coupling partners such as $ArSO_2Cl$, Pd(II)/Pd(IV) catalytic systems might be involved. Determining the reasons of the various reactivities and regioselectivities for these reactions could allow further improvement in terms of efficiency and reaction scope.

References

1. Nakamura N, Tajima Y, Sakai K (1982) Heterocycles 17:235
2. Akita Y, Inoue A, Yamamoto K, Ohta A, Kurihara T, Shimizu M (1985) Heterocycles 23:2327
3. Ohta A, Akita Y, Ohkuwa T, Chiba M, Fukunaga R, Miyafuji A, Nakata T, Tani N, Aoyagi Y (1990) Heterocycles 31:1951
4. Bellina F, Rossi R (2009) Tetrahedron 65:10269
5. Ackermann L (2009) Modern arylation methods. Wiley, Weinheim
6. Roger J, Gottumukkala AL, Doucet H (2010) ChemCatChem 2:20
7. Arockiam PB, Bruneau C, Dixneuf PH (2012) Chem Rev 112:5879

8. Wencel-Delord J, Droge T, Liu F, Glorius F (2011) Chem Soc Rev 40:4740
9. Deprez NR, Kalyani D, Krause A, Sanford MS (2006) J Am Chem Soc 128:4972
10. Okazawa T, Satoh T, Miura M, Nomura M (2002) J Am Chem Soc 124:5286
11. Hara O, Nakamura T, Sato F (2006) Heterocycles 68:1
12. Roger J, Doucet H (2008) Org Biomol Chem 6:169
13. Kozhushkov SI, Potukuchi HK, Ackermann L (2013) Catal Sci Technol 3:562
14. Biajoli AFP, da Penha ET, Correia CRD (2012) RSC Adv 2:11930
15. Colas C, Goeldner M (1999) Eur J Org Chem 1999:1357
16. Campeau L-C, Bertrand-Laperle M, Leclerc J-P, Villemure E, Gorelsky S, Fagnou K (2008) J Am Chem Soc 130:3276
17. Loukotova L, Yuan K, Doucet H (2014) ChemCatChem 6:1303
18. Yang SD, Sun CL, Fang Z, Li BJ, Li YZ, Shi ZJ (2008) Angew Chem 120:1495
19. Huang Y, Ma T, Huang P, Wu D, Lin Z, Cao R (2013) ChemCatChem 5:1877
20. Zhao J, Zhang Y, Cheng K (2008) J Org Chem 73:7428
21. Zhao B (2012) Org Biomol Chem 10:7108
22. Kirchberg S, Fröhlich R, Studer A (2009) Angew Chem Int Ed 48:4235
23. Liu B, Qin X, Li K, Li X, Guo Q, Lan J, You J (2010) Chem Eur J 16:11836
24. Ranjit S, Liu X (2011) Chem Eur J 17:1105
25. Liu W, Li Y, Wang Y, Kuang C (2013) Eur J Org Chem 2013:5272
26. Yang Y, Gong H, Kuang C (2013) Synthesis 45:1469
27. Kirchberg S, Vogler T, Studer A (2008) Synlett 2008:2841
28. Schnapperelle I, Breitenlechner S, Bach T (2011) Org Lett 13:3640
29. Schnapperelle I, Bach T (2013) ChemCatChem 5:3232
30. Schnapperelle I, Bach T (2014) Chem Eur J 20:9725
31. Kirchberg S, Tani S, Ueda K, Yamaguchi J, Studer A, Itami K (2011) Angew Chem Int Ed 50:2387
32. Yamaguchi K, Yamaguchi J, Studer A, Itami K (2012) Chem Sci 3:2165
33. Uehara TN, Yamaguchi J, Itami K (2013) Asian J Org Chem 2:938
34. Tani S, Uehara TN, Yamaguchi J, Itami K (2014) Chem Sci 5:123
35. Sekizawa H, Amaike K, Itoh Y, Suzuki T, Itami K, Yamaguchi J (2014) ACS Med Chem Lett 5:582
36. Lohrey L, Uehara TN, Tani S, Yamaguchi J, Humpf H-U, Itami K (2014) Eur J Org Chem 2014:3387
37. Yamaguchi K, Kondo H, Yamaguchi J, Itami K (2013) Chem Sci 4:3753
38. Gooßen LJ, Rodriguez N, Gooßen K (2008) Angew Chem Int Ed 47:3100
39. Cornella J, Larrosa I (2012) Synthesis 2012:653
40. Xie K, Yang Z, Zhou X, Li X, Wang S, Tan Z, An X, Guo C-C (2010) Org Lett 12:1564
41. Zhou J, Hu P, Zhang M, Huang S, Wang M, Su W (2010) Chem Eur J 16:5876
42. Hu P, Zhang M, Jie X, Su W (2012) Angew Chem Int Ed 51:227
43. Luo H-Q, Dong W, Loh T-P (2013) Tetrahedron Lett 54:2833
44. Pei K, Jie X, Zhao H, Su W (2014) Eur J Org Chem 2014:4230
45. Yang S, Li B, Wan X, Shi Z (2007) J Am Chem Soc 129:6066
46. Liang Z, Yao B, Zhang Y (2010) Org Lett 12:3185
47. Han W, Mayer P, Ofial AR (2011) Chem Eur J 17:6904
48. Funaki K, Sato T, Oi S (2012) Org Lett 14:6186
49. Kasahara A, Izumi T, Kudou N, Azami H, Yamamoto S (1988) Chem Ind 51
50. Miura M, Hashimoto H, Itoh K, Nomura M (1989) Tetrahedron Lett 30:975
51. Miura M, Hashimoto H, Itoh K, Nomura M (1990) J Chem Soc Perkin Trans 1:2207
52. Dubbaka SR, Vogel P (2003) J Am Chem Soc 125:15292
53. Labadie SS (1989) J Org Chem 54:2496
54. Dubbaka SR, Vogel P (2004) Org Lett 6:95
55. Reddy Dubbaka S, Vogel P (2004) Adv Synth Catal 346:1793
56. Dubbaka SR, Vogel P (2005) Chem Eur J 11:2633

57. Dubbaka SR, Vogel P (2006) Tetrahedron Lett 47:3345
58. Rao Volla CM, Vogel P (2008) Angew Chem Int Ed 47:1305
59. Johary NS, Owen LN (1955) J Chem Soc 1307
60. Barco A, Benetti S, Pollini GP, Taddia R (1974) Synthesis 1974:877
61. Fujita S (1982) Synthesis 1982:423
62. Kataoka T, Iwama T, Setta T, Takagi A (1998) Synthesis 1998:423
63. Blotny G (2003) Tetrahedron Lett 44:1499
64. Zhao X, Dimitrijević E, Dong VM (2009) J Am Chem Soc 131:3466
65. Zhang M, Zhang S, Liu M, Cheng J (2011) Chem Commun 47:11522
66. Jafarpour F, Olia MBA, Hazrati H (2013) Adv Synth Catal 355:3407
67. Yuan K, Doucet H (2014) Chem Sci 5:392
68. Yanagisawa S, Ueda K, Sekizawa H, Itami K (2009) J Am Chem Soc 131:14622
69. Ueda K, Yanagisawa S, Yamaguchi J, Itami K (2010) Angew Chem Int Ed 49:8946
70. Tang D-TD, Collins KD, Glorius F (2013) J Am Chem Soc 135:7450
71. Beladhria A, Yuan K, Ben Ammar H, Soulé J-F, Ben Salem R, Doucet H (2014) Synthesis 46:2515
72. Jin R, Yuan K, Chatelain E, Soulé J-F, Doucet H (2014) Adv Synth Catal 356:3831
73. Skhiri A, Beladhria A, Yuan K, Soulé J-F, Ben Salem R, Doucet H (2015) Pd-cataylsed direct arylation of heteroaromatics using (poly)halobenzene sulfonyl chlorides as coupling partners: one step access to (poly)halo-substituted bi(hetero)aryls. Eur J Org Chem 4428–4436. doi:10.1002\ejoc.201500354
74. Hfaiedh A, Yuan K, Ben Ammar H, Ben Hassine B, Soulé J-F, Doucet H (2015) ChemSusChem 8:1794
75. Zhang W, Liu F, Zhao B (2015) Appl Organomet Chem. doi:10.1002/aoc.3326
76. Liu LK, Chi Y, Jen K-Y (1980) J Org Chem 45:406
77. Chen R, Liu S, Liu X, Yang L, Deng G-J (2011) Org Biomol Chem 9:7675
78. Wang M, Li D, Zhou W, Wang L (2012) Tetrahedron 68:1926
79. Liu B, Guo Q, Cheng Y, Lan J, You J (2011) Chem Eur J 17:13415
80. Wang S, Liu W, Lin J, Jiang Y, Zhang Q, Zhong Y (2014) Synlett 25:586
81. Wu M, Luo J, Xiao F, Zhang S, Deng G-J, Luo H-A (2012) Adv Synth Catal 354:335
82. Miao T, Li P, Wang G-W, Wang L (2013) Chem Asian J 8:3185
83. Yu X, Li X, Wan B (2012) Org Biomol Chem 10:7479
84. Yuen OY, So CM, Wong WT, Kwong FY (2012) Synlett 23:2714
85. Ackermann L, Barfüsser S, Pospech J (2010) Org Lett 12:724
86. Sharma A, Vachhani D, Van der Eycken E (2012) Org Lett 14:1854

Top Organomet Chem (2016) 55: 119–136
DOI: 10.1007/3418_2015_128

Published online: 14 July 2015

Functionalization of [60]Fullerene via Palladium-Catalyzed C–H Bond Activation

Guan-Wu Wang

Abstract The palladium-catalyzed C–H bond activation strategy has been successfully applied to fullerene chemistry, and several types of [60]fullerene-fused heterocycles and carbocycles have been obtained. The synthesis of [60]fullerene-fused indolines, isoquinolinones, azepines, tetrahydroisoquinolines, tetrahydrobenzazepines, sultones, tetrahydrobenzooxepines/isochromans, and dihydrophenanthrenes has been achieved by the palladium-catalyzed reactions of [60] fullerene with anilides, benzamides, *N*-sulfonyl-2-aminobiaryls, *N*-benzyl sulfonamides, *N*-(2-arylethyl) sulfonamides, arylsulfonic acids, 2-phenylethyl/benzyl alcohols, and 2-arylbenzoic acids, respectively.

Keywords [60]Fullerene · Annulation · Carbocycle · C–H activation · Heterocycle · Palladium catalyst

Contents

G.-W. Wang (✉)
CAS Key Laboratory of Soft Matter Chemistry, iChEM (Collaborative Innovation Center of Chemistry for Energy Materials), National Laboratory for Physical Sciences at Microscale, and Department of Chemistry, University of Science and Technology of China, Hefei, Anhui 230026, P. R. China
e-mail: gwang@ustc.edu.cn

Abbreviations

BQ	*p*-Benzoquinone
Bs	Benzenesulfonyl
Cs	4-Chlorobenzenesulfonyl
equiv.	Equivalent(s)
Et	Ethyl
h	Hour(s)
Me	Methyl
MeCN	Acetonitrile
mL	Mililiter(s)
mol	Mole(s)
Ms	Methanesulfonyl
ODCB	*ortho*-Dichlorobenzene
Ph	Phenyl
Piv	Pivaloyl
PTSA	*p*-Toluenesulfonic acid
rt	Room temperature
TFA	Trifluoroacetic acid
Ts	*p*-Toluenesulfonyl

1 Introduction

Functionalization of fullerenes can not only retain the unique characteristics of pristine fullerenes but also modulate their properties by attaching different organic addends. Over the past two decades, various types of reactions have been developed to provide a diversity of fullerene derivatives [1]. Among them, transition-metal-mediated or -catalyzed reactions of [60]fullerene (C_{60}) have attracted increasing attention [2–4]. The functionalized fullerenes have huge potential applications in materials science, biology, and nanotechnology [5, 6]. On the other hand, the palladium-catalyzed C–H bond activation has emerged as one of the most important methodologies to construct C–C and C–X bonds in organic synthesis [7–9]. C–H bond activation enables the late-stage diversification of various kinds of organic scaffolds, ranging from relatively small molecules like drug candidates, luminescent compounds for optical applications, and photochromic molecules to complex polydisperse organic compounds such as metal–organic frameworks (MOFs) and polymers [10–12]. Recently, we have been interested in the functional group-directed sp^2 C–H activations. By employing $NHCOCH_3$, $CONHOCH_3$, $CONH_2$, and N=NAr as the directing groups, we have achieved the *ortho*-acetoxylation [13] and *ortho*-alkoxylation [14] of anilides, *ortho*-alkoxylation of *N*-methoxybenzamides [15], synthesis of phenanthridinones from *N*-methoxybenzamides and aryl iodides [16], synthesis of isoindolinones from *N*-methoxybenzamides and alkenes [17], *ortho*-arylation of benzamides [18], decarboxylative *ortho*-acylation of *O*-methyl ketoximes [19], and decarboxylative *ortho*-acylation of azobenzenes [20]. We have also successfully extended the palladium-catalyzed C–H bond activation strategy to fullerene chemistry. It is known that palladium-catalyzed reactions via non-C–H activation routes have been used for the functionalization of C_{60} [21–27]. Luh and coworkers first developed the palladium-catalyzed [3+2] cycloaddition of C_{60} [21, 22]. The Itami group then reported the palladium-catalyzed hydroarylation of C_{60} with boronic acids and subsequent cleavage of organo(hydro)fullerenes [23, 24], as well as the regioselective unsymmetrical tetraallylation of C_{60} [25, 26]. In this review article, we will focus on the progress in functionalization of C_{60} by the palladium-catalyzed and ligand-directed C–H activation protocols to construct C_{60}-fused heterocycles and carbocycles.

2 Palladium-Catalyzed Formation of [60]Fullerene-Fused Heterocycles

The functionalization of C_{60} via the Pd-catalyzed C–H bond activation usually requires a directing group to achieve high regioselectivity and efficiency. When the directing group contains a heteroatom, it would take part in the formation of palladacycle and the subsequent insertion of C_{60} to give a larger palladacycle intermediate. Reductive elimination of the intermediate generates the C_{60}-fused heterocycle. In this section, we will describe several Pd-catalyzed reactions of C_{60} with different substrates bearing directing groups with a nitrogen or oxygen atom, which involves the formation of the C–N or C–O bond in the C_{60}-fused heterocycles.

2.1 Palladium-Catalyzed Reaction of [60]Fullerene with Anilides

In 2009, the synthesis of [60]fulleroindolines through the Pd-catalyzed heteroannulation of C_{60} with *o*-iodoanilines was reported [28]. Nevertheless, halide by-products were generated, and a large number of *o*-iodoanilines were expensive or difficult to be prepared, thus limiting further application of the present reaction. We had investigated the *ortho*-acetoxylation [13] of anilides via the palladium-

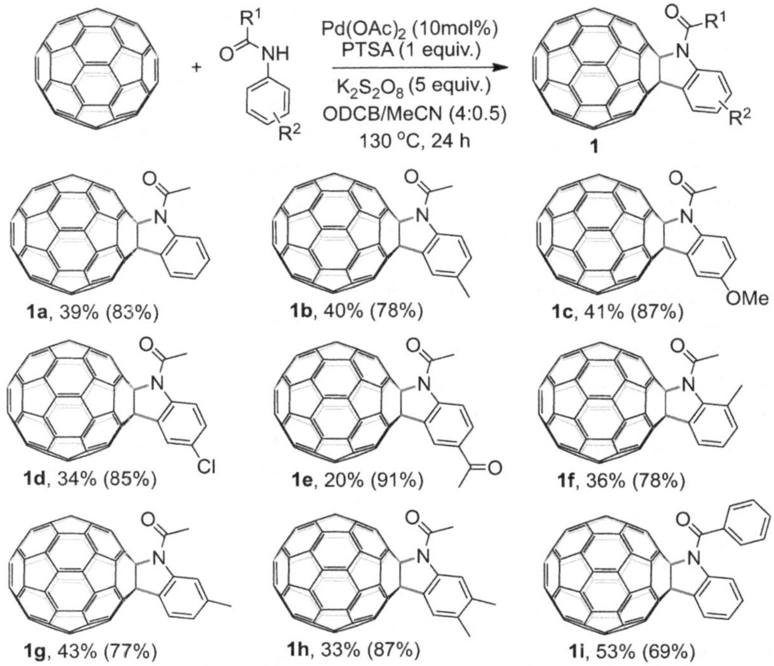

Scheme 1 Pd-catalyzed reaction of C_{60} with anilides

catalyzed C–H activation at that time. It is obvious that it would be highly desirable to utilize anilides instead of *o*-iodoanilines as starting material to functionalize C_{60} (Scheme 1) [29].

With the commercially available acetanilide as the model substrate to react with C_{60}, $K_2S_2O_8$ was found to be a better oxidant than Oxone, $Cu(OAc)_2$, and *p*-benzoquinone (BQ), and *p*-toluenesulfonic acid (PTSA) played a critical role for the success of current heteroannulation of C_{60}. It was noteworthy that a mixture of *ortho*-dichlorobenzene (ODCB, 4 mL) and acetonitrile (MeCN, 0.5 mL) was employed as the solvents. MeCN was added to increase the solubility of the employed inorganic salts.

As shown in Scheme 1, acetanilide and other acetanilides bearing electron-donating groups and weak electron-withdrawing group on the *para* position of the phenyl ring afforded products **1a–d** in 34–41% yields (78–87% yields based on consumed C_{60}). It should be pointed out that the yields in the parentheses of products **1a–d** as well as other products were calculated on the basis of consumed C_{60}. However, substrates with an electron-withdrawing ketone group and the *ortho*-substituent on the phenyl ring retarded the reaction obviously, giving the corresponding products **1e** and **1f** only in 20% and 36% yield, respectively, even by increasing the $Pd(OAc)_2$ loading to 50 mol% in pure ODCB. When anilides were substituted at the *meta* position, products **1g** and **1h** resulting from the reactions at

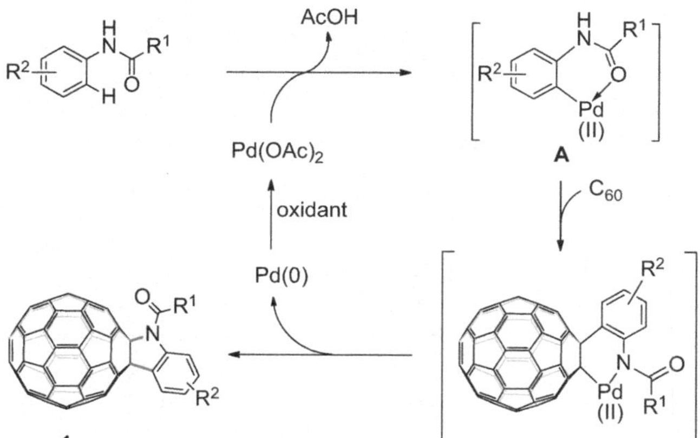

Scheme 2 Proposed reaction mechanism for the Pd-catalyzed reaction of C_{60} with anilides

the less sterically hindered positions were obtained regioselectively in good yields (33–43%). Intriguingly, *N*-benzoylated aniline was found to be the most effective substrate and provided product **1i** in 53% yield.

A plausible mechanism was proposed and is shown in Scheme 2 [29]. The reaction was supposed to be initiated by an amide-directed C–H deprotonation by AcO^- with the help of a Pd(II) species to give the intermediate **A**, followed by insertion of C_{60} into the arylpalladium bond to afford the intermediate **B**. Subsequent reductive elimination of the intermediate **B** produced fulleroindolines and Pd (0). The Pd(0) species was reoxidized to a Pd(II) species by the oxidant $K_2S_2O_8$ to complete the catalytic cycle. All the other annulation reactions of C_{60} to give [60] fullerene-fused heterocycles (vide infra) proceeded via a similar pathway and will not be described in details.

2.2 *Palladium-Catalyzed Reaction of [60]Fullerene with Benzamides*

In the early 2010, the palladium-catalyzed *ortho*-alkoxylation of *N*-methoxybenzamides was realized by using the CONHOMe group as a directing group [15]. Subsequently, the same directing group was utilized for the palladium-catalyzed synthesis of phenanthridinones from *N*-methoxybenzamides and aryl iodides [16] and the palladium-catalyzed synthesis of isoindolinones from *N*-methoxybenzamides and alkenes [17]. The simple amide $CONH_2$ group as the ligand was later also employed to direct the palladium-catalyzed *ortho*-arylation of benzamides [18]. Therefore, it was natural that the amide-directed C–H activation protocol was extended to fullerene chemistry. During our work on the palladium-catalyzed annulation of benzamides to C_{60}, Chuang and coworkers

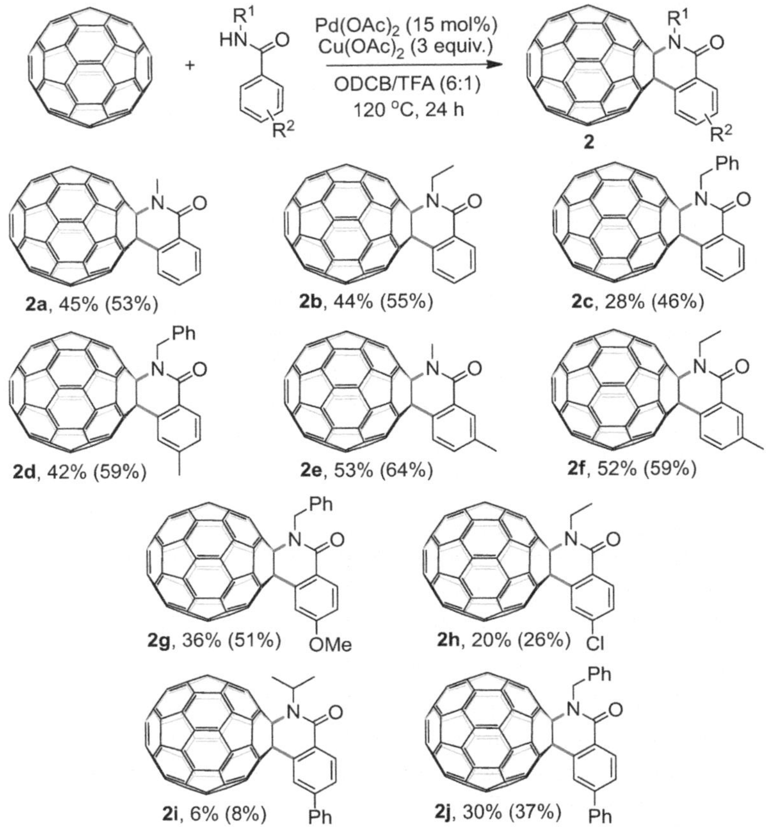

Scheme 3 Pd-catalyzed reaction of C_{60} with benzamides

independently discovered the same reaction. Thus, we jointly published these results (Scheme 3) [30].

Systematic screening of a range of oxidants and solvents for the Pd-catalyzed heteroannulation of C_{60} with *N*-methyl benzamide revealed that $Cu(OAc)_2$ and ODCB/TFA (6:1) performed best, giving fulleroisoquinolinone **2a** in 45% yield. Then a variety of substrates with either electron-donating or electron-withdrawing groups on their benzamide aryl rings were examined, and fulleroisoquinolinones **2b–j** were isolated in 6–53% yields (8–64% yield based on converted C_{60}). Substrates containing electron-donating groups generally afforded the corresponding fulleroisoquinolinones **2d–g** in good yields (36–53%). In addition, substrates with a *meta* substituent underwent regioselective C–H activations at their less hindered positions. In contrast, substrates bearing electron-withdrawing groups such as the chloro and phenyl units provided products **2h–j** in only moderate yields (6–30%). It should be noted that under the standard conditions, the reactions of amides bearing *N*-benzyl substituent with C_{60} yielded debenzylated products. Therefore, only

0.2 mL of TFA was used for these substrates to give the desired products **2c**, **2d**, **2g**, and **2j** in 28, 42, 36, and 30% yields, respectively. The extremely low yield of **2i** may be attributed to the bulkiness of the isopropyl group that made the formation of palladacycle difficult.

Similar to our previous conditions [29], we found that the combination of $K_2S_2O_8$ and mesitylenesulfonic acid (MesSA) was also applicable to the palladium-catalyzed reaction of C_{60} with benzamides.

2.3 *Palladium-Catalyzed Reaction of [60]Fullerene with N-Sulfonyl-2-aminobiaryls*

2-Aminobiaryls have been utilized to construct different sizable heterocycles. The Pd(II)-catalyzed cross-coupling of *N*-sulfonyl-2-aminobiaryls with alkenes to afford phenanthridine derivatives [31] and C–H bond activation/intramolecular amidation to synthesize carbazoles from 2-acetaminobiphenyls [32] have been demonstrated. Chuang and coworkers then developed the Pd(II)-catalyzed synthesis of [60]fulleroazepines from the reaction of C_{60} and *N*-sulfonyl-2-aminobiaryls through C–H bond activation and sequential C–C and C–N bond formation at room temperature (Scheme 4) [33].

Screening the Pd-catalyzed reaction of C_{60} with *N*-tosyl-2-aminobiphenyl showed that 10 mol% of $Pd(OAc)_2$, 3 equiv. of $KHSO_5$, and 1 equiv. of PSTA in 6.2 mL of ODCB/TFA (6:0.2) at ambient temperature were the optimal conditions. It was found that the new system using hybrid acids of PTSA and TFA was important for synthesis of the 7-membered-ring heterocycle under very mild conditions. This success was believed to be the unusual stability made by the hybrid acid system that stabilized the eight-membered palladacycle intermediates.

With the optimized results in hand, the scope and generality of the reaction by employing a variety of substrates containing electron-donating and electron-withdrawing groups on both the aromatic rings of 2-aminobiaryls were next investigated. The *N*-sulfonyl-2-aminobiaryls without substituent on any aryl rings underwent reaction with C_{60} smoothly to give **3a** and **3b** in 37% and 33% isolated yields, respectively. Substrates with electron-donating methyl groups on the aryl rings afforded **3c–g** in 31–33% yields. It was also observed that C–H activations occurred regioselectively at the less hindered and more electron-rich *para* positions relative to the methyl substituents to give **3c**, **3f**, and **3g**. Surprisingly, substrates bearing electron-withdrawing chloro group gave products **3h** and **3i** in excellent yields (40% and 51%). Products **3j–l** were isolated in lower chemical yields (19–22%) due to the formation of bis- and multi-addition products. It was suggested that the presence of an electron-donating methoxy group made the substrates more reactive. However, a relatively good yield (38%) could be obtained for **3m** with the additional presence of an electron-pulling chloro group. Substrates with a trifluoromethoxy (OCF_3) group afforded **3n–o** in 28–54% yields.

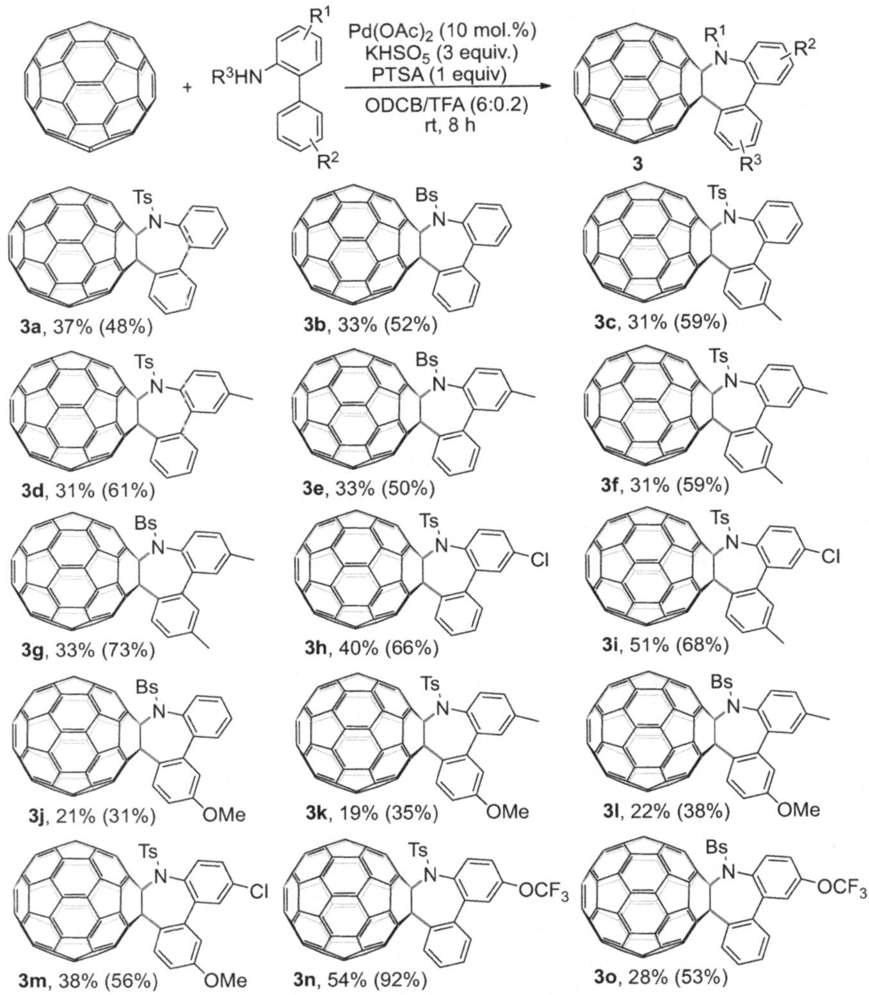

Scheme 4 Pd-catalyzed reaction of C_{60} with *N*-sulfonyl-2-aminobiaryls

2.4 Palladium-Catalyzed Reaction of [60]Fullerene with N-Benzyl Sulfonamides

It was previously reported that the amide-directed *ortho* C–H bond activation occurred selectively at the benzamide phenyl ring rather than at the phenyl ring of the *N*-benzyl moiety for the Pd-catalyzed reaction of C_{60} with *N*-benzyl benzamides [30]. Intriguingly, in efforts to extend *N*-benzyl benzamides to other substrates, we found that the usage of *N*-benzyl sulfonamides changed the site

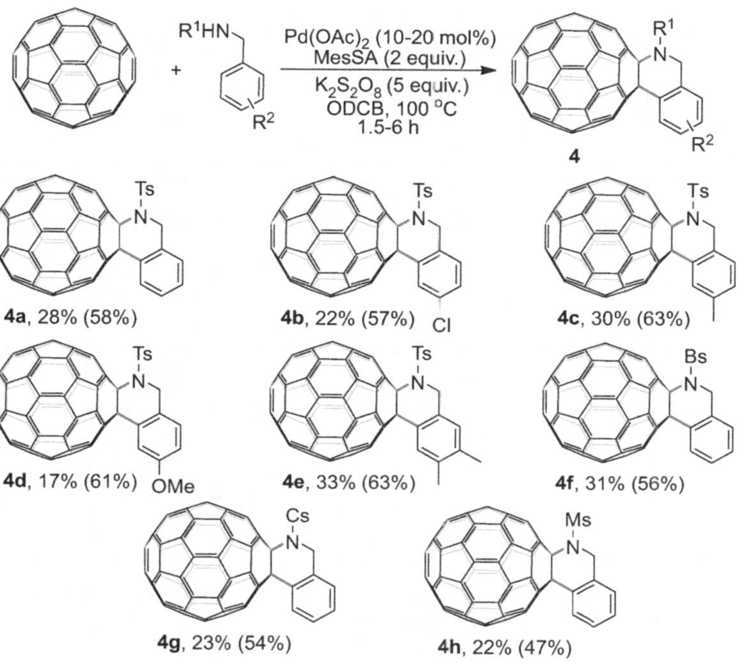

Scheme 5 Pd-catalyzed reaction of C_{60} with *N*-benzyl sulfonamides

selectivity of the *ortho* C–H activation to the phenyl ring of the *N*-benzyl moiety, which provided C_{60}-fused tetrahydroisoquinolines (Scheme 5) [34].

Initially, the Pd-catalyzed reaction of C_{60} with *N*-benzyl-4-toluenesulfonamide was selected as the model reaction for optimization study. It was found that $K_2S_2O_8$ was the best oxidant for the reaction at 100°C. Although both TFA (0.5 mL) and MesSA (2 equiv.) were effective additives to promote the reaction, MesSA was superior to TFA in terms of both the amount of the used acid and product yield based on converted C_{60}. A wide variety of *N*-benzyl sulfonamides with electron-withdrawing group (4-Cl) and electron-donating groups (4-Me, 4-MeO, 3,4-$(Me)_2$) on the benzylamine moiety could be used to afford products **4b–e** in 17–33% yields. It should be noted that the substrate with a 4-MeO group tended to generate more by-products under the employed standard conditions. As a result, the reaction temperature and the amount of MesSA were lowered to 70°C and 1 equiv., respectively, and the desired product **4d** could be obtained in 17% yield. Product **4e** was formed regioselectively from *N*-(3,4-dimethylbenzyl)-4-toluenesulfonamide in 33% yield due to steric hindrance. In addition, different functional groups attached to the nitrogen atom such as benzenesulfonyl (Bs), 4-chlorobenzenesulfonyl (Cs), and methanesulfonyl (Ms) worked well, and the corresponding products **4f–h** were obtained in 22–31% yields.

Intriguingly, the obtained C_{60}-fused tetrahydroisoquinolines could be transformed to C_{60}-fused indanes through a MesSA-promoted rearrangement.

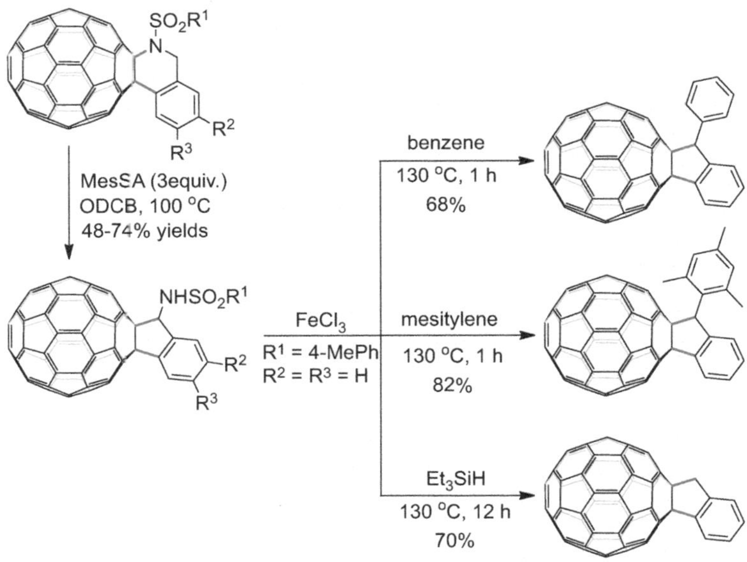

Scheme 6 MesSA-promoted rearrangement of C_{60}-fused tetrahydroisoquinolines to C_{60}-fused indanes and subsequent $FeCl_3$-promoted reactions

Furthermore, the sulfonamide group in C_{60}-fused indanes could be removed or replaced with an aryl group by the $FeCl_3$-promoted reduction and Friedel–Crafts-type reaction (Scheme 6) [34].

2.5 Palladium-Catalyzed Reaction of [60]Fullerene with N-(2-Arylethyl) Sulfonamides

2-Arylethylamines have been exploited to construct heterocycles via C–H activation reactions. Orito et al. reported the Pd-catalyzed direct aromatic carbonylation of 2-arylethylamines [35]. The Yu group developed the Pd-catalyzed intramolecular C–H aminations to prepare indolines from 2-arylethyl triflamides and 2-arylethyl 2-pyridylsulfonamides [36–38]. However, using 2-arylethylmine derivatives to construct seven-membered tetrahydrobenzazepines via C–H activation protocol had not been reported yet. We succeeded in the heteroannulation of C_{60} with various *N*-(2-arylethyl) sulfonamides to give the rare C_{60}-fused tetrahydrobenzazepines through the Pd-catalyzed C–H activation protocol (Scheme 7) [39].

The reaction of C_{60} with *N*-phenethyl-*p*-toluenesulfonamide as the model reaction was initially chosen to screen the optimal conditions. By employing the similar conditions for the Pd-catalyzed reaction of C_{60} with *N*-benzyl sulfonamides [34], that is, a combination of $K_2S_2O_8$ and MesSA, the model reaction gave a yield of only 6%. Further optimization revealed that $Cu(OAc)_2$ as the oxidant and ODCB/

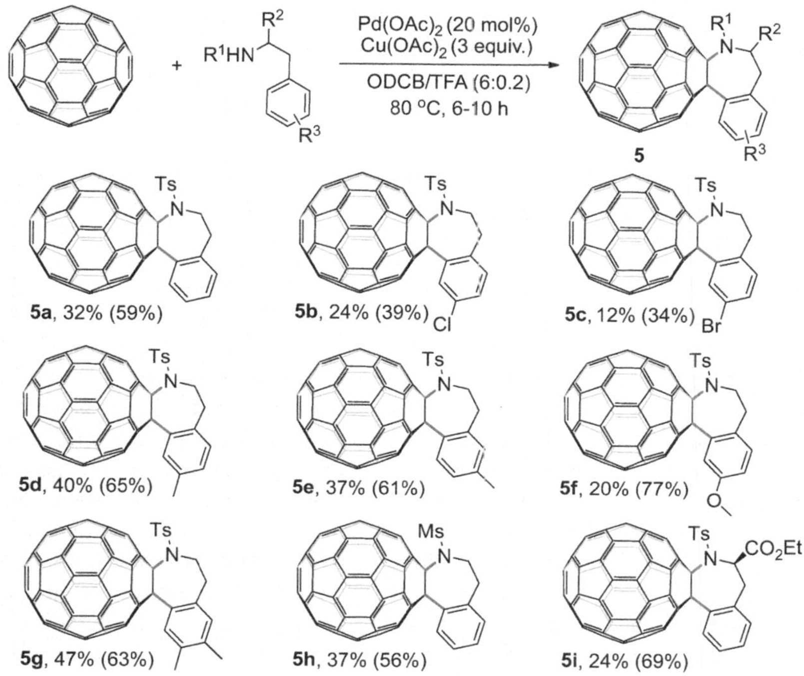

Scheme 7 Pd-catalyzed reaction of C_{60} with *N*-(2-arylethyl) sulfonamides

TFA (6.2 mL, v/v = 30:1) as the solvent were the optimal reaction conditions to provide product **5a** in 32% yield.

N-Phenethyl-*p*-toluenesulfonamide and other substrates with either electron-withdrawing or electron-donating groups on the 2-arylethylamine ring worked well and gave the desired products **5a–g** in 12–47% yields. Among them, substrates with an electron-withdrawing chloro or bromo group at the *para* position of the phenyl ring gave **5b** and **5c** in lower yields based on consumed C_{60} because they tended to generate some fullerene by-products. Furthermore, 2-phenethylamine with the Ms group attached to the nitrogen atom was also effective and gave **5h** in 37% yield. Interestingly, the tosylamide of L-phenylalanine was also reactive under our optimal conditions, and the novel C_{60}-fused amino acid derivative **5i** was obtained in 24% yield.

It is noteworthy that the synthesis of [60]fulleroazepines from rigid *N*-sulfonyl-2-aminobiaryls required a hybrid acid system (PTSA/TFA) [33]. In the present work, flexible *N*-(2-arylethyl) sulfonamides incorporating an alkyl chain, even the chiral amino acid moiety, were utilized as the substrates, and only TFA was required as the acid additive. In addition, $Pd(OAc)_2/Cu(OAc)_2$ in ODCB-TFA (7 mL, v/v = 6 : 1), which was similar to our system, was totally inert in Chuang's work. The rare C_{60}-fused seven-membered ring products were supposed to be generated by C–H activation via a hard-to-form eight-membered-ring palladacycle intermediate, which still remains as a great challenge.

2.6 *Palladium-Catalyzed Reaction of [60]Fullerene with Arylsulfonic Acids*

Carboxylic acids have been utilized to direct C–H bond activations, and C–H halogenation, arylation, alkylation, olefination, hydroxylation, and carboxylation reactions have been successfully realized. PTSA has been frequently added to promote Pd-catalyzed C–H activation reactions by increasing the electrophilicity of the Pd(II) center [29, 33, 40]. Just like carboxylic acids, it is possible that arylsulfonic acids can be functionalized via sulfonic acid group-directed C–H activation. However, such arylsulfonic acid derivatives had not been reported in the aforementioned reactions involving PTSA [29, 33, 40]. The electron-deficient aryl ring of arylsulfonic acids is less prone to the C–H activation step compared to that of aryl carboxylic acids. Nevertheless, we successfully achieved the synthesis of C_{60}-fused sultones by the Pd-catalyzed reaction of arylsulfonic acids with C_{60} via unprecedented sulfonic acid group-directed C–H bond activation (Scheme 8) [41].

Screening the reaction conditions showed that $K_2S_2O_8$ as the oxidant and ODCB (4 mL)/MeCN (0.5 mL) as the solvent at 140 °C for 24 h were the optimal conditions, similar to those for the previously reported Pd-catalyzed reaction of C_{60} with anilides [29].

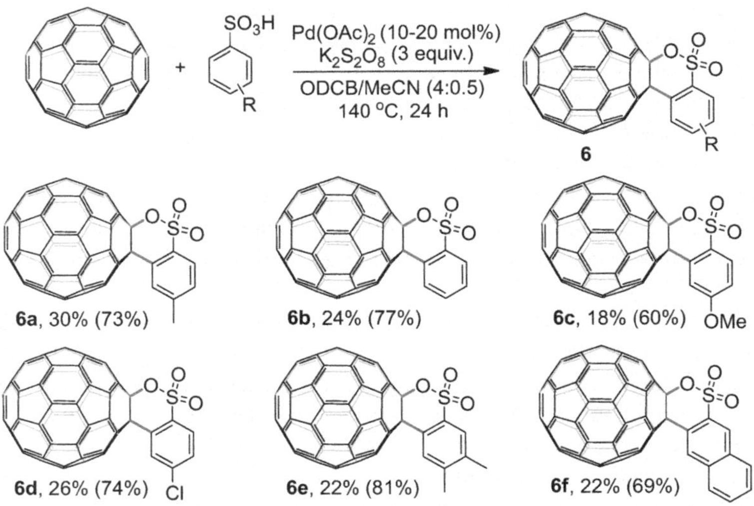

Scheme 8 Pd-catalyzed reaction of C_{60} with arylsulfonic acids

Arylsulfonic acids with both electron-donating and electron-withdrawing substituents on the phenyl ring could be smoothly transformed into the desired products **6a–e** in 18–30% yields. β-Naphthylsulfonic acid could also be employed and provided sultone **6f** in 22% yield. Although C_{60} could be successfully utilized in the Pd-catalyzed C–H activation of arylsulfonic acids, the attempts to functionalized PTSA with other olefins such as acrylates and styrenes failed under conditions previously employed for aromatic carboxylic acids or other combinations of $Pd(OAc)_2$, oxidants, and solvents, reflecting the unique property of C_{60}.

2.7 *Palladium-Catalyzed Reaction of [60]Fullerene with 2-Phenylethyl/Benzyl Alcohols*

The Pd-catalyzed hydroxyl-directed C–H activation reactions remain difficult to perform due to the possible oxidation, β-hydride elimination, and weak coordination of alcohols with Pd(II) [42–44]. Only three works on the Pd-catalyzed olefination [42], intramolecular cyclization [43], and carbonylation [44] of phenethyl alcohols had been reported before we investigated their application in fullerene chemistry. We recently achieved the hydroxyl-directed C–H activation/C–O cyclization reactions of phenethyl alcohols and benzyl alcohols with C_{60} to afford C_{60}-fused tetrahydrobenzooxepine and isochroman derivatives (Scheme 9) [45]. It should be noted that the Pd-catalyzed C–H activation of benzyl alcohols had no precedent.

Screening the $Pd(OAc)_2$-catalyzed reaction of representative 2-phenylethanol with C_{60} indicated that oxidant $K_2S_2O_8$ performed best and arylsulfonic acids such as PTSA and MesSA played critical role in the reaction. Unlike the previous Pd-catalyzed hydroxyl-directed C–H activations [26–28], our protocol does not require extra base additives as well as amino acid ligands. As shown in Scheme 9, phenylethyl alcohol gave the desired C_{60}-fused tetrahydrobenzooxepine **7a** in 40% yield. The methyl substituent at the benzylic position did not affect the reaction, and product **7b** was obtained in 42% yield. Substrates containing either an electron-donating group or electron-withdrawing group on the phenyl ring could also be used and provided products **7c–e** in 20–31% yields. In addition, it was found that benzyl alcohols were also compatible with the present reaction. Benzyl alcohol provided C_{60}-fused isochroman **7f** in 43% yield, showing a high efficiency of this C–H activation/C–O cyclization protocol. Nevertheless, 4-methylbenzyl alcohol and 3,4-dimethylbenzyl alcohol gave inferior results and afforded products **2g** and **2h** in 27% and 24% yields, respectively. A secondary alcohol could also be employed, and product **2i** was isolated in 20% yield.

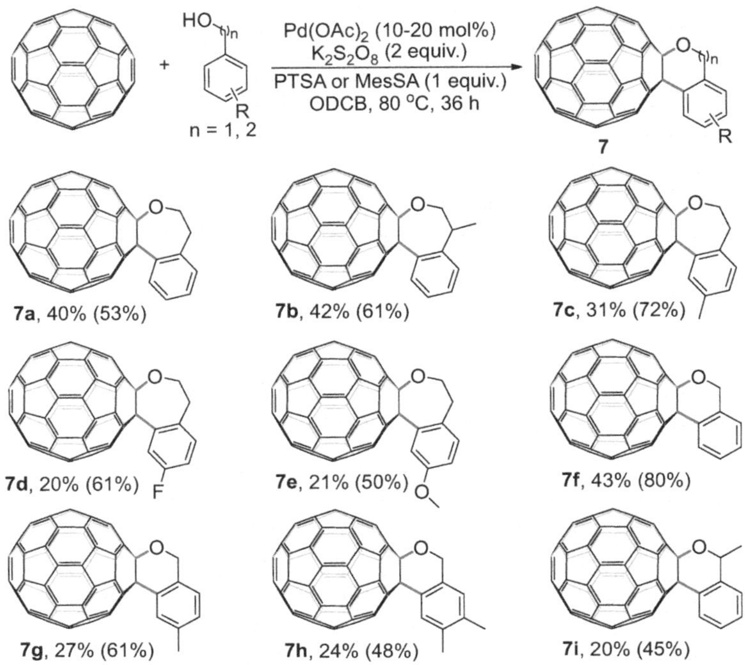

Scheme 9 Pd-catalyzed reaction of C_{60} with phenethyl and benzyl alcohols

3 Palladium-Catalyzed Reaction of [60]Fullerene with 2-Arylbenzoic Acids

The abovementioned palladium-catalyzed reactions of C_{60} via C–H activation involved directing groups with a nitrogen or oxygen atom, which participated in the formation of palladacycle intermediates and C_{60}-fused heterocycles. The attached heteroatoms in C_{60}-fused heterocycles would lower the LUMO energy levels of fullerene derivatives and thus are not beneficial to the improvement of open-circuit voltage (V_{oc}), a key factor for organic photovoltaic devices. Therefore, it is desirable to synthesize fullerene derivatives with two carbon atoms directly attached to the fullerene skeleton in order to achieve higher V_{oc}. The recent synthesis of phenanthrenes from 2-phenylbenzoic acids with alkynes through a decarboxylation/C–H activation sequence prompted us to investigate the palladium-catalyzed decarboxylative coupling of 2-arylbenzoic acids with C_{60} to obtain C_{60}-fused carbocycles (Scheme 10) [46].

Optimization of the reaction conditions revealed that $Cu(OAc)_2$ was the best oxidant and Lewis acid $ZnCl_2$ was crucial for the success of this formal [4+2] annulation reaction. It was believed that $ZnCl_2$ may facilitate the process of decarboxylation. Furthermore, the addition of K_2CO_3 and PivOH could improve the reaction efficiency.

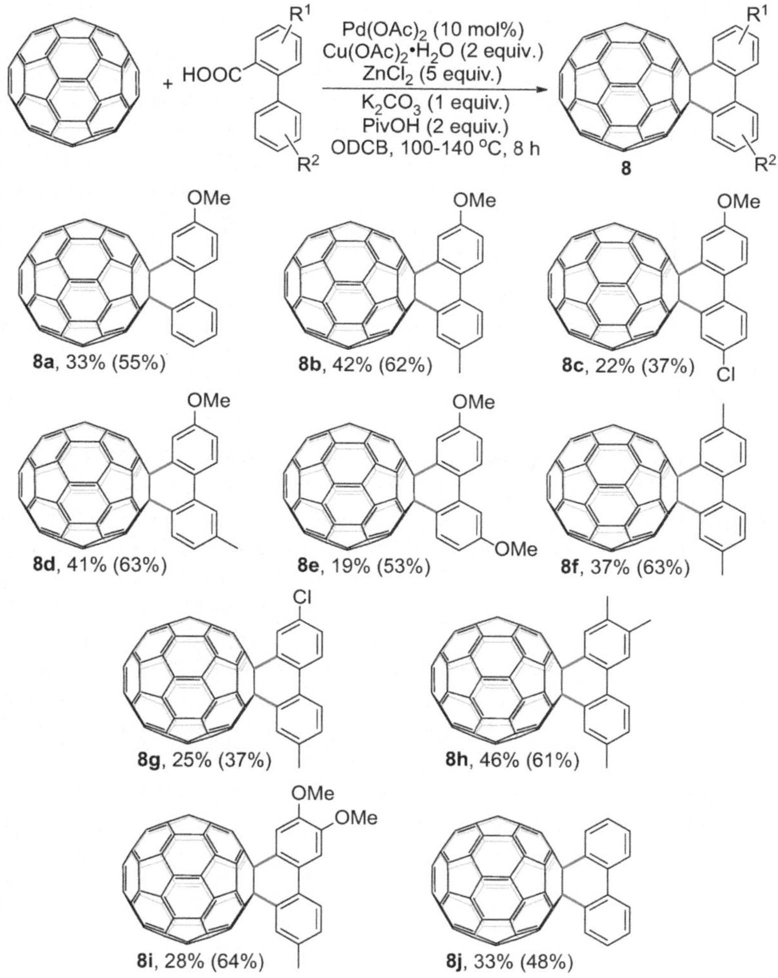

Scheme 10 Pd-catalyzed reaction of C_{60} with 2-arylbenzoic acids

4-Methoxy biphenyl-2-carboxylic acid with an electron-donating methyl group at the 4′-position exhibited good reactivity to afford **8b** in 42% yield. In comparison, only 22% yield was obtained for **8c** with an electron-withdrawing chloro group at the 4′-position. Similarly, the electronic effect was also very obvious when different substituent groups were situated at the 3′-position. The methyl group performed far better than the methoxy group, and products **8d** and **8e** were isolated in 41% and 19% yields, respectively. Furthermore, 4′-methyl-biphenyl-2-carboxylic acids with both electron-donating and electron-withdrawing substituents at the 4-position could be smoothly transformed into the desired products **8f–i** in 25–46% yields.

Finally, the simplest 2-phenylbenzoic acid could also be used to produce **8j** in 33% yield.

The cyclic voltammetry of **8a–j** showed that they had similar first reduction potentials to that of 6,6-phenyl-C_{61}-butyric acid methyl ester (PCBM), which has been widely used as an acceptor in organic photovoltaics. The LUMO levels of fullerene compounds are estimated by their onset reduction potentials (LUMO = $(E_1 + 4.8)$ eV). The high LUMO levels of the obtained C_{60}-fused carbocycles are expected to result in high V_{oc}, and thus, they may have potential application as acceptors in organic photovoltaic devices.

4 Conclusion

In summary, the palladium-catalyzed C–H bond activation protocols have been successfully exploited in the functionalization of C_{60}. Several types of substrates containing directing groups with a nitrogen or oxygen atom have been employed to form C_{60}-fused heterocycles. Anilides, benzamides, *N*-sulfonyl-2-aminobiaryls, *N*-benzyl sulfonamides, *N*-(2-arylethyl) sulfonamides, arylsulfonic acids, and 2-phenylethyl/benzyl alcohols have been employed to react with C_{60} to synthesize C_{60}-fused heterocycles. These C–H activation reactions involve palladacycle intermediates, of which ring size ranges from a 6-membered ring to a hard-to-form 8-memebered ring. The palladium-catalyzed decarboxylative coupling of 2-arylbenzoic acids with C_{60} to generate C_{60}-fused carbocycles has also been disclosed recently. It is expected that continual exploration of the C–H bond activation strategy to functionalize fullerenes will widen the substrate scope and produce more novel fullerene derivatives with different heterocycles and carbocycles.

References

1. Hirsch A, Brettreich M (2005) Fullerenes: chemistry and reactions. Wiley-VCH, Weinheim
2. Wang G-W, Li F-B (2007) Radical reactions of [60]fullerene mediated by manganese(III) acetate dihydrate. J Nanosci Nanotechnol 7:1162–1175
3. Matsuo Y, Nakamura E (2008) Selective multiaddition of organocopper reagents to fullerenes. Chem Rev 108:3016–3028
4. Wang G-W, Li F-B (2012) Transition metal salt-mediated radical reactions of [60]fullerene. Curr Org Chem 16:1109–1127
5. Nakamura E, Isobe H (2003) Functionalized fullerenes in water. The first 10 years of their chemistry, biology, and nanoscience. Acc Chem Res 36:807–815
6. Li C-Z, Yip H-L, Jen AK-Y (2012) Functional fullerenes for organic photovoltaics. J Mater Chem 22:4161–4177
7. Xu L-M, Li B-J, Yang Z, Shi Z-J (2010) Organopalladium(IV) chemistry. Chem Soc Rev 39:712–733

8. Engle KM, Mei T-S, Wasa M, Yu J-Q (2012) Weak coordination as a powerful means for developing broadly useful C-H functionalization reactions. Acc Chem Res 45:788–802
9. Topczewski JJ, Sanford MS (2015) Carbon–hydrogen (C–H) bond activation at Pd^{IV}: a frontier in C–H functionalization catalysis. Chem Sci 6:70–76
10. Wencel-Delord J, Glorius F (2013) C–H bond activation enables the rapid construction and late-stage diversification of functional molecules. Nat Chem 5:369–375
11. Beydoun K, Roger J, Boixel J, Le Bozec H, Guerchais V, Doucet H (2012) A straightforward access to photochromic diarylethene derivatives via palladium-catalysed direct heteroarylation of 1,2-dichloroperfluorocyclopentene. Chem Commun 48:11951–11953
12. Beydoun K, Zaarour M, Williams JAG, Roisnel T, Dorcet V, Planchat A, Boucekkine A, Jacquemin D, Doucet H, Guerchais V (2013) Palladium-catalyzed direct arylation of luminescent bis-cyclometalated iridium(III) complexes incorporating C^N- or O^O-coordinating thiophene-based ligands: an efficient method for color tuning. Inorg Chem 52:12416–12428
13. Wang G-W, Yuan T-T, Wu X-L (2008) Direct ortho-acetoxylation of anilides via palladium-catalyzed sp^2 C–H bond oxidative activation. J Org Chem 73:4717–4720
14. Jiang T-S, Wang G-W (2012) Palladium-catalyzed ortho-alkoxylation of anilides via C–H activation. J Org Chem 77:9504–9509
15. Wang G-W, Yuan T-T (2010) Palladium-catalyzed alkoxylation of *N*-methoxybenzamides via direct sp^2 C–H bond activation. J Org Chem 75:476–479
16. Wang G-W, Yuan T-T, Li D-D (2011) One-pot cascade C–C and C–N bond formation via palladium-catalyzed dual C–H activations: synthesis of phenanthridinones. Angew Chem Int Ed 50:1380–1383
17. Li D-D, Yuan T-T, Wang G-W (2011) Synthesis of isoindolinones via palladium-catalyzed C–H activation of *N*-methoxybenzamides. Chem Commun 47:12789–12791
18. Li D-D, Yuan T-T, Wang G-W (2012) Palladium-catalyzed ortho-arylation of benzamides via direct sp^2 C–H bond activation. J Org Chem 77:3341–3347
19. Cai H-T, Li D-D, Liu Z, Wang G-W (2013) Palladium-catalyzed decarboxylative ortho-acylation of *O*-methyl ketoximes via direct sp^2 C–H bond activation. Acta Chim Sinica 71:717–721
20. Li Z-Y, Li D-D, Wang G-W (2013) Palladium-catalyzed decarboxylative ortho acylation of azobenzenes with α-oxocarboxylic acids. J Org Chem 78:10414–10420
21. Shiu L-L, Lin T-I, Peng S-M, Her G-R, Ju D D, Lin S-K, Hwang J-H, Mou C. Y, Luh T-Y (1994) Palladium-catalysed [3+2] cycloaddition of trimethylenemethane (TMM) and fullerene. Observation of the room-temperature fluorescence spectrum of the TMM-C_{60} adduct. J Chem Soc Chem Commun 647–648
22. Shen CKF, Chien K-M, Liu T-Y, Lin T-I, Her G-R, Luh T-Y (1995) Palladium-catalyzed [3 +2] cycloaddition of 60-fullerene with cis-$HOCH_2CH{=}CHCH_2OCO_2Et$. Tetrahedron Lett 36:5383–5384
23. Mori S, Nambo M, Chi L-C, Bouffard J, Itami K (2008) A bench-stable Pd catalyst for the hydroarylation of fullerene with boronic acids. Org Lett 10:4609–4612
24. Nambo M, Itami K (2009) Palladium-catalyzed carbon–carbon bond formation and cleavage of organo(hydro)fullerenes. Chem Eur J 15:4760–4764
25. Nambo M, Wakamiya A, Yamaguchi S, Itami K (2009) Regioselective unsymmetrical tetraallylation of C_{60} through palladium catalysis. J Am Chem Soc 131:15112–15113
26. Nambo M, Wakamiya A, Itami K (2012) Palladium-catalyzed tetraallylation of C_{60} with allyl chloride and allylstannane: mechanism, regioselectivity, and enantioselectivity. Chem Sci 3:3474–3481
27. Lu S, Jin T, Bao M, Asiri AM, Yamamoto Y (2012) Palladium-catalyzed bisfunctionalization of active alkenes by β-acetonitrile-α-allyl addition: application to the synthesis of unsymmetric 1,4-di(organo)fullerene derivatives. Tetrahedron Lett 53:1210–1213
28. Zhu B, Wang G-W (2009) Synthesis of [60]fulleroindolines: palladium-catalyzed heteroannulations of [60]fullerene with *o*-iodoanilines. J Org Chem 74:4426–4428

29. Zhu B, Wang G-W (2009) Palladium-catalyzed heteroannulation of [60]fullerene with anilides via C−H bond activation. Org Lett 11:4334–4337
30. Chuang S-C, Rajeshkumar V, Cheng C-A, Deng J-C, Wang G-W (2011) Annulation of benzamides with [60]fullerene through palladium(II)-catalyzed C–H bond activation. J Org Chem 76:1599–1604
31. Miura M, Tsuda T, Satoh T, Pivsa-Art S, Nomura M (1998) Oxidative cross-coupling of *N*-(2′-phenylphenyl)benzene-sulfonamides or benzoic and naphthoic acids with alkenes using a palladium – copper catalyst system under air. J Org Chem 63:5211–5215
32. Tsang WCP, Zheng N, Buchwald SL (2005) Combined C–H functionalization/C–N bond formation route to carbazoles. J Am Chem Soc 127:14560–14561
33. Rajeshkumar V, Chan F-W, Chuang S-C (2012) Palladium-catalyzed and hybrid acids-assisted synthesis of [60]fulleroazepines in one pot under mild conditions: annulation of *N*-sulfonyl-2-aminobiaryls with [60]fullerene through sequential C–H bond activation, C–C and C–N bond formation. Adv Synth Catal 354:2473–2483
34. Su Y-T, Wang Y-L, Wang G-W (2012) Palladium-catalysed heteroannulation of [60]fullerene with *N*-benzyl sulfonamides and subsequent functionalisation. Chem Commun 48:8132–8134
35. Orito K, Horibata A, Nakamura T, Ushito H, Nagasaki H, Yuguchi M, Yamashita S, Tokuda M (2004) Preparation of benzolactams by $Pd(OAc)_2$-catalyzed direct aromatic carbonylation. J Am Chem Soc 126:14342–14343
36. Li J-J, Mei T-S, Yu J-Q (2008) Synthesis of indolines and tetrahydroisoquinolines from arylethylamines via Pd^{II}-catalyzed C–H activation reactions. Angew Chem Int Ed 47:6452–6455
37. Mei T-S, Wang X, Yu J-Q (2009) Pd(II)-catalyzed amination of C–H bonds using single-electron or two-electron oxidants. J Am Chem Soc 131:10806–10807
38. Mei T-S, Leow D, Xiao H, Laforteza BN, Yu J-Q (2013) Synthesis of indolines via Pd(II)-catalyzed amination of C–H bonds using $PhI(OAc)_2$ as the bystanding oxidant. Org Lett 15:3058–3061
39. Su Y-T, Wang Y-L, Wang G-W (2014) Palladium-catalyzed heteroannulation of [60]fullerene with *N*-(2-arylethyl) sulfonamides via C–H bond activation. Org Front Chem 1:689–693
40. Boele MDK, van Strijdonck GPF, de Vries AHM, Kamer PCJ, de Vries JG, Leeuwen PWNM (2002) Selective Pd-catalyzed oxidative coupling of anilides with olefins through C–H bond activation at room temperature. J Am Chem Soc 124:1586–1587
41. Li F, Liu T-X, Wang G-W (2012) Synthesis of [60]fullerene-fused sultones via sulfonic acid group-directed C–H bond activation. Org Lett 14:2176–2179
42. Lu Y, Wang D-H, Engle KM, Yu J-Q (2010) Pd(II)-catalyzed hydroxyl-directed C–H olefination enabled by monoprotected amino acid ligands. J Am Chem Soc 132:5916–5921
43. Wang X, Lu Y, Dai H-X, Yu J-Q (2010) Pd(II)-catalyzed hydroxyl-directed C–H activation/C–O cyclization: expedient construction of dihydrobenzofurans. J Am Chem Soc 132:12203–12205
44. Lu Y, Leow D, Wang X, Engle KM, Yu J-Q (2011) Hydroxyl-directed C–H carbonylation enabled by mono-N-protected amino acid ligands: an expedient route to 1-isochromanones. Chem Sci 2:967–971
45. Zhai W-Q, Peng R-F, Jin B, Wang G-W (2014) Synthesis of [60]fullerene-fused tetrahydrobenzooxepine and isochroman derivatives via hydroxyl-directed C–H activation/C–O cyclization. Org Lett 16:1638–1641
46. Zhou D-B, Wang G-W (2015) Palladium-catalyzed decarboxylative annulation of 2-arylbenzoic acids with [60]fullerene via C–H bond activation. Org Lett 17:1260–1263

Top Organomet Chem (2016) 55: 137–188
DOI: 10.1007/3418_2015_134

Published online: 19 August 2015

Ruthenium(II)-Catalysed Functionalisation of C–H Bonds with Alkenes: Alkenylation versus Alkylation

Christian Bruneau and Pierre H. Dixneuf

Dedicated to the memory of Guy Lavigne

Abstract This chapter describes the recent achievements since 2011 of ruthenium(II)-catalysed transformations of sp^2C–H bonds of a variety of functional arenes, heterocycles, alkenes and ferrocene derivatives on their reaction with alkenes, either via oxidative dehydrogenation to produce functional alkenes or via formal insertion of alkene into C–H bonds to afford alkylated products. The regioselectivity of *ortho* C–H bond alkenylation or alkylation is shown to be directed by both strongly and weakly coordinating functional groups from N-heterocycles and bidentate groups to carbonyl-containing groups such as ketone, ester, amide, carbamate and sulfonic acid derivatives. The ruthenium(II) catalysts often based on $[RuCl_2(p\text{-cymene})]_2$ or $RuCl_2(PPh_3)_3$ derivatives sometimes require the presence of a halide abstractor and an oxidant. The alkenylation of heterocycles will be shown to occur in the presence of base and their alkylation in the presence of proton source to give branched or linear alkylated isomers. Discussion of catalytic mechanism will involve the initial formation of a cyclometallate via C–H bond deprotonation and that of an intermediate resulting from alkene insertion into the Ru–C bond before its evolution to alkenylation or alkylation.

Keywords Alkenylation · Alkylation · Alkene hydroarylation · C–C bond cross coupling · Directing groups · Ruthenium(II) catalysis · sp^2C–H bond activation

---

C. Bruneau (✉) and P.H. Dixneuf
Institut des Sciences Chimiques de Rennes, UMR 6226 CNRS-Université de Rennes, "Organometallics: Materials and Catalysis" Centre for Catalysis and Green Chemistry, Campus de Beaulieu, 35042 Rennes, France
e-mail: christian.bruneau@univ-rennes1.fr; pierre.dixneuf@univ-rennes1.fr

Contents

1 Introduction

Metal-catalysed regioselective functionalisation of C–H bonds for the cross-coupling formation of C–C and C–heteroatom bonds has brought a revolution in synthetic methods, by providing atom- and step-economical processes, for the production of complex pharmaceutical molecules, industrial intermediates or molecular materials [1–8].

Besides successful C–C bond cross couplings from sp^2C–H bonds promoted by palladium [7, 9–14] and rhodium [14–20] catalysts, the Murai group since 1993 has shown that lower-cost ruthenium(0) catalysts offer the direct access to functional arenes, heterocycles and alkenes, via initial directed ruthenium(0) insertion into the sp^2C–H bond [21–26].

More recently, stable ruthenium(II) catalysts have been shown to promote the directed sp^2C–H bond activation and functionalisation, initiated by the pioneer work since 2001 of Oi and Inoue [27, 28]. Some applications, even in water as solvent, have been presented in several reviews especially for arylations and heteroarylations of (hetero)aromatic compounds [29–35]. It was actually shown that the ruthenium(II) activation of arene sp^2C–H bond is *ortho*-directed by the functional group and easily takes place at room temperature via C–H bond deprotonation by cooperative action of the Ru(II) site and carbonate [36] or external carboxylate [37, 38], initially leading to a cyclometallated intermediate on coordination of the functional group and C–Ru bond formation.

The easy ruthenium(II) activation of C–H bond by carboxylate deprotonation has been applied to the oxidative dehydrogenation of two sp^2C–H bonds as an alternative route to polyfunctional alkenes, by cross coupling of a (hetero)arene and an alkene, a reaction which was previously discovered using Pd(II) catalysts and oxidant by Moritani and Fujiwara [1, 39, 40]. A general method of ruthenium(II)-catalysed alkenylation of arene and heterocycle directly on reaction with alkenes was demonstrated only recently in 2011. First, Satoh and Miura showed the alkenylation of heteroarylcarboxylic acids [41], Ackermann alkenylated arylcarboxylic acids in water, [42] Bruneau and Dixneuf performed the alkenylation of heteroaryl arenes with catalytic amount of Cu(II) [43] and Jeganmohan demonstrated that ketone could direct alkenylation [44].

Since 2011, many applications of this Ru(II)-catalysed *alkenylation* reaction have shown that a variety of weakly bonding functional groups (FG) could direct this *ortho* C–H bond functionalisation with alkenes of arene and heterocycle derivatives, thus offering a low-cost, general route to multifunctional alkenes, some of them were profitably used for sequential oxa- or aza-Michael addition (Eq. 1).

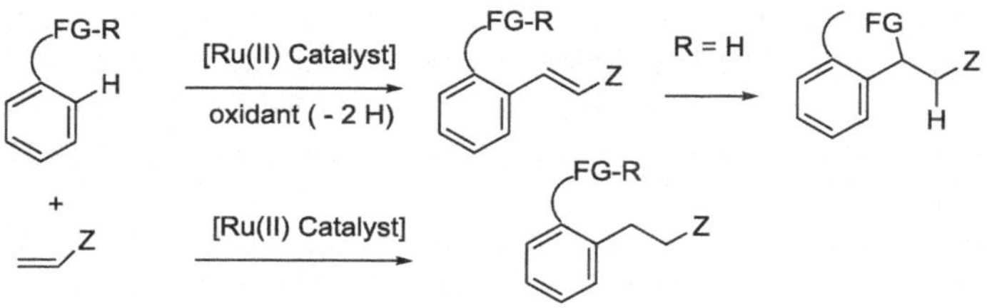

(1)

(2)

More importantly, the Ru(II)-catalysed reaction of functional (hetero)arenes with alkenes can also lead to the formal *alkylation* at the *ortho*-position of the functional directing group. It was first observed for the hydroarylation of ethylene with Ru(II) catalyst by Gunnoe et al. [45]. This Ru(II)-catalysed alkylation corresponds to the formal insertion of the alkene into the *ortho* C–H bond (Eq. 2) as it was previously observed to be the main product in the Murai reaction but with Ru(0) catalyst [21, 23]. Thus, it becomes crucial to understand how to orientate the reaction of alkenes with (hetero)arenes towards alkylation or alkenylation with ruthenium(II) catalytic systems and explore the role of additional oxidant. Only a few examples of ruthenium(II)-catalysed alkenylations have appeared in recent reviews [32, 46–49].

The objective of this chapter is to show the regioselective *alkenylation or alkylation*, on reaction with alkenes of functional arenes, heteroarenes and alkenes promoted by simple ruthenium(II) catalysts, as new routes to promote sp^2C–H bond activation for C–C cross coupling. This chapter will cover the literature since the first direct alkenylations in 2011 till the end of 2014 even if a few 2015 crucial reports will be cited. This chapter will provide opportunities to discuss the ruthenium(II) sp^2C–H bond activation mechanism leading to cyclometallate intermediates and their further functionalisation of the (C–H) carbon leading to C–C cross couplings.

2 Ruthenium(II)-Catalysed Alkenylation of sp²C–H Bonds with Alkenylboronic Acids

Ruthenium-catalysed alkenylation of sp²C–H bonds directly with alkenes has been initially performed by the Murai group mostly using ruthenium(0) catalyst precursors such as $RuH_2(CO)(PPh_3)_3$ or $Ru_3(CO)_{12}$ [21–26]. However, it was often in competition with preferential alkylation by formal insertion of alkene into the same *ortho* C–H bond [26, 50–52] as also shown by Genet with $[RuCl_2(p\text{-cymene})]_2/MO_2CH$ catalyst [53, 54]. It is noteworthy that Chatani and Kakiuchi have recently succeeded to perform *ortho*-alkenylation of arenes, linked to a variety of heterocycles as directing groups, with another Ru(0) species Ru(COD)(COT), and this catalyst allows alkenylation with functional alkenes such as alkenyl esters [55] and alkenyl carbonates [56].

Ruthenium-catalysed regioselective alkenylations of aromatic sp²C–H bonds have been easily performed first with alkenylboron derivatives and since 2011 more generally simply by oxidative dehydrogenation with alkenes with ruthenium(II) catalysts. These two approaches will be successively presented.

Alkenylarenes have been produced by reaction of arylboronic acids with alkenes in the presence of ruthenium(II) catalysts with a Cu(II) oxidant by Brown [57, 58] (Eq. 3). This reaction corresponds to the functionalisation of alkene C–H bonds, and by contrast to palladium catalysts, it tolerates arene carbon–halide bonds.

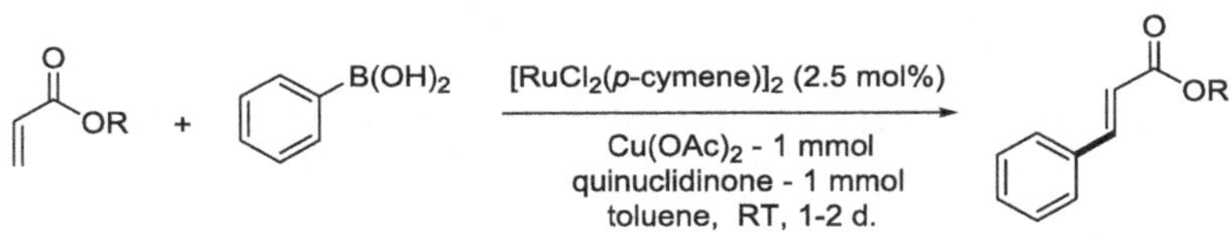

(3)

Chatani and Kakiuchi have demonstrated that a ruthenium(II) cyclometallate catalysed the selective monoalkenylation of aromatic ketones, containing a tolerated ether group, at the *ortho* C–H bond of the ketone group. A related cyclometallate was demonstrated to be formed on reaction of the ketone with the Ru(0) species arising from $RuH_2(CO)(PPh_3)_3$ (Eq. 4) [59].

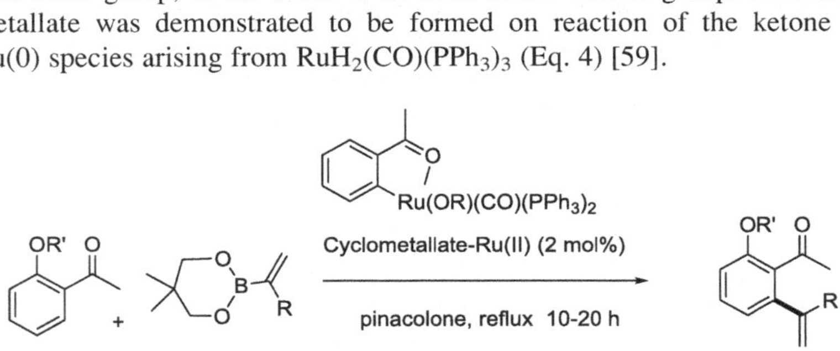

(4)

Recently, Jeganmohan has developed the *ortho*-alkenylation of *N*-alkylbenzamides with alkenylboronic acids in the presence of a simple $[RuCl_2(p\text{-cymene})]_2/AgSbF_6$ catalytic system, with 1 equiv. of Ag_2O in THF. The reaction leads, via a five-membered Ru(II) cyclometallate, to the clean functionalisation of *ortho* C–H bond via C–C bond cross coupling (Eq. 5) [60]. Ag_2O accelerates the transmetallation of the alkenyl group from boron to the ruthenium(II) centre.

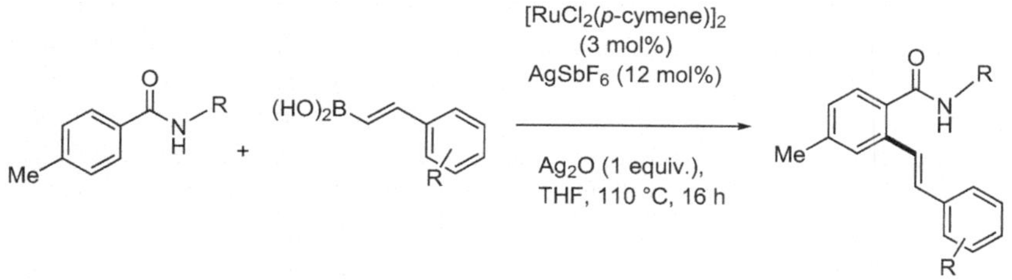

(5)

3 Alkenylation of sp^{2}C–H Bonds with Alkenes and Ruthenium(II) Catalysts

The synthesis of functional olefins has made a tremendous improvement with the discovery of the Mizoroki–Heck reaction by combination of an alkene with an aromatic halide, catalysed by palladium(0) catalysts [61]. The Fujiwara–Moritani alkenylation has made another decisive step for the formation of Heck-type products, but via the cross oxidative dehydrogenation of alkene and (hetero)aromatic C–H bonds, catalysed by palladium(II) catalysts in the presence of an oxidant [39]. These discoveries have motivated the search for simple and cheaper catalysts based on ruthenium(II) precursors for the dehydrogenative coupling of an alkene with arene or heterocycle sp^{2}C–H bonds.

The first example of alkenylation of an arene with ruthenium (II or III) catalysts was actually presented by Milstein in 2001 [62], for the oxidative coupling of acrylate with an arene without directing group. The reaction was catalysed by either $RuCl_3.xH_2O$, $[RuCl_2(CO)_3]_2$ or $[RuCl_2(C_6H_6)]_2$ but under weak pressure of carbon monoxide and using oxygen as the only oxidant. Ruthenium(0) catalyst such as $Ru_3(CO)_{12}$ was less efficient (Eq. 6).

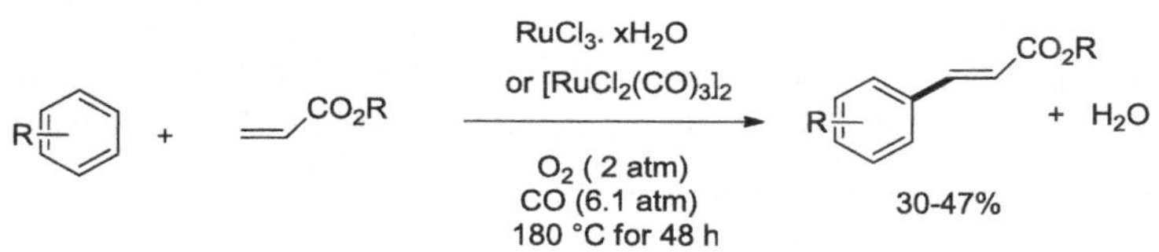

(6)

Then the cross coupling of two sp^{2}C–H bonds between an alkene and an arene has been achieved first by Yi in 2009 using of Ru(II) hydride precursor and recently since 2011 by a variety of groups using the simple $[RuCl_2(p\text{-cymene})]_2$ catalyst precursor with an oxidant. These two processes will be successively presented.

3.1 Ruthenium Hydride-Catalysed Alkenylation of Arenes with Functional Alkenes

Chae S. Yi has shown that the ruthenium(II) hydride catalyst [RuH(CO)(PCy_3)(C_6H_6)][BF_4] in the presence of a strong acid $HBF_4.OEt_2$ could perform the alkenylation at the *ortho*-position of an aromatic ketone or amide with a rather good selectivity. He first succeeded the alkenylation with styrene of an aromatic ketone at the *ortho* C–H bond [63]. The directed activation of the *ortho* C–H bond is demonstrated by the alkenylation of the naphthylketone. The reaction is believed to generate a metallacycle which inserts the alkene and leads either to the alkenylated ketone by β-elimination or to indene after carbonyl insertion into the Ru–C bond (Eq. 7) [63].

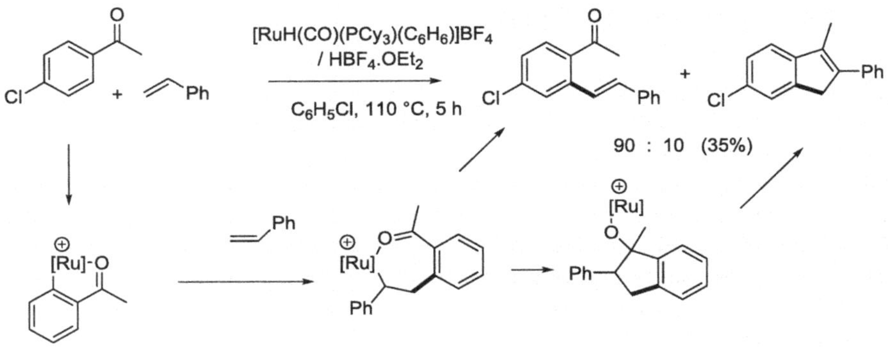

(7)

Chae S. Yi also explored, in the presence of the same [RuH(CO)(PCy_3)(C_6H_6)][BF_4] catalyst, the reaction of arylamides with cyclic and disubstituted alkenes under mild conditions. Cyclopentene and cycloheptene led to the preferential alkenylation at the *ortho*-position of the $CONR_2$ or CONHR directing group, with a small amount of the alkylated product corresponding to the formal alkene insertion into the *ortho* C–H bond (Eq. 8) [64]. The alkenylation product may arise from the initial formation of a cyclometallate by *ortho* C–H bond cleavage as for acetophenone, followed by alkene double bond insertion into Ru–C bond followed by β-elimination.

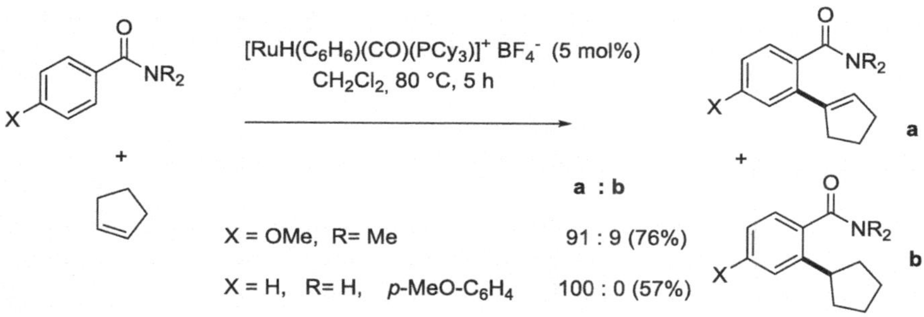

(8)

The same catalyst also promotes the reaction of arylamides with 1,1-disubstituted terminal alkenes to preferentially afford the *ortho*-alkylated products at the expenses of the *ortho*-alkenylated arylamides [64].

3.2 Dehydrogenative Alkenylation of Arenes with Ruthenium(II) Catalysts According to Directing Groups

The high potential of the dehydrogenative cross coupling of alkenes with (hetero) aromatic C–H bonds has initiated efforts to open synthetic routes to multifunctional alkenes using stable and low-cost catalysts. The understanding of the mechanism of activation of sp^2C–H bond with ruthenium(II) catalysts [37, 38], via C–H bond deprotonation with the help of carbonate or especially carboxylates, which can take place at room temperature has attracted interest to attempt further oxidative couplings with alkenes. Whereas the first attempts of ruthenium(II)-catalysed reactions of ethylene with arenes have led to hydroarylation of ethylene, e.g. the alkylation of benzene with ethylene [45], alkenylations with functional olefins of arenes and heteroarenes with ruthenium(II) catalysts but in the presence of an oxidant have emerged since 2011 and were shown to be regioselectively directed by a variety of functional groups including weakly coordinating ones.

The first examples of ruthenium(II)-catalysed alkenylation with alkenes were shown within a few months of 2011 first by Satoh and Miura with (hetero)aromatic carboxylic acids [41], then by Ackermann who could perform successive alkenylation of carboxylic acids and oxa-Michael reaction in water [42], then Bruneau and Dixneuf who used nitrogen-containing heterocycle groups to direct alkenylation with catalytic amount of Cu(II) in air [43] and Jeganmohan who showed that a simple weakly bonding ketone could direct the activation/alkenylation at *ortho*-position [44]. Since 2012, a general method of ruthenium(II)-catalysed alkenylation has been developed, via ruthenium(II)-catalysed dehydrogenative cross coupling of functional arenes or heterocycles and alkenes which will be presented now. It will be shown that a variety of functional groups direct the C–H bond alkenylation such as carboxylate, heterocycles, ketone, aldehyde, amides, esters, carbamates, free amine or alcohol and sulfonic acid derivatives.

3.2.1 Alkenylation with Carboxylic Acid as a Directing Group

Carboxylate groups have been shown to easily coordinate to Ru(II) complexes without decarboxylation, and they lead to weakly bonded ligands that easily dissociate from the Ru(II) centre and deprotonate neighbour C–H bonds [37, 65]. Satoh and Miura reported in 2011 the first example of ruthenium(II)-catalysed alkenylation with acrylates. The selected substrates were *heterocycles* such as benzothiophene, benzofuran, pyrrole and indole derivatives, but containing a carboxylate group as a directing group. The reaction was performed with $[RuCl_2(p\text{-cymene})]_2$ catalyst and required a stoichiometric amount of $Cu(OAc)_2.H_2O$ as oxidant with LiOAc as a base (Eq. 9) [41]. It was demonstrated before that carboxylic acids did favour the C–H bond activation by Ru(II) systems, but the tremendous advantage of ruthenium(II) catalysts was that they tolerated this functional group, whereas useful palladium(II) catalysts for similar reaction undergo decarboxylation of the aromatic carboxylate directing group [66]. The thiophene-3-carboxylic acid could lead to its dialkenylation at both C2 and C4 positions thus demonstrating the directing ability of the carboxylate directing group [41].

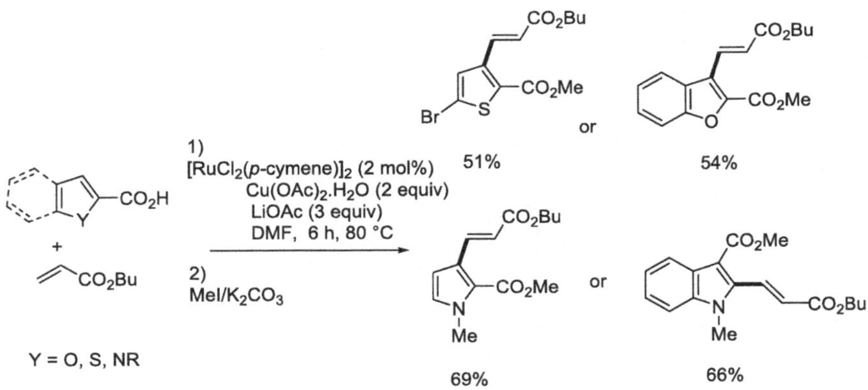

(9)

Ackermann showed that the carboxylic group could direct the alkenylation with Ru(II) catalyst at the *ortho*-position of benzoic acid derivatives with alkyl acrylates or acrylonitrile with 2 equiv. of oxidant $Cu(OAc)_2.H_2O$ (Eq. 10) [42]. This reaction took place efficiently in *water as solvent*, and the alkenylation was followed by oxa-Michael addition, thus leading to a variety of lactones. The reaction with Ru(II) catalysts was shown to tolerate a bromide linked to the arene. This constitutes a new example of the stability and efficiency in water of ruthenium(II) catalysts for functionalisation of sp^2C–H bonds [33, 67].

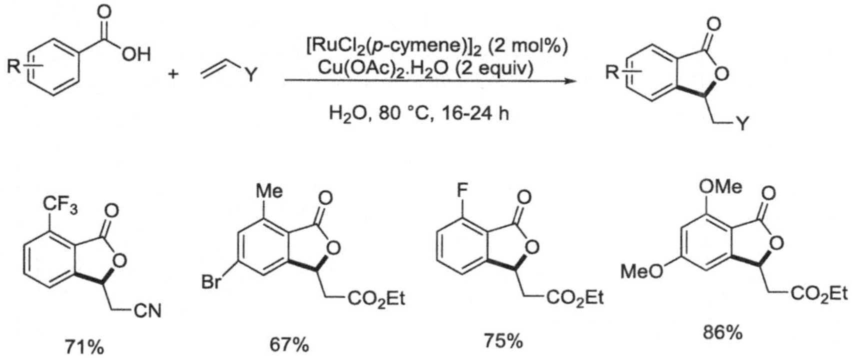

(10)

3.2.2 Heterocycles as Directing Groups for Alkenylation

Evidence for Directed C–H Bond Alkenylation

Nitrogen-containing heterocycles are known to give strong N–Ru bond with Ru(II) complexes, and several of them can be isolated [65]. Then, oxidative alkenylation of arenes, directed by a heterocycle in *N*-arylpyrazoles, with nonactivated alkenes such as styrene was shown by Bruneau and Dixneuf (Eq. 11a) [43]. The reaction was performed by $Ru(OAc)_2$(*p*-cymene) catalyst, but with only catalytic amount of $Cu(OAc)_2.H_2O$ (20 mol%) in air at 100°C. A key of the success was the use of a simple $Ru(OAc)_2$(*p*-cymene) catalyst in acetic acid which together allows easy C–H bond cleavage [37, 38].

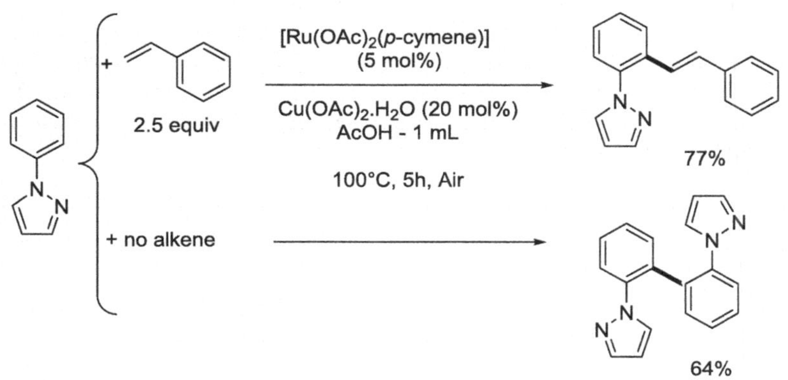

(11a)

(11b)

It is noteworthy that in the absence of styrene, the catalytic system performed easy *ortho* C–H bond cleavage of *N*-arylpyrazole and *ortho* C–C homocoupling leading to bidentate ligand (Eq. 11b) [43].

This alkenylation of arylpyrazoles can be applied to electrophilic acrylamides or acrylates under similar conditions, but the reaction is slower. However, good yields can be obtained using 1 equiv. of $Cu(OAc)_2.H_2O$ in acetic acid. The *ortho*-dialkenylation with acrylate of N-*p*-methoxy-substituted phenylpyrazole can also be achieved (Eq. 12) [43].

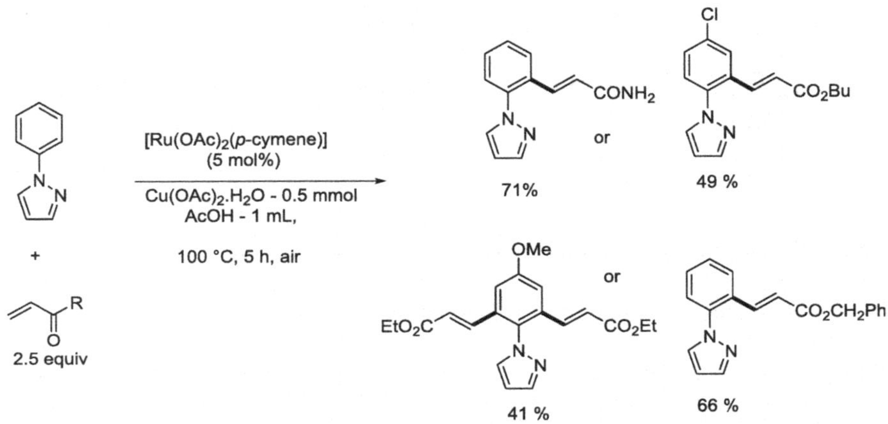

(12)

The selective monoalkenylation of 1-phenylpyrazoles has also been performed by Satoh and Miura with $[RuCl_2(p\text{-cymene})]_2$ catalyst and 2 equiv. of $Cu(OAc)_2.H_2O$ in DMF under nitrogen atmosphere at 100°C. The reaction was shown to tolerate various substituents on the arene ring such as Cl, Me, CO_2R and CN groups [68].

The alkenylation directed by pyrazole has been applied to illustrate the successive functionalisations of two *ortho* C–H bonds of the *N*-phenylpyrazole aryl group. After selective monoarylation of *N*-phenylpyrazole with ruthenium(II) catalyst in water, the new *N*-phenylpyrazole derivative was *ortho*-alkenylated with *n*-butyl acrylate with catalyst $Ru(OAc)_2(p\text{-cymene})$ in acetic acid using 1 equiv. of oxidant $Cu(OAc)_2.H_2O$ (Eq. 13) [69].

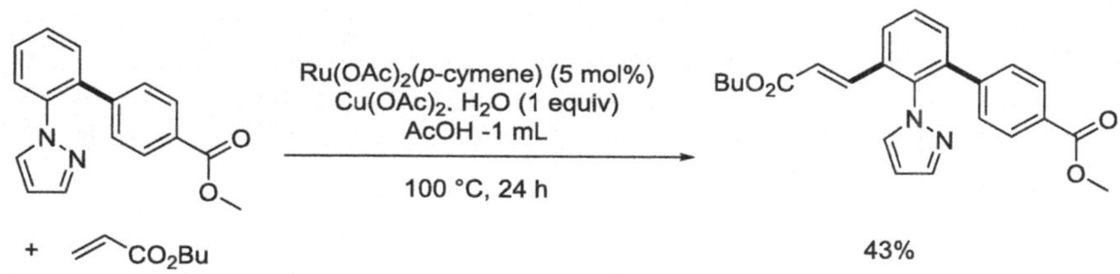

(13)

The oxazoline ring has been shown to be an efficient directing group for *ortho*-alkenylation of aryl C–H bonds with acrylates, acrylamides in toluene. However, the usual carboxylic acids such as AcOH or $PhCO_2H$ were not efficient partners, and the 1,1′-binaphthyl-2,2′-diyl hydrogen phosphonate (HMPAH) appeared to be an excellent Ru(II) partner for a reaction performed in simple ethanol as solvent, with less than 1 equiv. of $Cu(OAc)_2.H_2O$ under air (Eq. 14) [70].

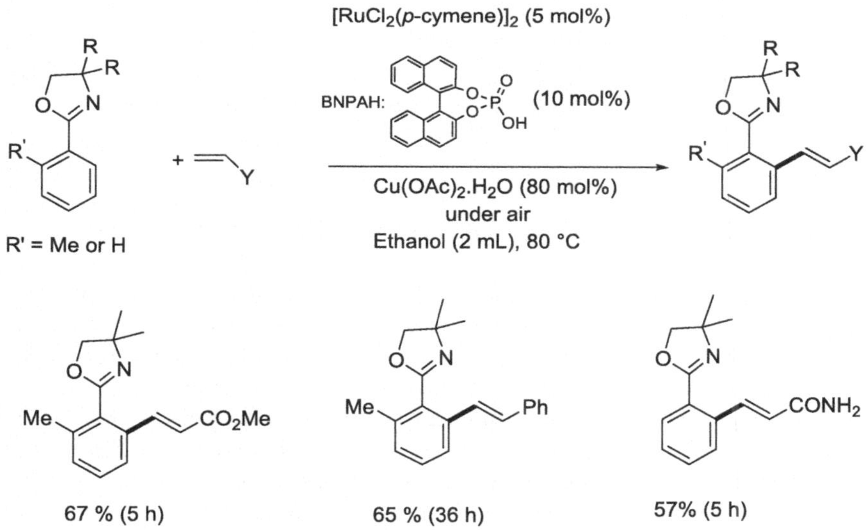

(14)

This transformation shows that a variety of C–H bond deprotonation partners have to be selected according to both the substrate and directing group nature, and phosphate derivatives such as $(alkylO)_2P(O)OK$ are now commonly used for catalytic monoarylation of arene C–H bonds directed by a tetrazole unit for the production of industrial intermediates [71].

Azole has also been used as a heterocycle directing group in the ruthenium(II) alkenylation of phenylazoles by Miura [72]. The catalytic system is based on Jeganmohan's catalyst discovered for alkenylation of aromatic ketones [44], with $[RuCl_2(p\text{-cymene})]_2/4AgSbF_6$ also with 2 equiv. of $Cu(OAc)_2.H_2O$ in *t*-AmOH solvent at 100°C under nitrogen (Eq. 15a) and under conditions presented later with amides as directing groups. The reaction with acrylate of phenylbenzothiazole with only $[RuCl_2(p\text{-cymene})]_2$ led to 20% only of the alkenylated product [68]. By contract with $AgSbF_6$, the same catalyst afforded 61% of the *ortho*-alkenylated product (Eq. 15b) [72]. $AgSbF_6$ favours halide abstraction and formation of Ru(II)–OAc bond.

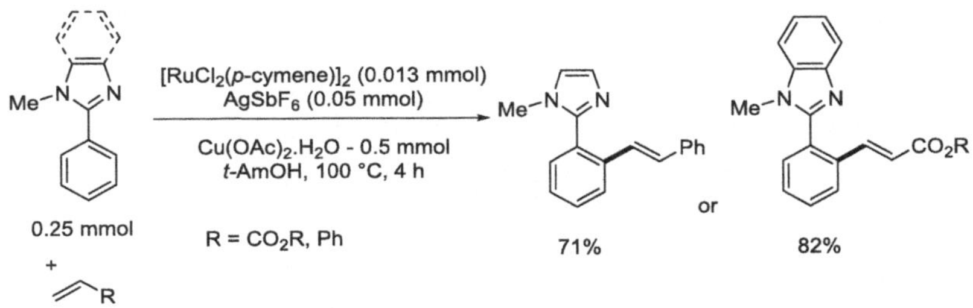

(15a)

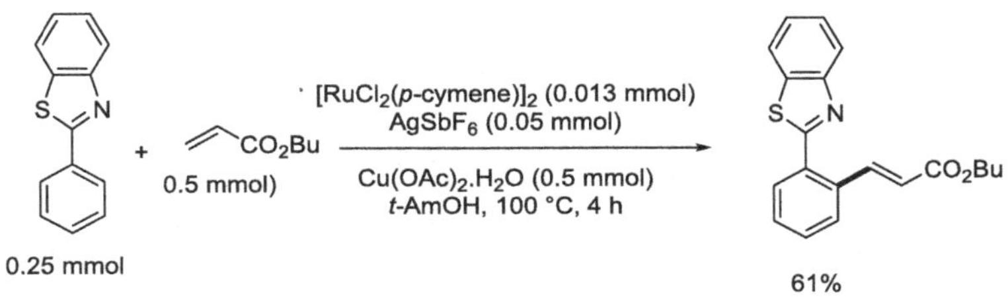

(15b)

First Approach of Mechanism for C–H Bond Activation and Alkenylation

At this stage, a mechanism for the alkenylation of functional arenes can be proposed as exemplified by that of phenyloxazoline (Eq. 14). It is based on the first step of the mechanism established with the help of kinetic studies for C–H bond activation and arylation at 27°C of functional arene: 2-phenylpyridine [37], 2-phenyl-2-oxazoline and 1-phenylpyrazole [38], using $Ru(OAc)_2(p\text{-cymene})$ catalyst (Scheme 1). The Ru(II)–OAc bond of $Ru(OAc)_2(p\text{-cymene})$ is weak [65], and acetate dissociation from an 18-electron complex takes place first to allow the coordination of the heterocycle nitrogen. The formation of the cyclometallate **A** is first produced by acetate C–H bond deprotonation and accelerated by additional KOAc, thus supporting an intermolecular deprotonation of C–H bond by external acetate. This cyclometallate formation is strongly accelerated by the freed AcOH, likely to keep the Ru(II) site coordinatively unsaturated, and this phenomenon reveals an *autocatalytic process*. This will be reflected in most C–H bond activation/deprotonation processes observed with Ru(II) catalysts, and the crucial role of carboxylic acids and carboxylates partners will be shown. More importantly, the kinetic studies of C–H bond activation process with phenylpyridine by $Ru(OAc)_2(p\text{-cymene})$ and $Pd(OAc)_2$, leading in both cases to a cyclometallate complex, show that the reaction of $Pd(OAc)_2$ is faster than with the ruthenium(II) catalyst, but that it is not affected at all by the addition of acetic acid or acetate [38]. Thus, $Pd(OAc)_2$ proceeds via an

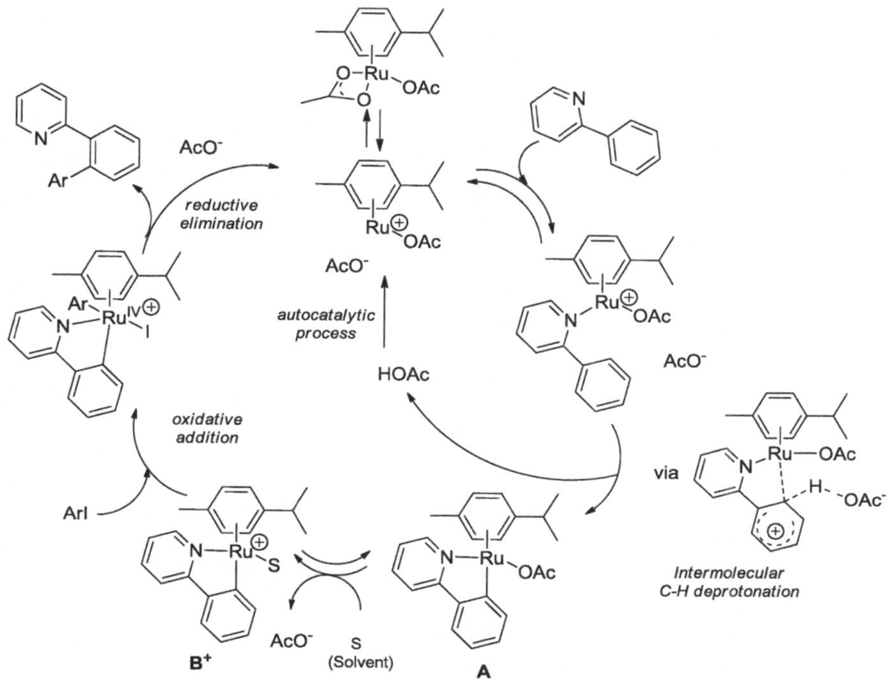

Scheme 1 Proposed mechanism for Ru(II)-OAc assisted C–H bond activation and arylation

intramolecular non-autocatalysed concerted metallation–deprotonation (CMD) mechanism, whereas $Ru(OAc)_2L_n$ catalysts proceed via an *intermolecular deprotonation favoured by carboxylate or phosphate via an autocatalytic process* [37, 38].

Thus, for Ru(II)-catalysed alkenylation of functional arenes such as phenyloxazoline (see (Eq. 14) [70]), the first step of C–H bond activation is expected to be analogous to that observed for arylation of phenylpyridine (Scheme 2). The reaction of $[RuCl_2(p\text{-cymene})]_2$ catalyst and $Cu(OAc)_2.H_2O$ gives easily $Ru(OAc)_2(p\text{-cymene})$ which, in the presence of the phosphonate BNPAH, can lead to a ruthenium(II) cation, with counter-anion (A^-) = (AcO^- or $BNPA^-$). The coordinatively unsaturated Ru(II) centre is able to coordinate the functional group and favours the interaction of the ruthenium(II) site with the *ortho* (C–H) carbon **C**, increasing its C–H bond acidity. The C–H bond deprotonation by external $BNPA^-$ is expected to afford the five-membered cyclometallate **D**, which via coordinative insertion of alkene leads to cyclometallate **E**. When the N–Ru bond is stable, then classical β-elimination cannot take place due to the rigidity of the seven-membered cyclometallate **E**, avoiding the coplanarity of both Ru–C and Cβ–H bonds. Then, the deprotonation of a β-hydrogen in *anti*-position to the Ru–C bond should take place to give the alkenylated product and a Ru(0) species $Ru(X)Ln^-$. The latter is oxidised into Ru(II) species with the Cu(II) salt in the presence of air. If the Ru–N

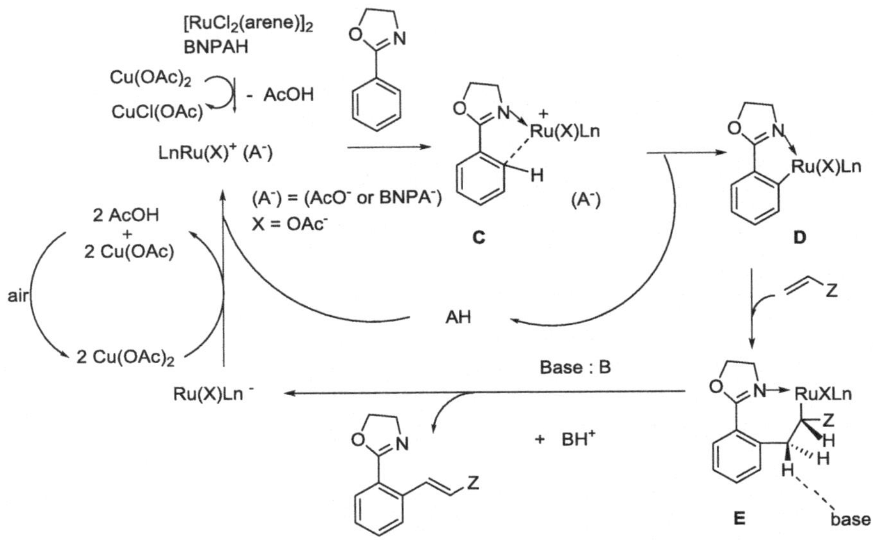

Scheme 2 Possible mechanism for alkenylation of ortho C–H bond of phenyloxazoline

bond can easily dissociate, then β-elimination takes place to generate a Ru(H)(X)Ln species leading, on reaction with $Cu(OAc)_2$, to the catalytic species.

3.2.3 Ketone as a Directing Group for Alkenylation

With the help of ruthenium(0) catalysts, the Murai group showed that ketone groups could direct *ortho*-arylation, with arylboronates, of aromatics, and with alkenes, they direct alkylation preferentially to alkenylation [21–26]. However, ketones with Ru(II) catalysts could not direct arylation with aryl halides or Fujiwara–Moritani alkenylation, whereas their corresponding imines could make direct arylation possible with aryl halides [28, 73–76].

A very useful modification of the Ru(II) catalysts was brought by Jeganmohan et al. at the end of 2011 which allowed the selective, efficient dehydrogenative alkenylation of aromatic ketones (Eq. 16) [44]. Thus, by reacting the catalyst $[RuCl_2(p\text{-cymene})]_2$ with more than 4 equiv. of $AgSbF_6$ to abstract the chlorides from the ruthenium(II) complex, in the presence of $Cu(OAc)_2.H_2O$ as an acetate provider and as oxidant, they succeeded to perform the *ortho*-alkenylation of a large variety of substituted aryl methyl ketones (X=F, I, MeO, CO_2Me) (Eq. 16). The reaction of benzoketone can lead either to *ortho*-mono- or dialkenylation.

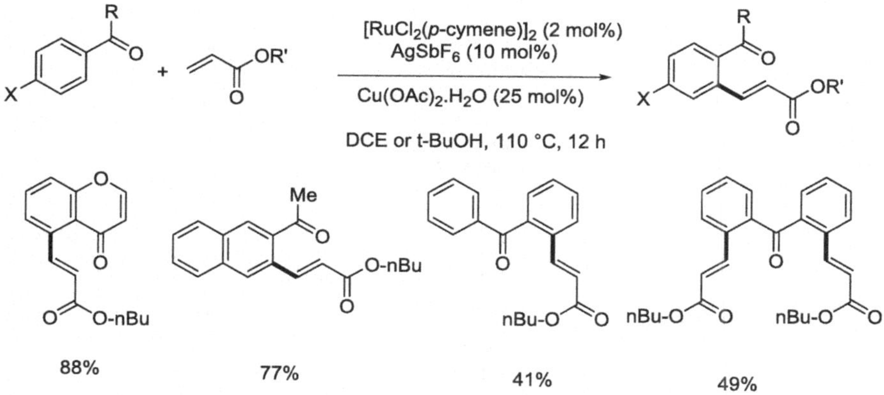

(16)

This Jeganmohan catalytic system, based on the abstraction of chlorides by Ag^+ SbF_6^- salt, which favours the formation of Ru–OAc bond in the presence of $Cu(OAc)_2$, has contributed to further establish general methods of alkenylation of a variety of arylketones, but also of arenes and heteroarenes containing a carbonyl bond for weak interaction with the Ru(II) centre, such as formyl, ester, amide, amidine and carbamate directing groups of which applications will be presented later.

The ketone group has also been used to directly alkenylate ferrocene derivatives which could not be arylated before. This functionalisation directed to *ortho*-position of the ketone would open the direct access to compounds with planar chirality. Using a similar catalytic system as Jeganmohan for alkenylation of aromatic ketones (Eq. 16) [44], Singh and Dixneuf have reported the catalytic alkenylation with acrylates of ferrocenyl ketones, with COMe and COEt directing group, using 8 mol% of $[RuCl_2(p\text{-cymene})]_2$, 40 mol% of $AgSbF_6$ and 2 equiv. of $Cu(OAc)_2.H_2O$ in DCE. The C2-alkenylation product was selectively produced at 80–100°C but was obtained in moderate yields (<30%) (Eq. 17a) [77].

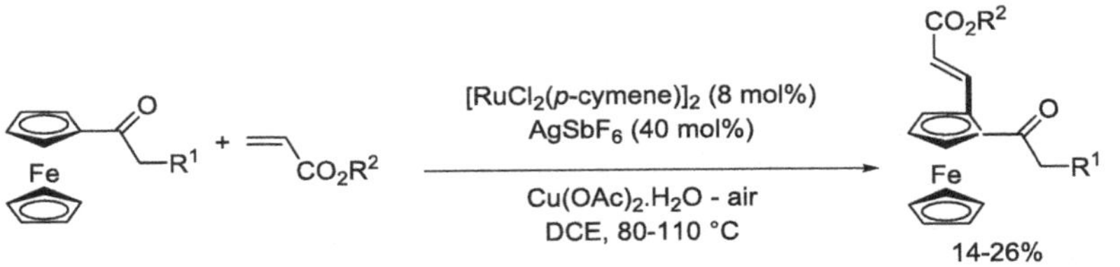

(17a)

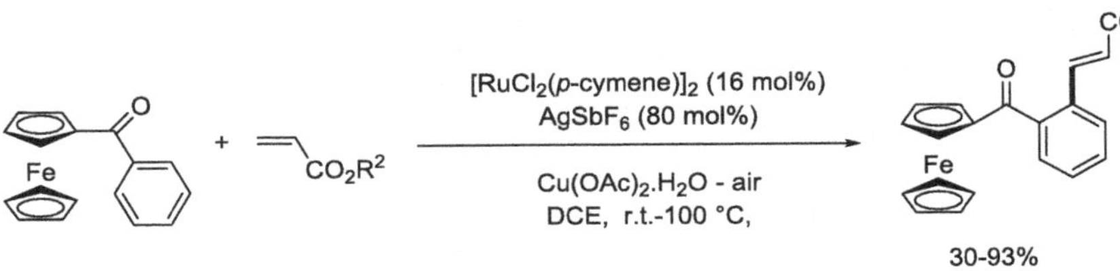

(17b)

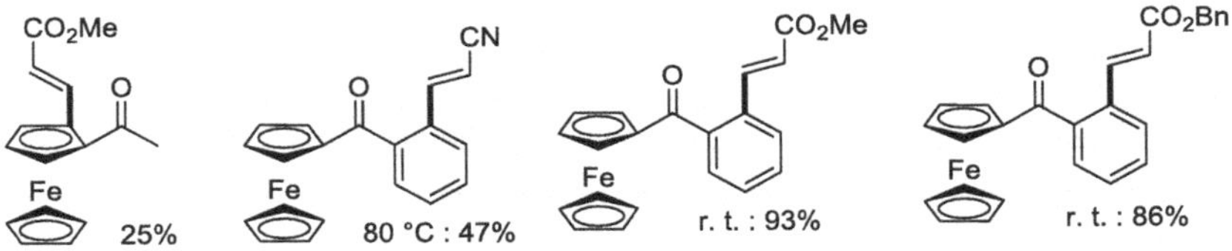

Then, in order to evaluate the easiness of activation and functionalisation of (Fc)C–H versus (Ph)C–H bonds, the reaction was applied to ferrocenyl phenyl ketone. Surprisingly in that case, the alkenylation selectively took place by functionalisation of the *ortho* (Ph)C–H bond only and not at the electron-rich ferrocenyl cyclopentadienyl group. With acrylates, alkenylation occurred *at room temperature* and in quantitative yields (Eq. 17b) [77]. The same alkenylation was performed at the phenyl group with acrylonitrile and styrene derivatives but at 80–100°C. These reactions show that the FcCO acyl group actually favours *ortho* aryl C–H bond activation for alkenylation.

3.2.4 Formyl as a Directing Group for Alkenylation

Soon after, in 2012, Jeganmohan used his catalytic system $RuCl_2$(*p*-cymene)]$_2$/6 $AgSbF_6$ with $Cu(OAc)_2.H_2O$ under air to show that the formyl group also directed the alkenylation at its *ortho*-position of aromatic aldehydes. He showed that the electron-withdrawing substituents such as Cl, CN and CO_2Me on the aryl groups disfavour the reaction; by contrast, electron-rich substituents such as OMe, Me, NMe_2 and dioxole led to good yields (Eq. 18) [78]. The directing ability is likely due to the weak coordination of the carbonyl oxygen to the Ru(II) site. It is noteworthy that with Ru(0) catalysts, the Murai alkylation with olefins of aromatic aldehydes was disfavoured. The formation of the alkenylated products is suggested to take place via initial activation with a Ru(II)–OAc species leading to the five-membered cyclometallate by *ortho* C–H bond cleavage.

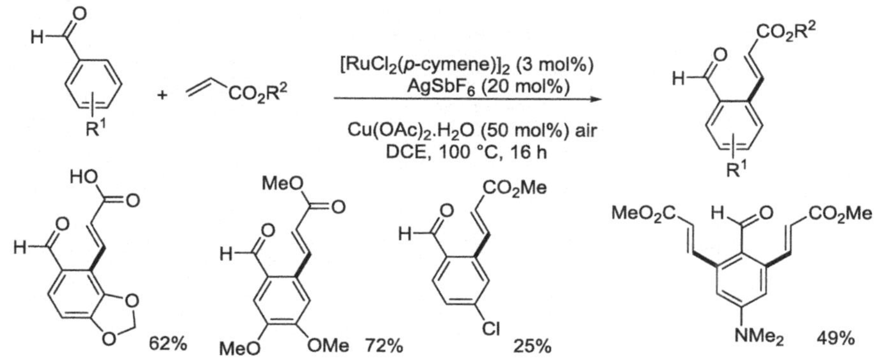

(18)

Jeganmohan showed the interest of the produced *ortho*-alkenylated aromatic aldehydes, as under irradiation in benzene they afforded the four-membered cyclic ketones or the polysubstituted isochromanone derivatives (Eq. 19) [78].

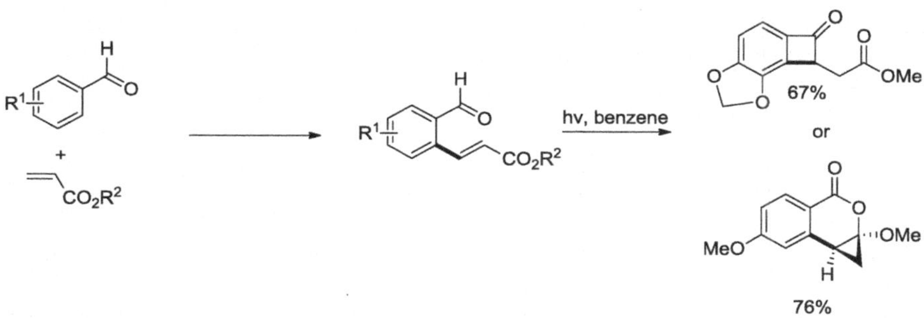

(19)

3.2.5 Ester as a Directing Group for Alkenylation

As for ketone and formyl groups, it was shown soon after first by Jeganmohan [79] and by Ackermann [80] that the weakly coordinating ester function could also direct the alkenylation of arenes with acrylates on action of the catalyst $[RuCl_2(p\text{-cymene})]_2$/n $AgSbF_6$. Under air atmosphere, $Cu(OAc)_2.H_2O$ in 1,2-dichloroethane could be used this time in a catalytic amount of 30 mol%. A variety of aromatic and heteroaromatic esters could be alkenylated by acrylates (Eq. 20) [79, 80]. It is shown in the two reports, especially by cross experiments, that the reaction is favoured by electron-rich substituents on the aryl groups. The reaction tolerates the presence of iodide substituent.

The reaction appears quite general and can be applied to thiophene, furan and indole derivatives, but an electron-withdrawing group on aromatic esters disfavours this cross-coupling reaction (Eq. 20) [79, 80].

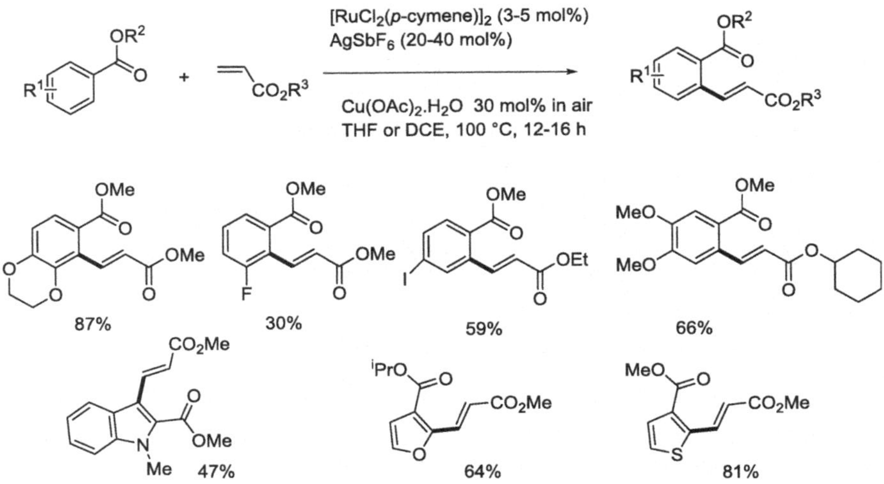

(20)

It is noteworthy that Loh used the ester group of acrylates to direct the alkene C–H bond activation to give functional dienes in the presence of another alkene (see Sect. 3.4) [81].

3.2.6 Amides CONHR and NHCOR as Directing Groups for Alkenylation

At the same time as the alkenylation of *N*-phenylpyrazoles, Miura et al. [68] have also found in 2011 that an amide group CONHR could direct the *ortho*-alkenylation of a phenyl ring with $[RuCl_2(p\text{-cymene})]_2$ catalyst and 2 equiv. of $Cu(OAc)_2.H_2O$ in *o*-xylene. The reaction of benzylanilide and *n*-butyl acrylate led first to *ortho*-alkenylation followed by intramolecular aza-Michael addition and the production of the lactam (Eq. 21) [68]. Thus, they showed for the first time that both *ortho* C–H and N–H bonds could be easily functionalised with ruthenium(II) catalyst.

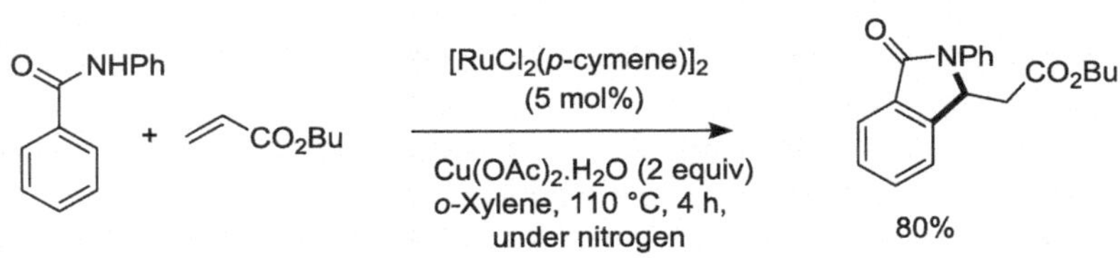

(21)

Bin Li and Baiquan Wang, in the beginning of 2012, reported that the amide group CONH(OMe) could direct *ortho*-alkenylation and behave as an oxidising reagent. They showed that $[RuCl_2(p\text{-cymene})]_2$, without $Cu(OAc)_2$ this time, allowed the general catalysed oxidative alkenylation with acrylates, simply in the presence of NaOAc (30 mol%) in methanol at 60°C (Eq. 22a) [82].

The similar reaction with unactivated alkenes, with styrene and norbornadiene, but in CF_3CH_2OH as solvent, allows the formation of 3,4-dihydroisoquinolinones via initial *ortho*-alkenylation of *N*-methoxybenzamides followed by intramolecular nucleophilic addition of the nitrogen to the formed C=C bond (Eq. 22b) [82].

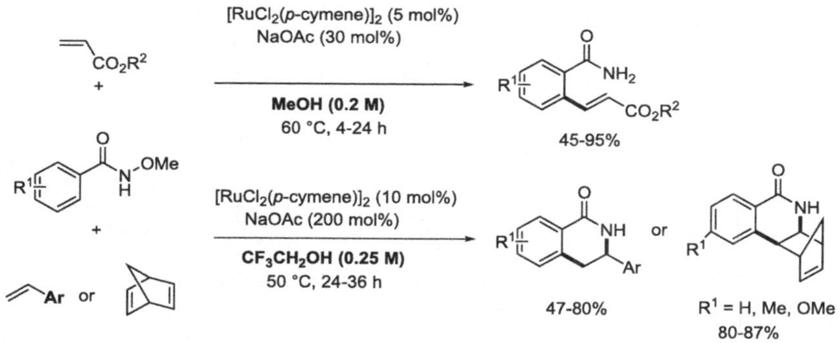

The authors proposed a realistic mechanism based on initial formation of $Ru(OAc)_2$(*p*-cymene) catalyst shown in Scheme 3 [82]. The cyclometallated intermediate **F** is likely to be formed from *N*-methoxybenzamide on acetate deprotonation of both *ortho* C–H and N–H bonds. The intermediate **G** should result from coordinative insertion of the alkene. The alkenylated product is produced via β-elimination from the intermediate **G**, or by deprotonation with copper freed acetate, in the case of acrylate in methanol. By contrast, the alkyl and aryl group in **G** arising from styrene or norbornadiene in CF_3CH_2OH should favour reductive elimination. These processes lead to the release of MeOH with the regeneration of the $Ru(OAc)_2$(*p*-cymene) catalyst.

Miura and co-workers, as they did for the alkenylation of phenylazole and phenylthiazole (Eq. 15) [72], used the $[RuCl_2(p\text{-cymene})]_2/4\ AgSbF_6$ catalytic system and 2 equiv. of $Cu(OAc)_2.H_2O$ under nitrogen, in *t*-AmOH for the regioselective *ortho*-alkenylation with alkyl acrylates of *N*,*N*-dialkylbenzamides (Eq. 23) [72].

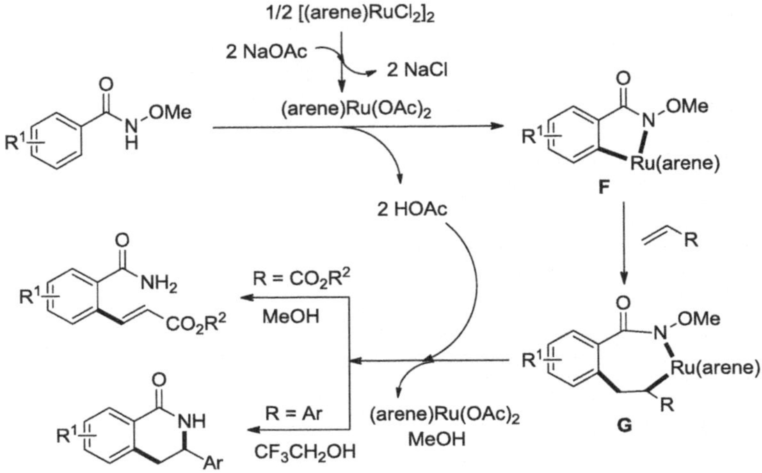

Scheme 3 Proposed mechanism for alkenylation of *N*-methoxybenzamides

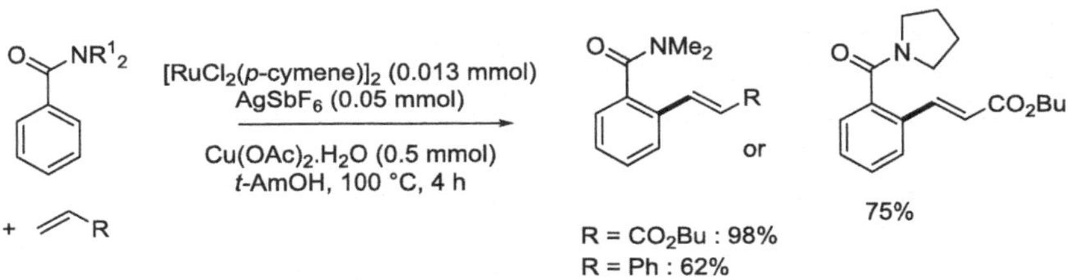

(23)

All the previous ruthenium(II)-catalysed directed alkenylations with alkenes of functional arenes were performed in organic solvents except that of arylcarboxylic acids performed in water as soon as 2011 by Ackermann's group (Eq. 10) [42]. Ackermann's group succeeded to perform this catalytic monoalkenylation of benzamides *in water* using the $[RuCl_2(p\text{-cymene})]_2$ catalyst with the non-coordinating salt KPF_6 (20 mol%) in the presence of $Cu(OAc)_2.H_2O$ as the oxidant. The *N*-(pentafluorophenyl)benzamides were first monoalkenylated and further led to intramolecular aza-Michael addition to form the expected lactams (Eq. 24) [83].

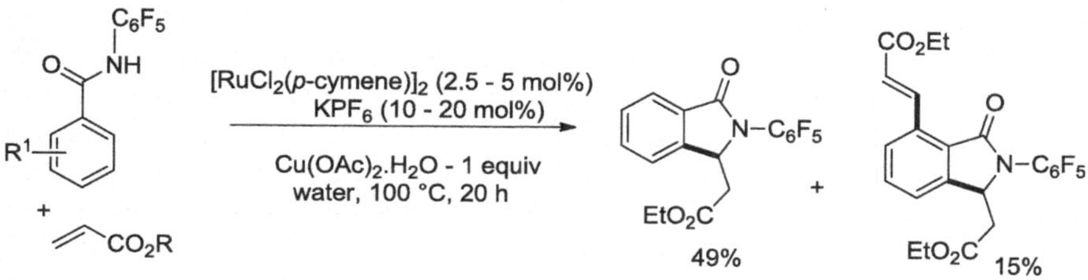

(24)

They showed that the NHCOR group of anilides, more generally than the CONHR group, directed the *ortho*-alkenylation with acrylates and *in water as solvent*. This catalytic system $[RuCl_2(p\text{-cymene})]_2/KPF_6$ (20 mol%) in the presence of $Cu(OAc)_2.H_2O$ is more active in water than in DMF, NMP, *t*-AmOH. The reaction is general and favoured by electron-rich anilides (Eq. 25) [83].

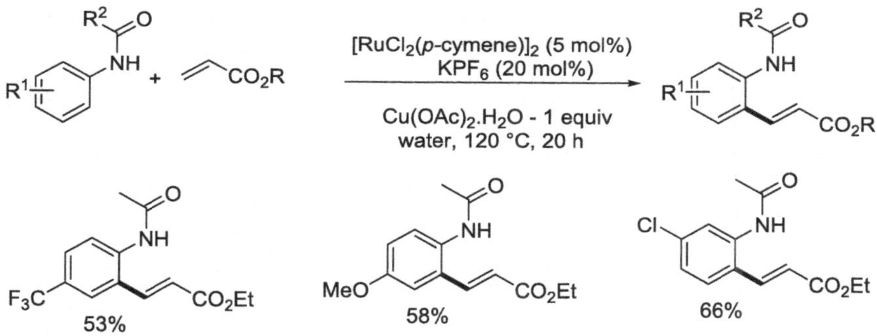

(25)

The alkenylation of N-benzoyl aniline took place only at the *ortho* C–H bond of the aromatic ring linked to the amide carbonyl showing the preferential activation/alkenylation by the -CONHPh than the -NHCOPh group. It is consistent with the activation of the more acidic C–H bond of the phenyl linked to the carbonyl group (Eq. 26) [83].

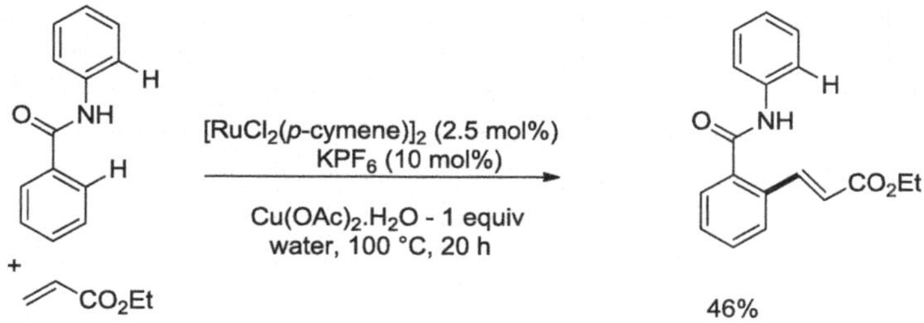

(26)

Ackermann also demonstrated that the CONHR group could efficiently direct the alkenylation of heterocycles such as indole, thiophene, benzothiophene and benzofuran [83].

The amide group $CONR_2$ has first been used in 2013 by Huanfeng Jiang to direct the ruthenium(II)-catalysed *ortho*-alkylation of arenes and heterocycles with allylic alcohols [84] (see alkylation part of Sect. 4.1). Now Jeganmohan has profitably used the amide group CONHR to direct *ortho*-alkenylation of arenes, directly with allylic alcohols, followed by isoindolinone formation via intramolecular Michael addition. The alkenylated intermediate offers the direct access to a variety of isoindolinones (Eq. 27) [85]. The catalyst is similar to that used for acrylates, $[RuCl_2(p\text{-cymene})]_2$/4 $AgSbF_6$, that is only effective with $Cu(OAc)_2.H_2O$ (2 equiv). With the $CONH^tBu$ group, a mixture of isoindolinones and *ortho*-alkenylated product with a 1:0.4 ratio was obtained in 45% yield, suggesting an initial formation of alkenylated intermediate.

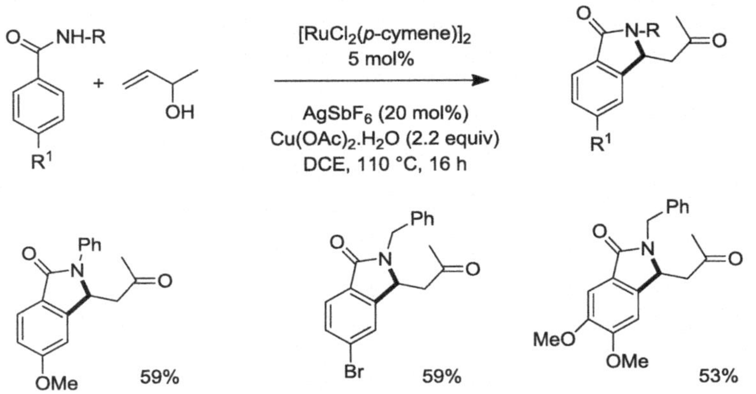

(27)

The reaction occurs for a variety of allylic alcohols in DCE at 110°C. It is shown that under the catalytic conditions, (1) allylic alcohol is dehydrogenated into the α,β-unsaturated ketone and (2) the unsaturated ketone with arylamide leads to isoindolinone. Thus, the suggested mechanism of the reaction is presented on Scheme 4. It involves the formation of the cyclometallate **H** that was isolated in one example: the insertion of the α,β-unsaturated ketone, arising from allylic alcohol, leads to the cyclometallate **I** and then to the alkenylated intermediate **J**. The latter double bond is activated by the ruthenium(II) catalyst to form the isoindolinone via intramolecular addition. It is also shown that the *ortho*-alkylated product **K** can be obtained in small quantity in one example and that *in the presence of AcOH its formation is increased*. This formation of **K** is expected to result from protonation of the inserted product Ru–C bond of **I**, before β-elimination.

It is noteworthy that Loh used the $CONR_2$ amide group of acrylamides to direct this alkene C–H bond activation towards alkenylation in the synthesis of (*Z*,*E*)-dienamides (see Sect. 3.4) [86].

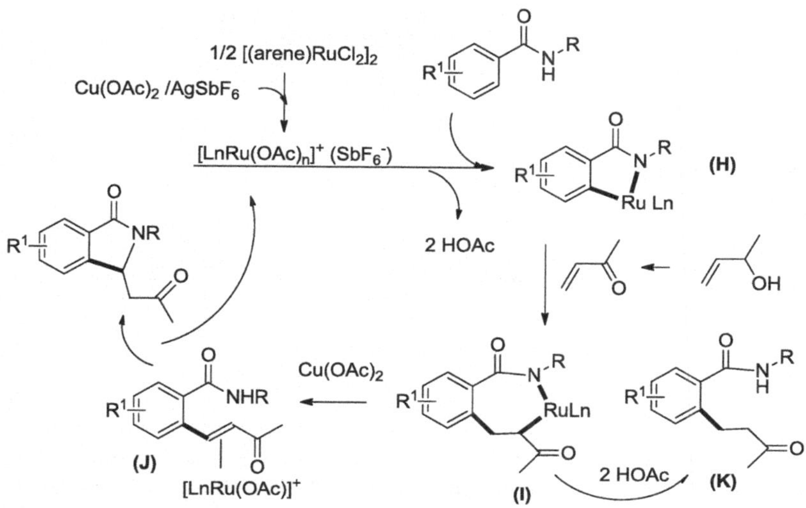

Scheme 4 Proposed mechanism for alkenylation of benzamides with allyl alcohol

3.2.7 Amidines as Directing Groups

One of the *ortho* C–H bonds of benzylamidines was efficiently and selectively activated by ruthenium catalyst generated from $[RuCl_2(p\text{-cymene})]_2$ in the presence of silver acetate and copper acetate, owing to the excellent directing properties of the amidine functionality [87]. The dehydrogenative coupling with acrylates took place at 100–120°C to give 3-alkylidene-1-iminoisoindolines. Two types of products were formed depending on the steric hindrance of the *N*-substituent of the benzamidine. With the less sterically hindered *iso*- and *n*-butyl or benzylic groups, the amino group was involved in the C–N bond formation (Eq. 28). A variety of functional groups on the benzyl group of benzamidine were tolerated in this reaction.

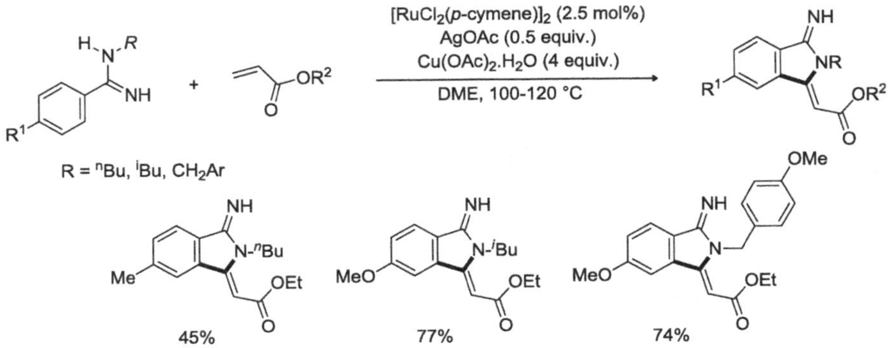

22

(28)

By contrast with the bulky *t*-butyl group, the imino group was incorporated in the five-membered ring of the isoindoline (Eq. 29). Mechanistic studies indicated that the transformation resulted from a sequence of oxidative alkenylation followed by intramolecular aza-Michael reaction and then by copper-catalysed dehydrogenation to generate the acrylic system (Eq. 29).

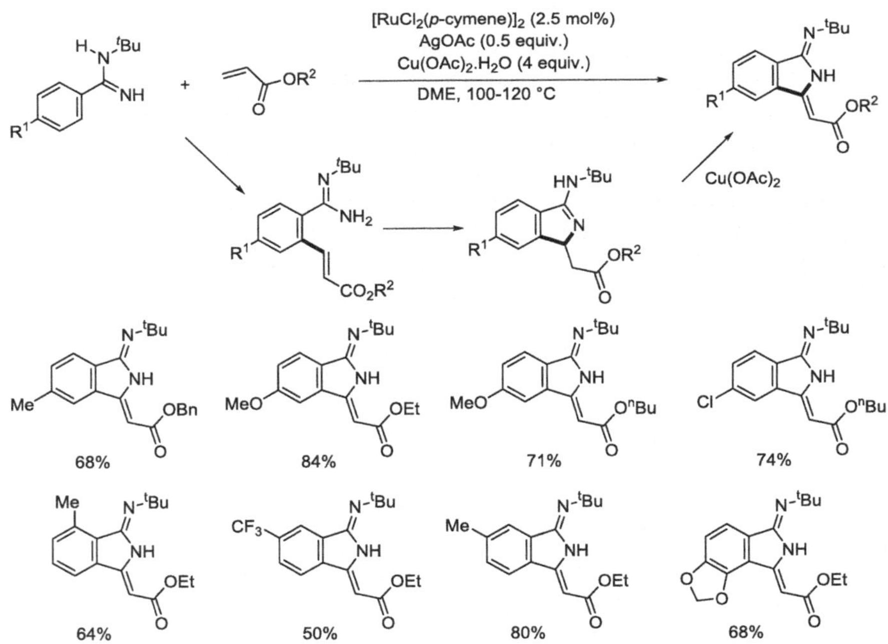

(29)

3.2.8 Free Amine as a Directing Group for Alkenylation

The coordinating ability of the free primary amino group of benzylamines was shown by Miura and Satoh to be strong enough to direct the *ortho* C–H bond activation of the phenyl group and to allow alkenylation. α,α-Disubstituted benzylamines, in the presence of catalytic amounts of $[Cp^*RhCl_2]_2$, or with the low-cost $[RuCl_2(p\text{-cymene})]_2$ and $AgSbF_6$, and 2 equiv. of copper(II) diacetate, led to the direct coupling with electron-deficient acrylates and the selective formation of isoindolines *at room temperature* in dioxane (Eq. 30) [88]. The reaction is supposed to produce first an *ortho*-alkenylated benzylamine, which is prone to cyclise via aza-Michael addition. The formation of the five-membered ring was favoured when the starting benzylamine was disubstituted at the benzylic position, whereas the α-unsubstituted benzylamine only gave aza-Michael addition to the electron-deficient olefin. This is the first example of ruthenium(II)-catalysed C–H bond functionalisation directed by a free NH_2 group. An intermediate situation leading to a mixture of isoindoline and secondary linear benzylamine was observed when acrylonitrile was used as coupling partner. This catalytic system based on ruthenium was not active for the alkenylation with styrene. On the other hand, a catalytic system based on $[Cp^*RhCl_2]/AgSbF_6$ and $Cu(OAc)_2$ was also able to achieve this alkenylation/cyclisation at 80°C from acrylate and also the simple *ortho*-alkenylation from styrene.

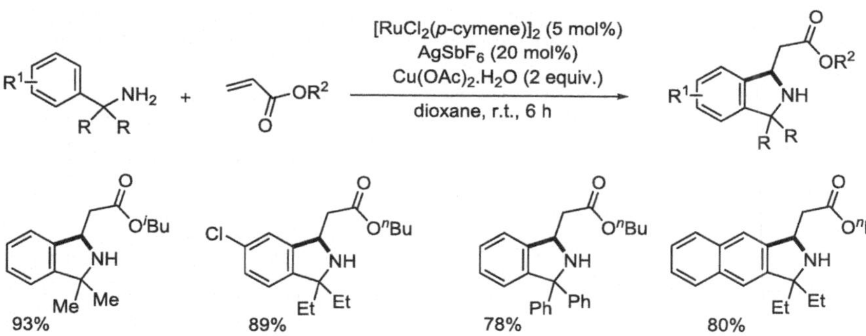

(30)

The mechanism is suggested to involve the initial formation of a Ru(II)–OAc species leading to a five-membered cyclometallate, the regioselective coordinative insertion of the acrylate double bond into the Ru–C bond, the β-elimination with metal hydride formation and the intramolecular Michael addition. Reoxidation of the metal hydride species with $Cu(OAc)_2$ is proposed to release the Ru(II) catalyst (Scheme 5) [88].

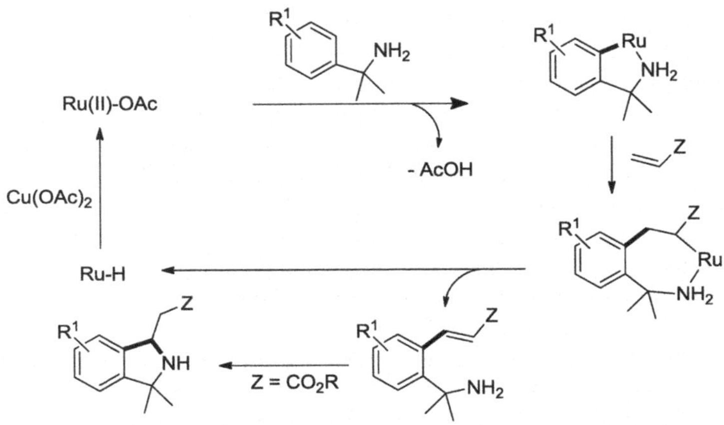

Scheme 5 Proposed mechanism for the alkenylation of benzylamines directed by free amine

3.2.9 Free Hydroxy Directing Group for Alkenylation

The directing ability of a hydroxy group has been revealed by Lam during the alkenylation of 2-aryl-3-hydroxy-2-cyclohexenones with acrylate, acrylonitrile or acrylamide using simply the catalyst $[RuCl_2(p\text{-cymene})]_2$ with 2 equiv. of $Cu(OAc)_2.H_2O$. The reaction leads to the formation of benzopyrans via oxa-Michael-type addition of the enol oxygen atom to the alkenylated intermediate (Eq. 31) [89].

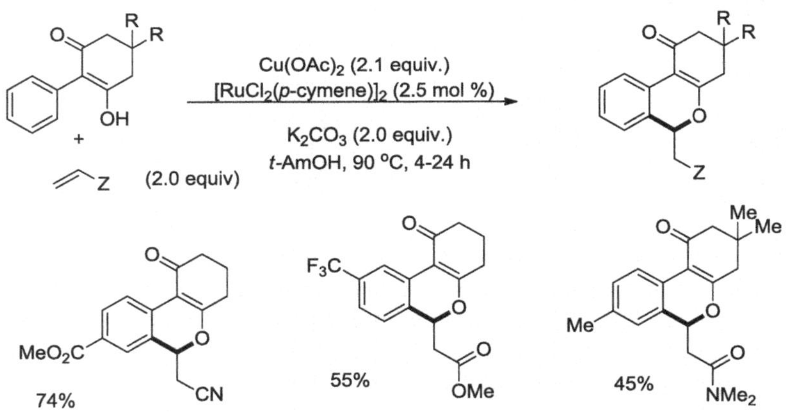

(31)

3.2.10 Carbamate, Carboxylate and 2-Pyridyl as Phenol Protecting and Directing Groups for Alkenylation

The hydroxy group of phenols does not direct ruthenium(II) C–H bond activation, but their protecting groups such as carbamates or carboxylates and the strongly coordinating 2-pyridyl group have been shown to direct the alkenylation of thus protected phenols, and the further deprotection allows to reach *ortho*-alkenylated phenols. Ackermann first [90] and then Jeganmohan [91] and Baiquan Wang [92] have shown that carbamate derivatives of phenols easily direct *ortho*-alkenylation with acrylates in a general way (Eq. 32), using the usual catalyst $[RuCl_2(p\text{-cymene})]_2$/4 $AgSbF_6$ with 2 equiv. of $Cu(OAc)_2.H_2O$ or better only 30 mol% but under air in 1,2-dimethoxyethane (Eq. 32) [91].

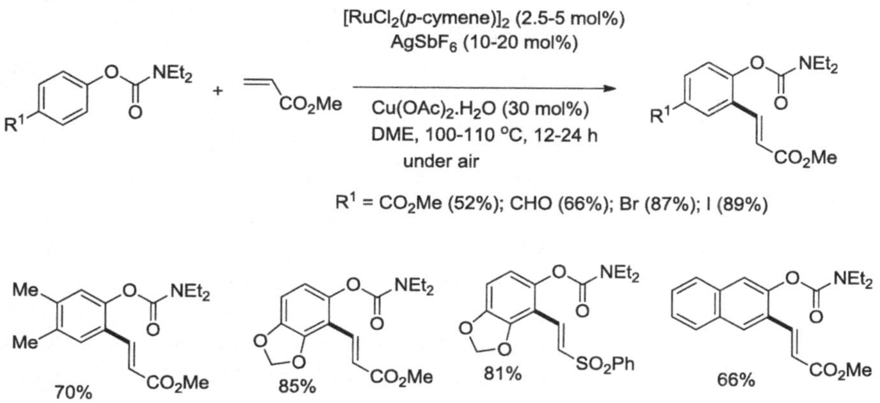

(32)

The reaction was successful with less reactive alkenes such as styrenes, alkenyl sulfone or acrylonitrile. With a weakly coordinating functional directing group at the *para*-position of carbamate, it was revealed that the carbamate is a stronger directing group than the CO_2Me or CHO directing group (Eq. 32) [91]. The carbamate directing group was easily removed and led to *ortho*-alkenylated phenol using $LiOH{\cdot}H_2O$ in THF/MeOH/H_2O (4:1:1) solvent at 80°C [91] or NaOH in EtOH at 80°C [90, 92].

Carboxylate groups have also been shown to direct *ortho*-alkenylation of their phenol derivatives, with alkyl acrylate. Thus, sesamol acetate and pivalate, in the presence of similar ruthenium–silver catalyst, lead to regioselective alkenylation at the C–H bond between both carboxylate and alkoxy groups (Eq. 33) [91].

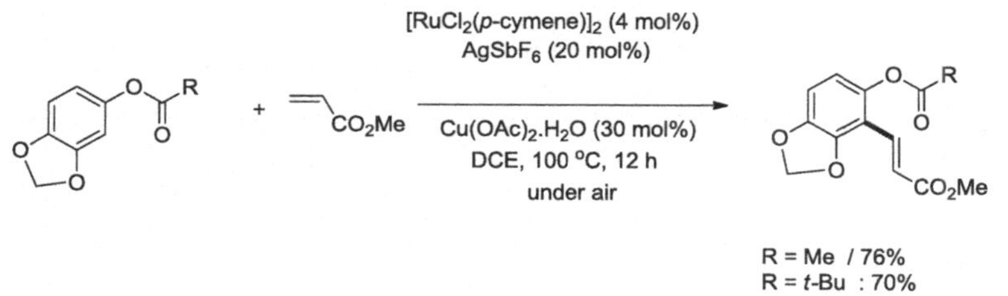

(33)

Ackermann et al. have shown that removable 2-pyridyl group of O-protecting phenols could direct *ortho*-alkenylation of the phenol derivatives, after he demonstrated their *ortho* C–H bond activation for C–C cross-coupling arylation [93].

A similar Ru(II)/$AgSbF_6$ catalytic system was used for general alkenylation with alkyl acrylates and with $Cu(OAc)_2.H_2O$ in aerobic atmosphere (Eq. 34) [94].

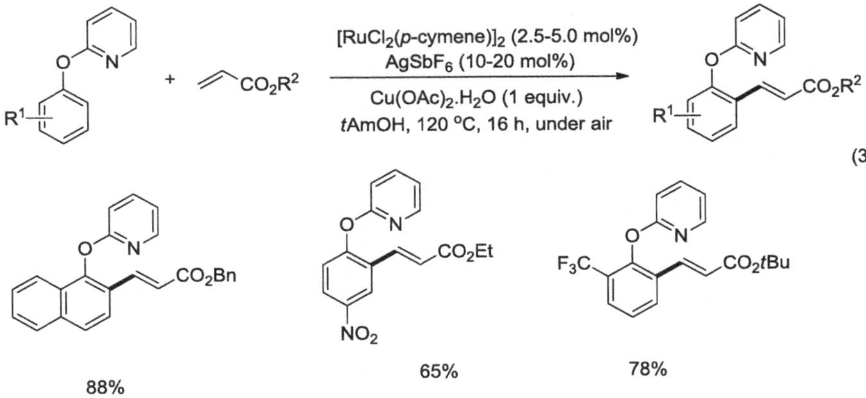

(34)

It is noteworthy that the intermediate is a six-membered cyclometallate that is usually more difficult to generate than the previous five-membered metallacycles. The 2-pyridyl protecting group is easily removed from alkenylated products with MeOTf and sodium in methanol without modification of phenol and alkene functional group.

This 2-pyridyl directing/protecting group has been used by Ackermann to functionalise heterocycles with alkenes. *N*-Methylindole containing an O-2-pyridyl group at C_3 led to alkenylation at C_2 only (Eq. 35a) [94]. The strong directing power of the O-2-pyridyl group was revealed by the reaction of acrylate with thiophene containing an O-Py group at C_2, which led to selective alkenylation at C_3 position and not at C_5 (Eq. 35b) [94].

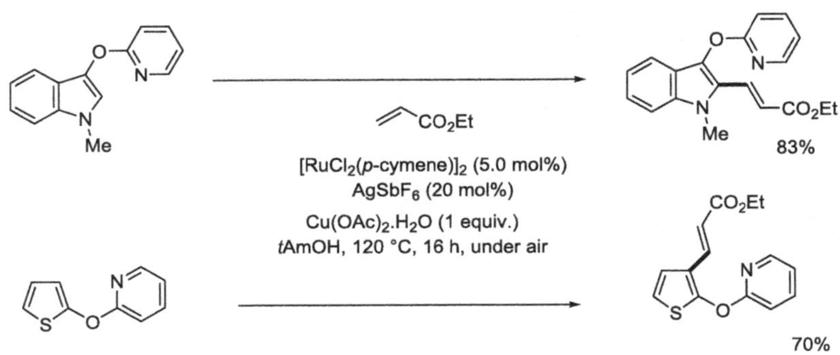

(35a)

(35b)

3.2.11 Azoxy Directing Group for Alkenylation

The azoxy group was shown by Wang to be another efficient directing group for sp^2C–H bond activation. The activation of the *ortho* sp^2C–H bond of azoxybenzenes has been completed with high selectivity in the presence of a new type of C–H bond activation ruthenium catalyst, a Ru(III) precursor $[Cp^*RuCl_2]_n$ associated to $AgSbF_6$ and $Cu(OAc)_2$, and alkenylations have been achieved with electron-deficient olefins such as acrylates and vinylphosphonates at 110°C in DCE (Eq. 36) [95]. Competition experiments have revealed that azoxybenzenes substituted by a methoxy electron-donating group reacted faster than by a neutral or electron-deficient analogue. With this azoxy directing group, the formation of cyclic products was not possible.

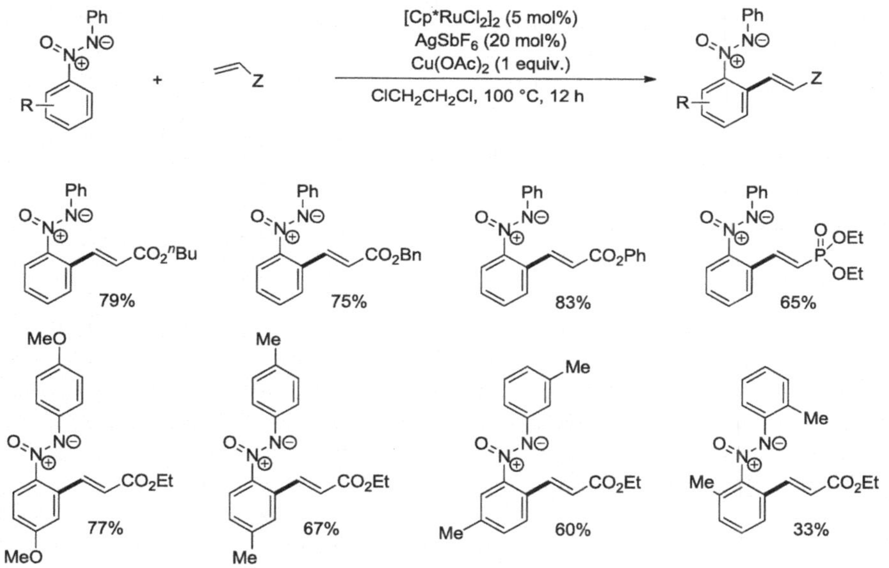

(36)

3.2.12 Sulfonic Acid, Sulfonamide and Sulfoximine Directing Groups for Alkenylation

Ackermann showed that the sulfonyl moiety arising either from *sulfonic acid or sulfonyl chloride* generated a five-membered bidentate *ortho*-phenylsulfonate ligand coordinated to the ruthenium(II) centre. Subsequent regioselective insertion of an alkene into the C–Ru bond followed by a type of β–H elimination process led to *ortho*-alkenylated sulfonic acids (Eq. 37) [96]. Interestingly, not only electron-poor olefins were suitable for this reaction, such as acrylates, α,β-unsaturated ketones, vinylsulfones or vinylphosphonates, but styrenes could also be efficiently used. Electron-rich and electron-deficient, as well as hindered *ortho*-substituted, sulfonic acids were reactive. This allowed the 2,6-bis(alkenylation) to take place in the presence of a threefold excess of alkene.

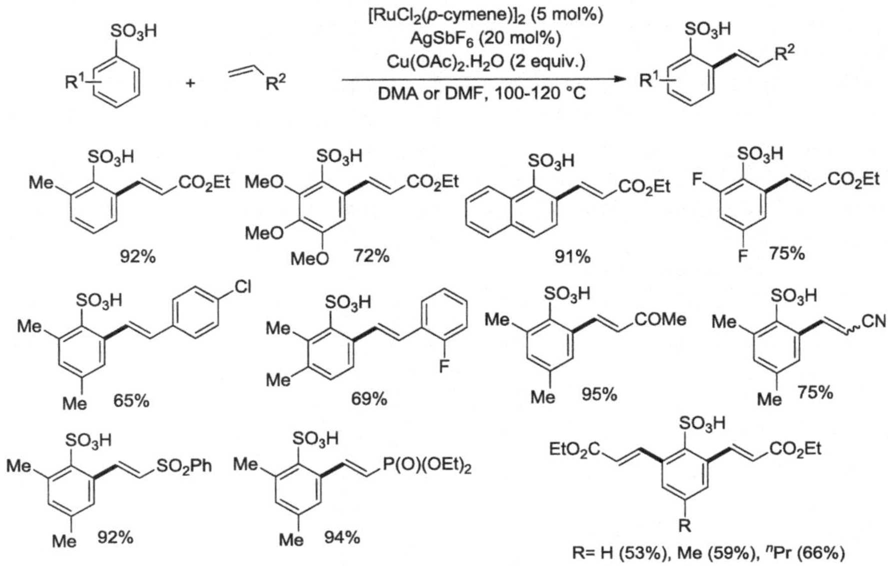

(37)

Sulfonamides, which also contain a NH in β-position with respect to the aromatic ring, have been shown by Ackermann to be excellent directing groups for alkenylation reaction with acrylates promoted by the same type of in situ generated ruthenium carboxylate catalyst from $[RuCl_2(p\text{-cymene})]_2$ with 2 equiv. of $Cu(OAc)_2.H_2O$. The alkenylation is followed by intramolecular aza-Michael addition leading to cyclisation into sultams in good yields (Eq. 38) [96].

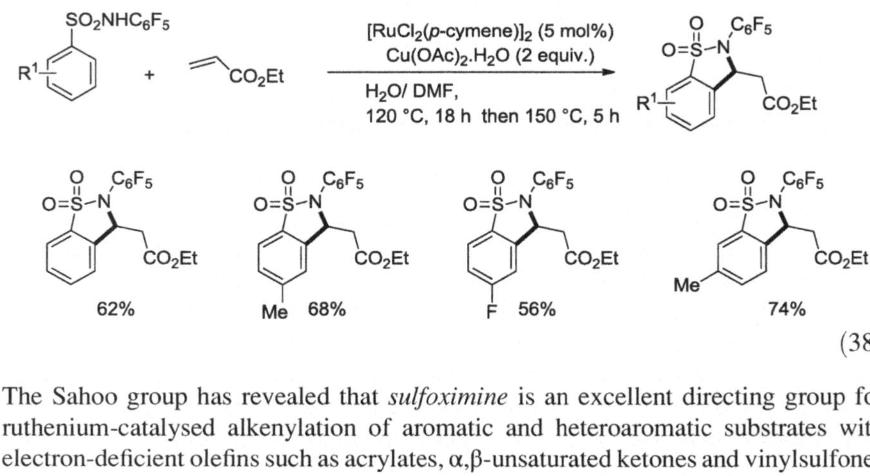

(38)

The Sahoo group has revealed that *sulfoximine* is an excellent directing group for ruthenium-catalysed alkenylation of aromatic and heteroaromatic substrates with electron-deficient olefins such as acrylates, α,β-unsaturated ketones and vinylsulfones

[97]. A variety of substituted arenes bearing either electron-donating or electron-attracting groups have been alkenylated in the presence of $[RuCl_2(p\text{-cymene})]_2$/ $AgBF_4$ and $Cu(OAc)_2$ as catalyst precursor (Eq. 39) [97].

It is noteworthy that heteroaromatic substrates including thiophenes, indoles and benzofurans equipped with a *sulfoximine* at carbon C_2 were alkenylated at C_3 with excellent regioselectivity and yields. It is demonstrated that the *sulfoximine prevails over the ester* as directing group for this alkenylation (Eq. 39) [97].

This directing group could be removed under basic treatment with NaOH in $MeOH/H_2O$ at 60°C to give the corresponding aromatic carboxylic acid and regenerate methyl phenyl sulfoximine that could be recycled to prepare the substrates.

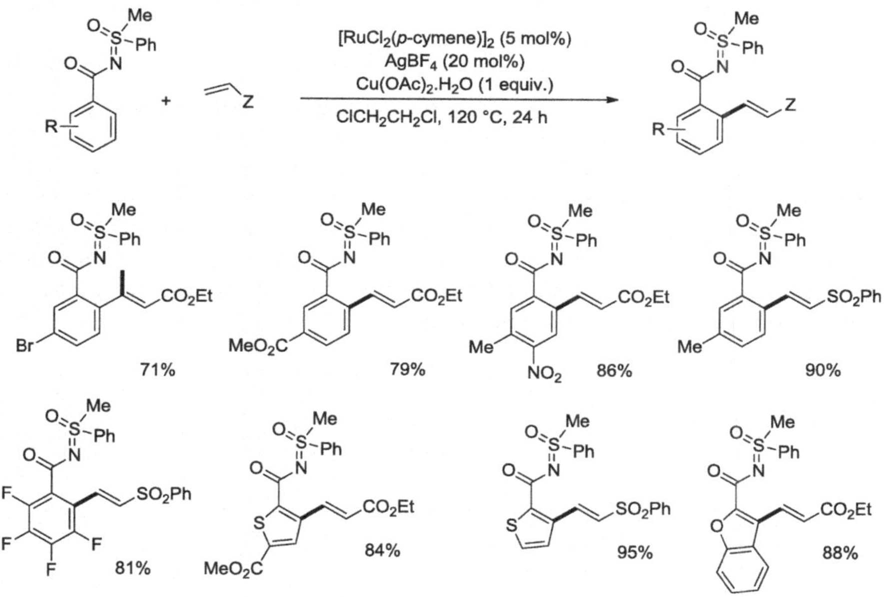

(39)

3.2.13 Nitrile as a Directing and Reactive Group for Alkenylation of Arenes

Jeganmohan has just shown that a nitrile group directly linked to a phenyl ring could direct *ortho*-alkenylation by acrylate, thus by π-interaction with the ruthenium(II) catalyst. The success was brought by the cooperative action of both $AgSbF_6$ (20 mol%) and AgOAc (2 equiv.) in addition to $[RuCl_2(p\text{-cymene})]_2$ (5 mol%) in dry acetic acid. Thus, a large variety of *ortho*-alkenylated arylnitriles were directly obtained in excellent yields on reaction with acrylates or alkenyl sulfones (Eq. 40) [98]. The proposed mechanism involves the formation of a cyclometallate containing a CN group π-bonded to the Ru(II) centre.

This reaction can be easily extended to heteroaromatic nitriles such as 2-nitrile and 3-nitrile thiophenes [98].

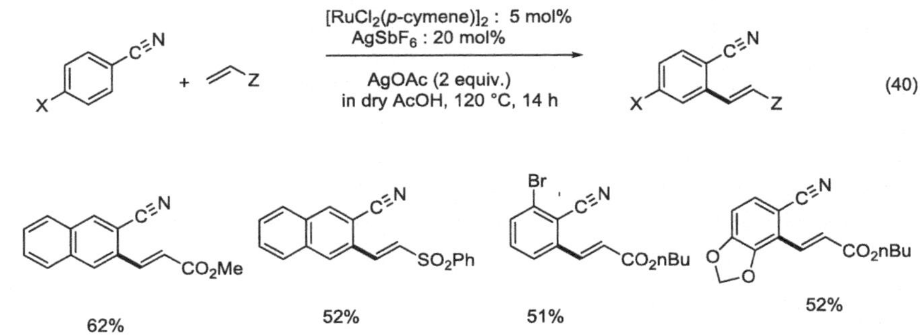

Previously, Jeganmohan also showed that the nitrile group could orientate alkenylation at the *ortho*-position of functional arenes but that in the presence of water the nitrile is hydrolysed into amide and can lead to the synthesis of (*Z*)-3-methylene-isoindilin-1-ones (Eq. 41) [99]. These cascade transformations were promoted by $[RuCl_2(p\text{-cymene})]_2$ catalyst with $AgSbF_6$ (20 mol%) but with $Cu(OAc)_2.H_2O$ (2 equiv.) in acetic acid. Thus, the only difference with the sole alkenylation step in dry medium (Eq. 40) is the introduction of $Cu(OAc)_2.H_2O$, instead of AgOAc, as an oxidant and acetate provider in the presence of water.

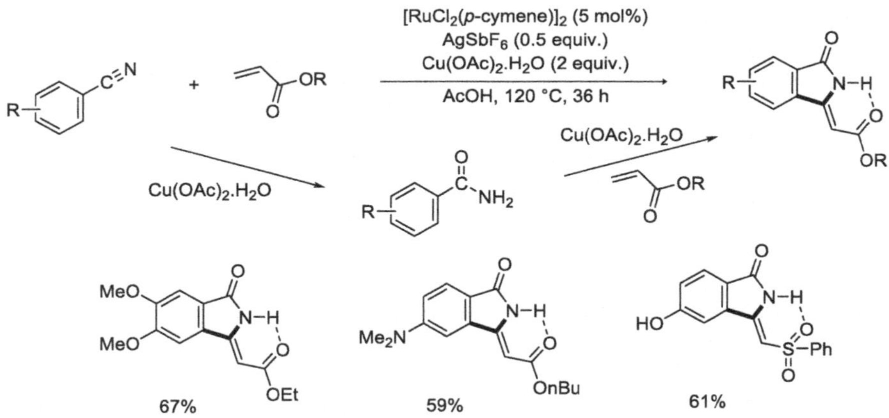

(41)

The mechanism is based on initial hydrolysis of the nitrile group into the amide group $CONH_2$ which becomes the directing group for *ortho*-alkenylation, followed by Ru(II)-assisted addition of the amide NH_2 group to the Ru(II)-coordinated double bond and finally β-elimination [99].

3.2.14 Enolisable Ketone as a Directing and Reactive Group in Alkenylation of C–H Bonds

Greaney has recently shown that ruthenium(II)-catalysed alkenylation of arylacetophenone C–H bonds with acrylates could further lead via cascade transformations to synthesis of heterocycles. First 1-indanones were selectively obtained as a major *trans*-isomer using $[RuCl_2(p\text{-cymene})]_2$ (5 mol%)/$AgBF_4$ (20 mol%) with $Cu(OAc)_2.H_2O$ (2 equiv.). The reaction is demonstrated to lead first to *ortho*-alkenylation and then the presence of $Cu(OAc)_2.H_2O$, and thus, free acetate allows the enolate formation and its Michael addition to the acrylic double bond (Eq. 42) [100].

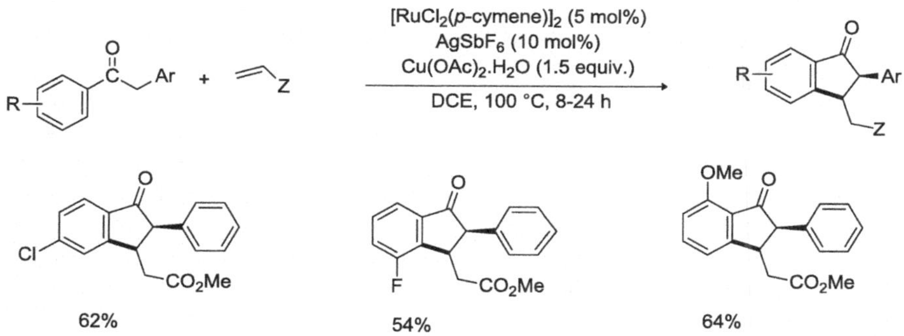

(42)

By using diarylated acetophenones, the similar catalytic reaction led to the formation of tetracyclic compounds: indenoindenes (Eq. 43) [100]. This transformation is proposed to take place via radical formation promoted by $Cu(OAc)_2.H_2O$ [100].

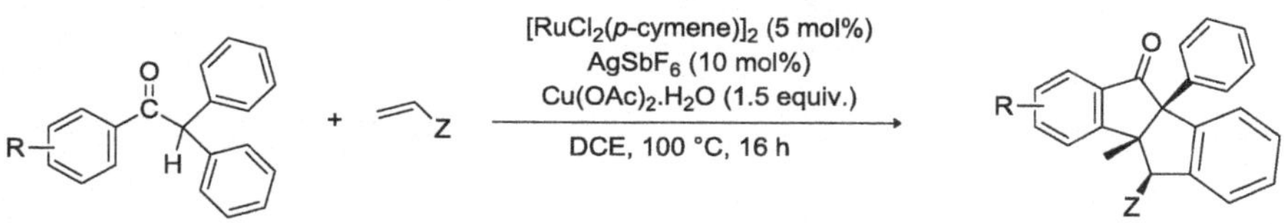

(43)

3.3 *Alkenylation of Heterocycles with Ruthenium(II) Catalysts and Copper(II)*

3.3.1 The First Examples

Several examples of ruthenium(II)-catalysed alkenylation of heterocycles have just been presented above. The first example of alkenylation of heterocycles with functional alkenes was performed by Miura and Satoh using the carboxylic acid directing group with $[RuCl_2(arene)]_2$ catalyst with stoichiometric amount of

$Cu(OAc)_2$ as oxidant. They especially showed that Ru(II) catalyst did not decarboxylate heterocycles such as furan, thiophene or indole derivatives as did previously Pd catalysts (Eq. 9) [41].

Then, both Jeganmohan [79] and Ackermann [80], by showing the directing ability of the ester group for ruthenium-catalysed alkenylation, also reported a few examples of direct alkenylation of thiophene, furan and indole derivatives (Eq. 20) [79, 80].

Ackermann et al. also showed that the O-2-pyridyl group linked to thiophene or indole, in the presence of $[RuCl_2(arene)]_2$/4 $AgSbF_6$ with 1 equiv. of oxidant $Cu(OAc)_2$, could direct alkenylation of heterocycles (Eq. 35) [94].

3.3.2 The Directing Ability of N-$CONR_2$ Groups of N-Containing Heterocycles

These pioneer examples of Ru(II)-catalysed alkenylation of heterocycles were directed by functional groups (FG): CO_2H, CO_2R and O-2-Py. In 2013, Prabhu demonstrated that a *N*-benzoyl group of indole could direct the alkenylation by acrylate at C_2 using the usual Ru(II)/$AgSbF_6$ catalytic system in the presence of $Cu(OAc)_2$ as oxidant (Eq. 44) [101]. It is noteworthy that with Pd(II) catalyst the alkenylation of indole takes place at C_3 and that the amide carbonyl does not direct alkenylation at *ortho*-positions of the phenyl group of the benzoyl or of the indole. The reaction tolerates a Cl or a Br group on the indole.

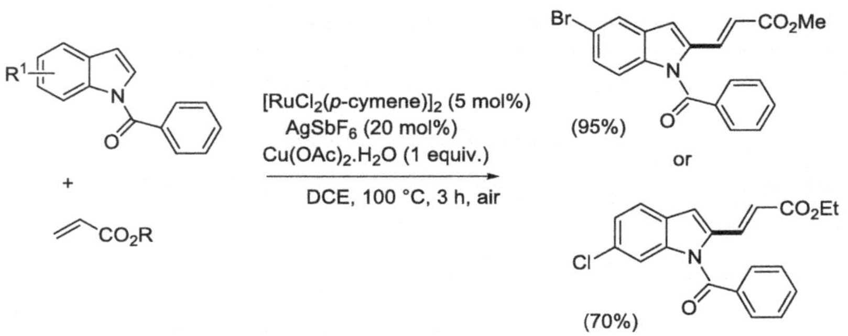

(44)

At the same time, by using the same type of Ru(II) catalytic system, Li and Wang have performed the alkenylation of indole containing a *N*,*N*-dimethyl amide group with a variety of functional alkenes: acrylate, styrene, *E*-methyl crotonate, vinylphosphonate and vinylsulfone. Cyclopentenone by contrast led to the addition product featuring a formal alkylation by alkene (Eq. 45) [102]. Competition experiments show that an electron-donating group on indole arene favours the reaction. This $CONMe_2$ directing group has the advantage of being easily removed with 6 equiv. of *t*BuOK at room temperature or with NaOH at 80°C in ethanol.

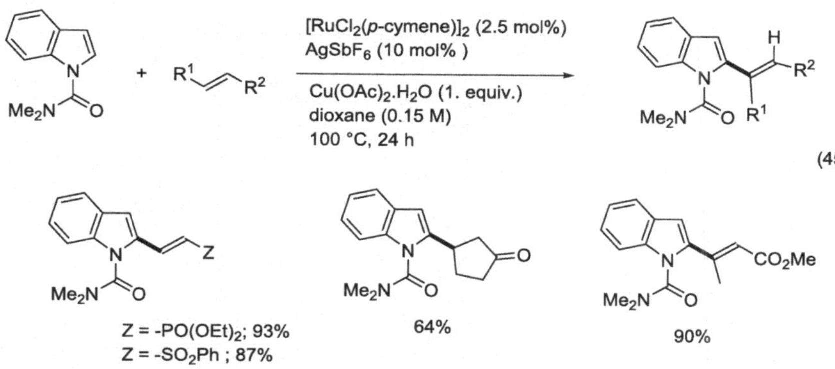

(45)

Analogously, under similar conditions with the same catalyst, the reaction of acrylate with N-protected dimethylamidopyrrole shows that the alkenylation successively takes place at C_2 and then C_5 positions thus leading to unsymmetrically dialkenylated pyrroles (Eq. 46) [102].

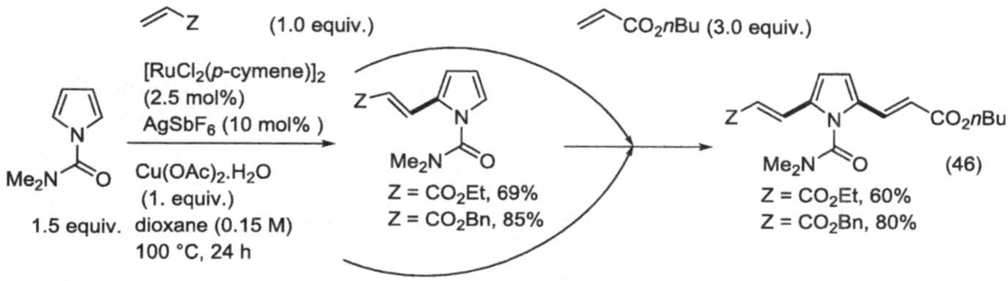

(46)

Song has also performed the direct alkenylation of indoles containing a removable *N,N*-dimethylcarbamoyl group, with acrylate derivatives, but also with unsaturated esters, phosphonates or sulfonyl derivatives and functional styrenes. The reaction is performed with $[RuCl_2(p\text{-cymene})]_2$/4 $AgSbF_6$ catalyst and catalytic amount (only 20 mol%) of $Cu(OAc)_2.H_2O$ in the presence of air. The profitable influence of NaOAc is surprisingly shown.

The direct alkenylation always takes place at the C_2–H carbon atom of the heterocycle, not at the *ortho* phenyl C–H bond adjacent to the Me_2NCO-N directing group (Eq. 47) [103].

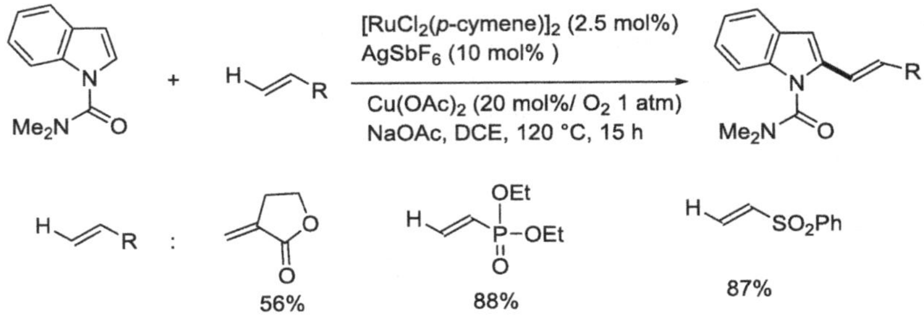

(47)

Under the same conditions, monoalkenylation or dialkenylation with acrylates of pyrrole can be selectively performed. More importantly, monoalkenylation of carbazole was achieved with either alkyl acrylates or styrene (Eq. 48) [103].

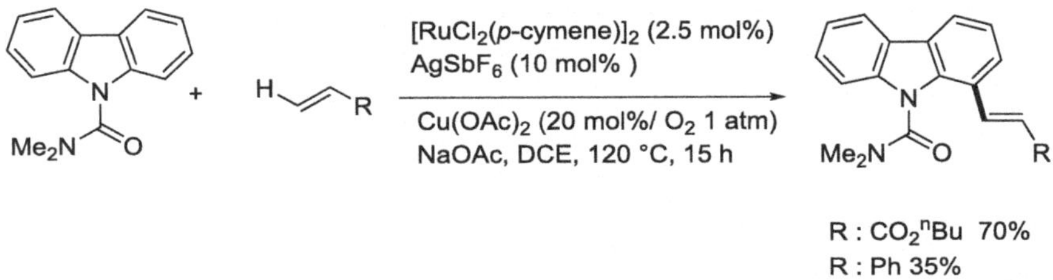

(48)

3.3.3 The Formyl Directing Group Ability in Heterocycles

Whereas most of the previous modifications of heterocycles took place at the heterocycle ring, in 2013, Prabhu showed the selective functionalisation of the C_4–H bond with no alkenylation at the C_2 position of *N*-protected indoles by introducing a formyl directing group at the C_3 position and by using the usual Ru (II)/$AgSbF_6$ catalytic system with $Cu(OAc)_2.H_2O$, (Eq. 49) [104]. These experiments, compared to the previous alkenylation of *N*-amide indole (Eq. 45), confirm that the *N*-amide group really directs alkenylation of indole at C_2 position. A cross experiment between *N*-benzoyl indole and the C_3-formyl *N*-amide-indole (Eq. 44) showed the preferential formation of the C_4-alkenyl C_3-formyl indole [104].

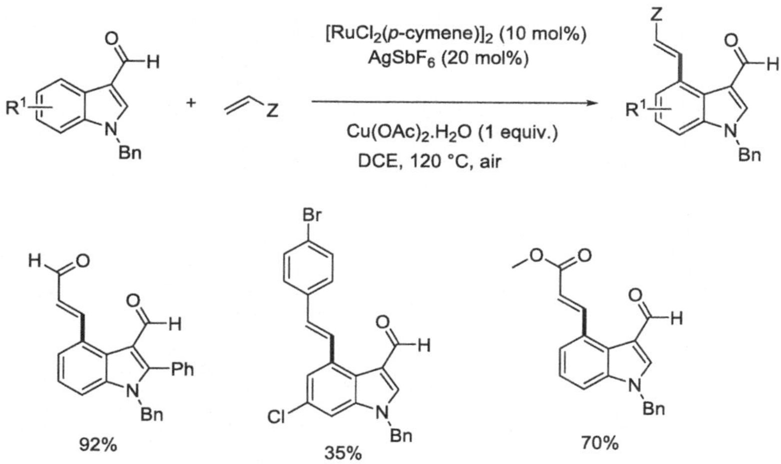

(49)

3.3.4 The Kommagalla–Ramana Alkenylation of Benzofurans: Evidence for Alkenylation Versus Alkylation

Alkenylation of furan and benzofuran derivatives with alkenes has been performed in several occasions with the Ru(II) catalyst and a Cu(II) salt as an oxidant, as for the C_3-alkenylation directed by C_2-CO_2Me group (Eq. 9) [41], or C_2-alkenylation directed by a C_3-CO_2Me group (Eq. 20) [79]. Kommagalla and Ramana have recently demonstrated crucial aspects of Ru(II)-catalysed alkenylation of benzofuran C–H bonds directly with alkene. They showed that changing the directing group to C_2-2-pyridyl group also favoured C_3–H bond activation but, more importantly, that only the presence or absence of base can tune the *alkenylation versus alkylation* of the C–H bond (Eq. 50) [105].

Thus, 2-pyridyl C_2-substituted benzofuran and methyl acrylate with $RuCl_2(PPh_3)_3$ (5 mol%)/AgOAc (30 mol%) catalyst react in toluene at 140°C for 24 h and lead to unique *alkylation* at C_3 (Eq. 50a) [105], thus corresponding to a formal insertion of alkene double bond into the C_3–H bond. More importantly under the same conditions except the addition of 3 equiv. of K_2CO_3, only the *C_3-alkenylation* was observed (Eq. 50b), thus demonstrating that carbonate is crucial for alkenylation and that the AgOAc acetate alone is not performing alkenylation. However, when $Cu(OAc)_2$ was used instead of AgOAc without K_2CO_3, the C_3-alkenyl (43%) and C_3-alkylated (31%) products were obtained, showing that $Cu(OAc)_2$ acetate is also operating as a base. These simple results bring crucial information for the first time to understand the mechanism.

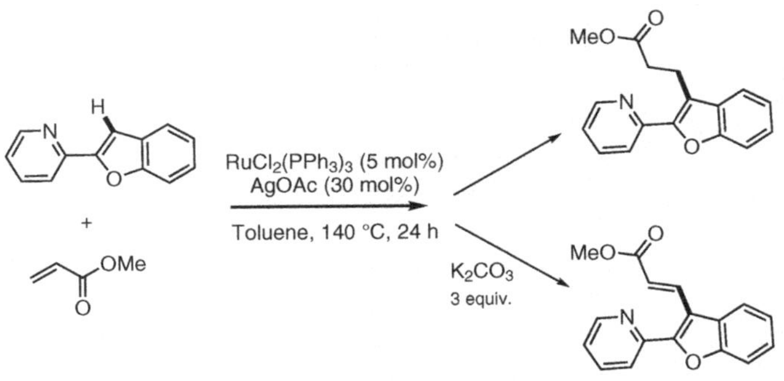

(50a)

(50b)

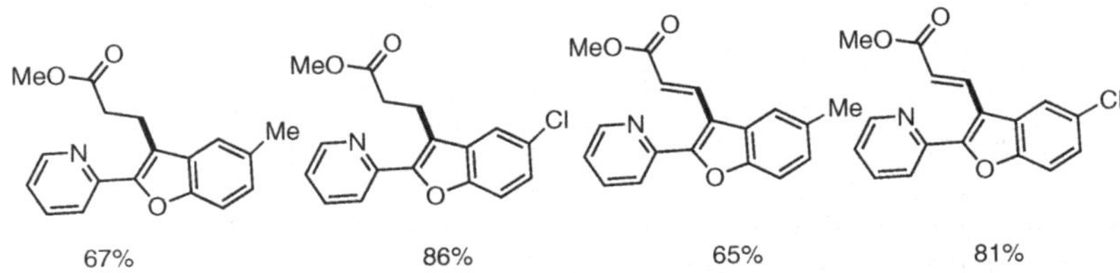

Kommagalla and Ramana [105] have proposed that the five-membered cyclometallate **L** is first produced via C–H bond deprotonation with AgOAc acetate, followed by coordinative insertion of the alkene (Scheme 6). The addition of the nucleophilic (Ru–C) carbon to the alkene electrophilic carbon to form the seven-membered cyclometallate **M** is thus expected. As the pyridine is strongly coordinating the Ru(II) site, further classical β-elimination is not possible from this rigid ruthenacycle, as the Ru–Cα and Cβ–H bonds cannot be coplanar and in *syn* position. Thus, AcOH protonation of the Ru–Cα bond takes place to form the alkylated product (Scheme 6).

In the presence of carbonate, this seven-membered cyclometallate **M** is deprotonated at the Cβ–H bond, at the *anti*-position of the Ru–Cα bond to give the alkenylated product and a Ru(0) species that can be reoxidised into a Ru(II) species with the available silver salt in the presence of acetic acid (Scheme 6). This mechanism brings light to the general Ru(II)-catalysed reaction of the C–H bond with alkene leading to either alkenylation or alkylation.

Each time a strong rigid metallacycle is formed via alkene insertion, the β-elimination will not take place and thus lead to the alkylated product when acid such as AcOH is present. By contrast with weakly bonding directing group (DG), the dissociation of this DG group may allow β-elimination and formation of the LnRu(H)(OAc) intermediate, even in the absence of a base. However, in the presence of large amount of base, the Cβ–H bond deprotonation should be favoured.

It is noteworthy that recently the nitrile ability to direct alkenylation of aromatic nitriles with ruthenium catalyst (Eq. 40) has been used by Jeganmohan to reach

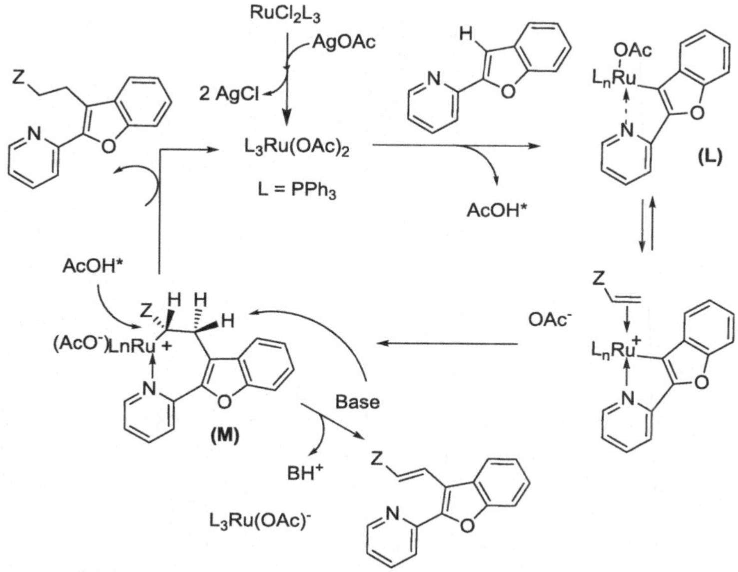

Scheme 6 Proposed mechanism for the alkenylation versus alkylation with alkenes

mono C_3-alkenylation of 2-nitrile thiophene, whereas 3-nitrile thiophene led easily to C_2- and C_4-dialkenylation [98].

3.4 Alkenylation of Alkene C–H Bonds with Ruthenium(II) and Copper(II)

The ruthenium(II) alkenylation of functional alkenes has been performed for the first time by Loh [86]. He demonstrated the successful synthesis of (*Z,E*)-dienamides upon alkenylation of acrylamides with electron-deficient alkenes in the presence of $[RuCl_2(p\text{-cymene})]_2$ (2.5 mol%) and 20 mol% of KPF_6, with 2 equiv. of oxidant $Cu(OAc)_2.H_2O$ and a small amount of acetic acid. Thus, substituted acrylamides reacted with various functional alkenes containing the CO_2R, CONHBn, $PO(OEt)_2$, CN, SO_2Ph or 4-Cl phenyl group to lead to the (*Z,E*)-dienamides, and the reaction of *N*-aryl methacrylamides with acrylate gave the desired products in 31–39% yield (Eq. 51) [86].

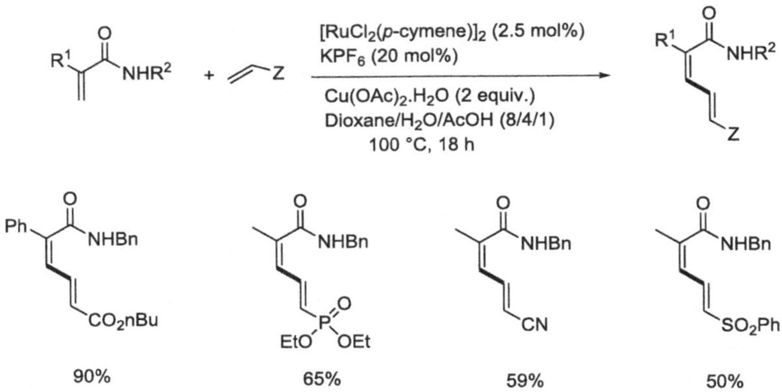

(51)

Loh also showed the cross coupling of two acrylate derivatives leading stereoselectively to (*Z,E*)-muconates which was achieved at 135°C with a closely related catalytic system involving an additional amount of $AgSbF_6$ (20 mol%) as additive. The self-alkenylation of unsubstituted acrylates was possible, but due to steric hindrance, 2-substituted propenoic acid esters were not self-alkenylated, which made the heterocoupling of diversely substituted acrylate derivatives with simple acrylates very effective in the presence of an excess of the latter (Eq. 52) [81].

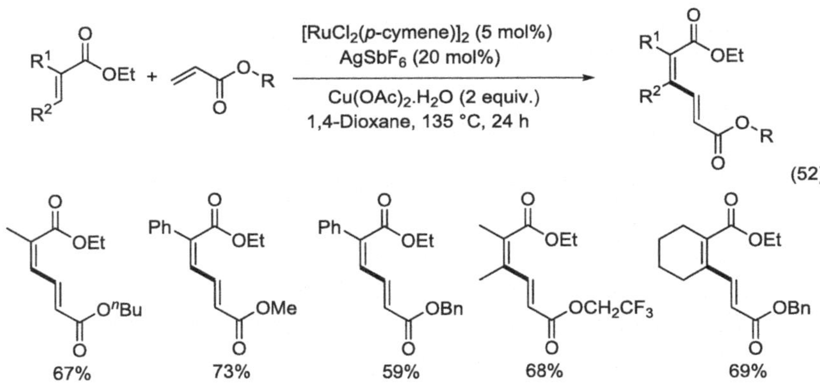

(52)

4 Alkylation with Alkenes of Arene and Heterocycle sp^{2}C–H Bonds with Ruthenium(II) Catalysts

4.1 Alkylation of Arenes with Alkenes

The formal insertion of an alkene into a sp^{2}C–H bond of arene, or hydroarylation of olefins, corresponds to an atom economy reaction. This alkylation of functional arenes by alkenes, rather than alkenylation, was first observed as the preferential reaction with Ru(0) catalysts in the Murai reaction [21–26]. It involves the oxidative addition of the arene C–H bond to Ru(0) species. Gunnoe has reported first that ruthenium(II) catalysts promote the addition of benzene C–H bond to ethylene C=C bond to form ethylbenzene. The ruthenium(II) catalysts used are of type TpRu-R(L)(NCMe) (Tp=hydrotris(pyrazolyl)borate) (Eq. 53) [45, 106–110].

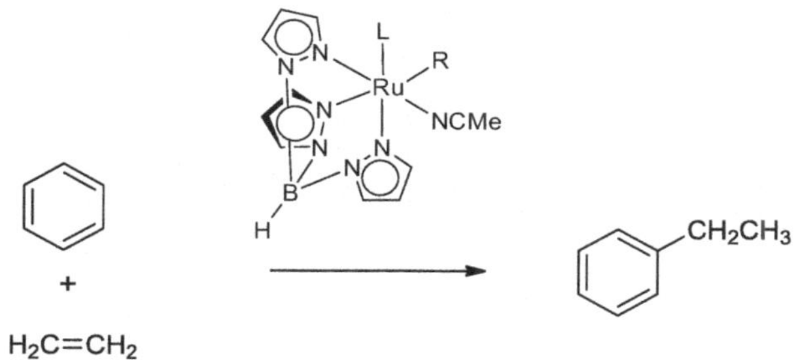

(53)

Kinetic studies and calculations have shown that the mechanism does not involve an oxidative addition of C–H bond and a ruthenium(IV) species, but undergoes the formation of Ru(II)–C_6H_5 species resulting from benzene coordination to the TpRu-R(L) intermediate and alkane R–H elimination. The insertion of ethylene into the Ru(II)–C_6H_5 bond leads to a Ru(II)–$CH_2CH_2C_6H_5$ species, which reacts with benzene to produce ethylbenzene (Scheme 7) [106–111].

The rate-determining step, favoured by electron-donating ligands L, is the C–H bond activation of the coordinated benzene by TpRu-R(L). By contrast, the withdrawing ligand L=CO decreases the energy barrier of the olefin insertion [111].

The catalyst TpRu-Me(L)(NCMe) similarly promotes the alkylation of heterocycles, thiophene and furan, via the formation of isolated intermediates TpRu(CO)(NCMe)(2-thienyl) and TpRu(CO)(NCMe)(2-furyl). The TpRu(CO)(NCMe)(2-thienyl) does catalyse the insertion of ethylene into the C_2–H bond of thiophene [112].

Ackermann has shown as early as 2008 the direct ruthenium(II)-catalysed intermolecular hydroarylation of highly strained methylenecyclopropanes using the $[RuCl_2(cod)]_n$ catalyst in the presence of the bulky electron-donating monophosphine XPhos. The hydroarylation of phenyl-substituted methylenecyclopropane and bicyclopropylidene was produced at the *ortho* C–H bond of phenylpyridine. The hydroarylation leads to the formation of *cis*-disubstituted cyclopropanes (Eq. 54) [113].

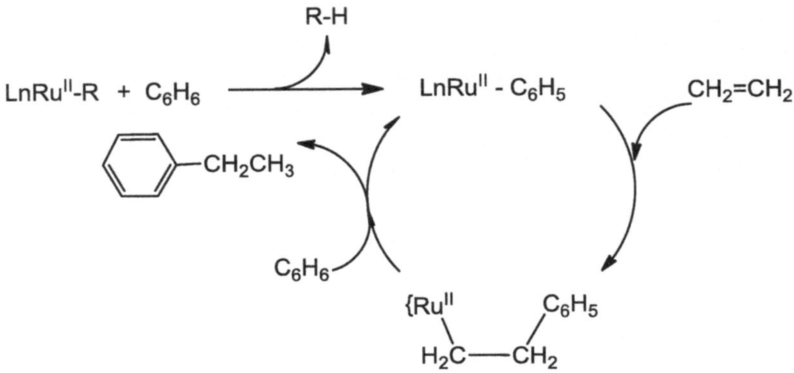

Scheme 7 Proposed mechanism for the TpRu-R(L) catalyzed hydroarylation of ethylene

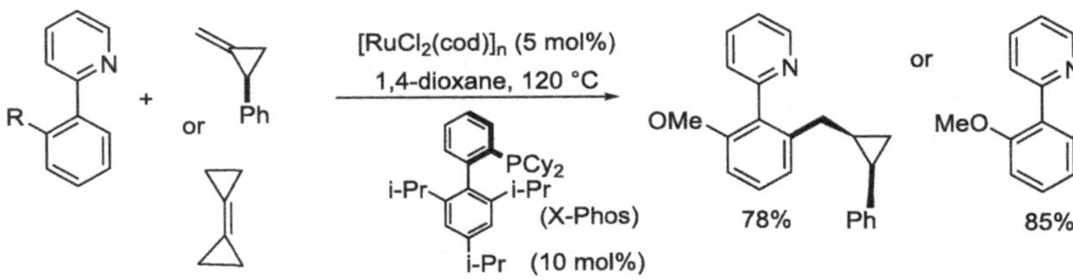

(54)

Recently, Ackermann extended this ruthenium(II)-catalysed hydroarylation of methylenecyclopropanes to a variety of *ortho* C–H bonds of 2-arylheterocycles directed by pyridine, oxazoline, diazole, pyrimidine, etc., using the same catalytic system [$RuCl_2$(cod)]$_2$/(XPhos). In some examples, an *anti*-Markovnikov addition was observed, or a ring opening occurred leading to Diels–Alder products [114].

The use of C_6D_5-Py showed a partial deuterium retention in the resulting alkyl group, and the authors suggest an oxidative addition of the C–H bond to the ruthenium(II) site, alkene insertion into the Ru–H bond and C–C bond formation via reductive elimination. However, this mechanism taking place previously with Ru(0)-catalysed Murai-type reaction [21–26] may not take place with a ruthenium(II) species even enriched by basic phosphine, and an alternative mechanism can be suggested later (Scheme 8).

Ackermann has now shown that a catalyst containing carboxylate ligand Ru(O_2CMes)$_2$(*p*-cymene) with 20 mol% of KO_2CMes is more efficient than the previous [$RuCl_2$(cod)]$_2$/phosphine catalyst to perform the same reaction (Eq. 55) [115]. This reaction is now highly regioselective and stereoselective without ring opening and can be applied to alkylation of cyclohexene C–H bond directed by a 2-pyridyl group.

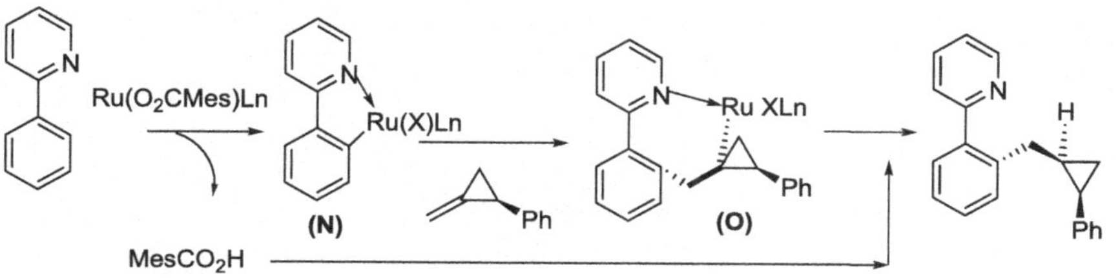

Scheme 8 Alternative mechanism for the alkylation of phenylpyridine with methylidenecyclopropane

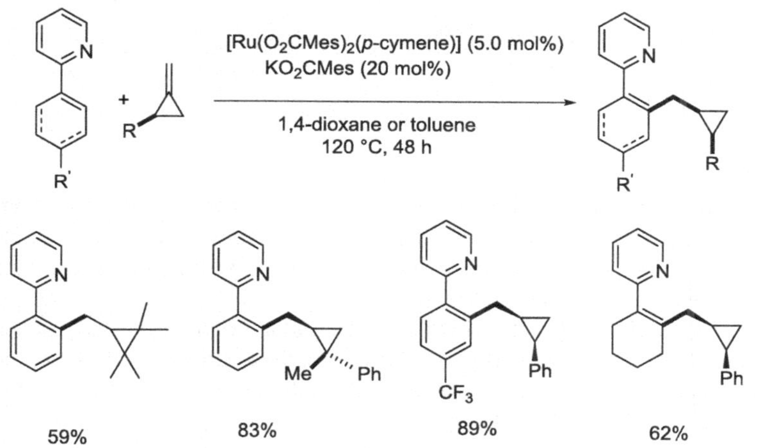

(55)

The authors suggest the initial insertion of the Ru(II) centre into the C–H bond and insertion of the alkene bond into the Ru–H bond [114, 115]; however another possible mechanism (Scheme 8) involves the classical Ru(II) activation of the *ortho* C–H bond via C–H deprotonation with carboxylate and cyclometallate **N** formation from phenylpyridine and formation of carboxylic acid [37]. The insertion of the C=C bond into the Ru–C bond corresponds to the *cis*-addition of the Ru–C bond to the less hindered face of the alkene, opposite to the substituent, to give the intermediate **O**. The protonation of the Ru–C(cyclopropyl) bond by the freed $MesCO_2H$ acid affords the alkylated product (Scheme 8). It is noteworthy that this alkylation occurs with strongly coordinating directing group pyridine, e.g. when rigid metallacycles are disfavouring β-elimination to form the corresponding alkene (see Scheme 6).

With the in situ generated carboxylate–Ru(II) catalyst $Ru(O_2CMes)_2$(*p*-cymene) from $[RuCl_2(p\text{-cymene})]_2$ and KO_2CMes, the alkylation of arene C–H bonds with unactivated and unstrained alkenes was performed by Ackermann in toluene and even in H_2O. Unactivated alkenes with a distant functional group such as ether, ketone, hydroxyl, ester and fluorine groups led to *ortho*-monoalkylated products (Eq. 56) [116]. The *ortho* C–H bond activation is directed by strongly coordinating nitrogen-containing heterocycles which also disfavour the alkenylation from the seven-membered cyclometallate. No alkenylation was observed.

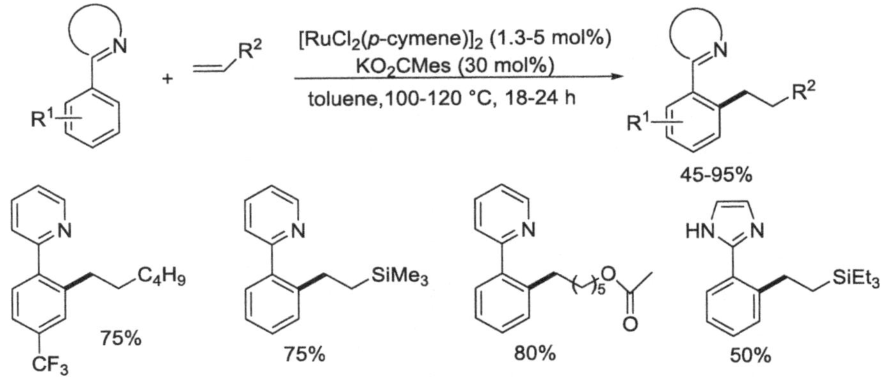

(56)

Peris has recently prepared mono- and di-*N*-heterocyclic carbenes based on pyrene and their related ruthenium(II) complexes and evaluated them for catalytic arylation and alkylation with alkenes of phenylpyridine. With nonactivated alkenes in the presence of KO_2CMes, both catalysts lead to only mono *ortho*-alkylation and show the same activity (Eq. 57) [117], with similar efficiency as the previous system $Ru(O_2CMes)_2$(*p*-cymene) [116]. However, these catalysts allow two cross *ortho* C–H bond functionalisations sequentially: alkylation with alkene and then arylation with phenyl bromide [117].

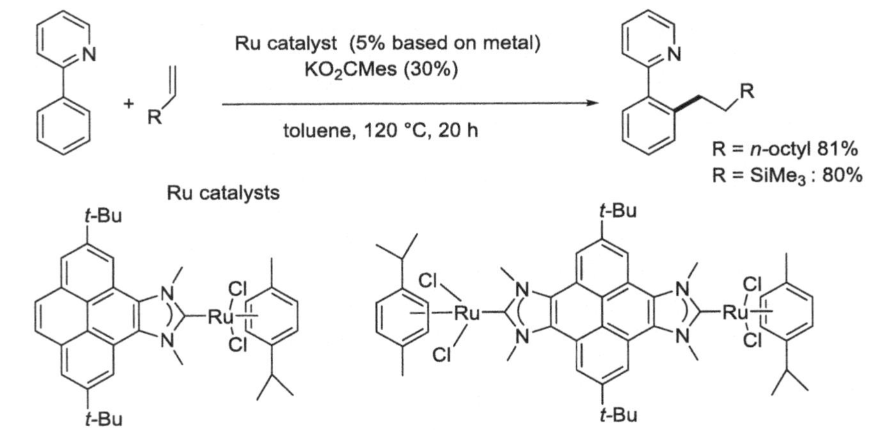

(57)

In 2013, Huanfeng Jiang used the amide group $CONR_2$ to direct the ruthenium(II)-catalysed *ortho*-alkylation of arenes on reaction of arylamides but with *allylic alcohols* [84]. The reaction was performed with $[RuCl_2(p\text{-cymene})]_2$ catalyst in the presence of $AgSbF_6$ and 1 equiv. of $Cu(OAc)_2.H_2O$ but also in the presence of 2 equiv. of AcOH. Thus, the reaction allows the selective formation of β-aryl ketones (Eq. 58) [84] and tolerates arene functional groups such as Br, CN, NO_2 and F. The alkylation occurs at the less sterically hindered *ortho* C–H bond, as shown from *meta*-substituted arylamide. It can be extended to thiophene derivative containing the C_2-$CONMe_2$ group directing the alkylation only at C_3. Jiang also showed that this reaction can be catalysed with $[RhCl_2Cp^*]_2$ with $AgSbF_6$ (7.5 mol %) and 1 equiv. of $Cu(OAc)_2.H_2O$, which allowed *ortho*-dialkylation of arylamides in *t*-butanol/water solvent, without the addition of AcOH [84]. In that case, the $CONR_2$ group is a weakly coordinating group, but the reaction is made in the presence of AcOH which favours both C–Ru bond cleavage and alkylation.

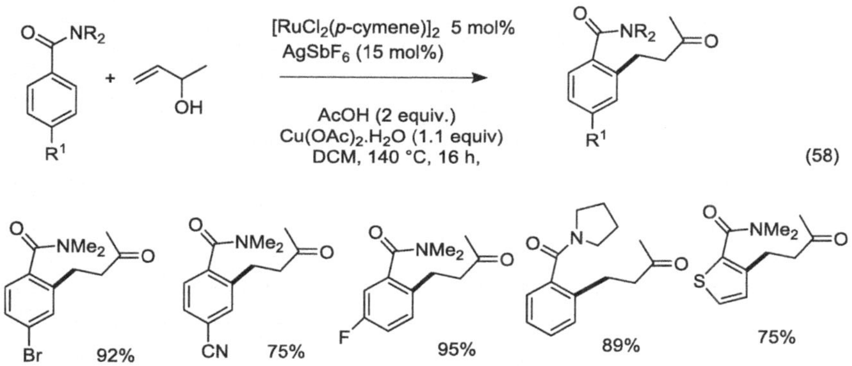

(58)

The mechanism is not described, but a realistic proposal can be made based on the recent reported ruthenium(II)-catalysed *ortho*-alkenylation, by Jeganmohan (Eq. 27) [85] of aryl-CONHR derivatives also with allylic alcohols, followed by isoindolinone formation and the recent demonstration by Kommagalla and Ramana on the *alkenylation versus alkylation* of benzofuran (Eq. 50) [105]. Thus, Jeganmohan showed that allyl alcohol is first dehydrogenated into unsaturated ketone leading to alkylation under acidic conditions, and Kommagalla and Ramana demonstrated that alkenylation takes place in the presence of base. Thus, it is expected that the Ru(II) catalyst generates the unsaturated ketone by hydrogen transfer, the enone double bond inserts into the Ru–C bond of the cyclometallate intermediate to give a seven-membered cyclometallate for which the β-elimination is disfavoured and, thus, protonation with AcOH of the Ru–C bond is expected to give the alkylated product (see Scheme 6 mechanism).

In 2013, Chatani has reported the ruthenium(II)-catalysed *ortho* C–H bond alkylation of aromatic amides with a variety of α,β-unsaturated ketones using the 8-aminoquinoline bidentate directing group. The best catalyst was based on $RuCl_2(PPh_3)_3$/NaOAc system (Eq. 59) [118]. The reaction of 2-substituted aromatic amides afforded monoalkylated products, whereas for *para*-substituted arylamides, the *ortho*-dialkylated amides were preferentially produced.

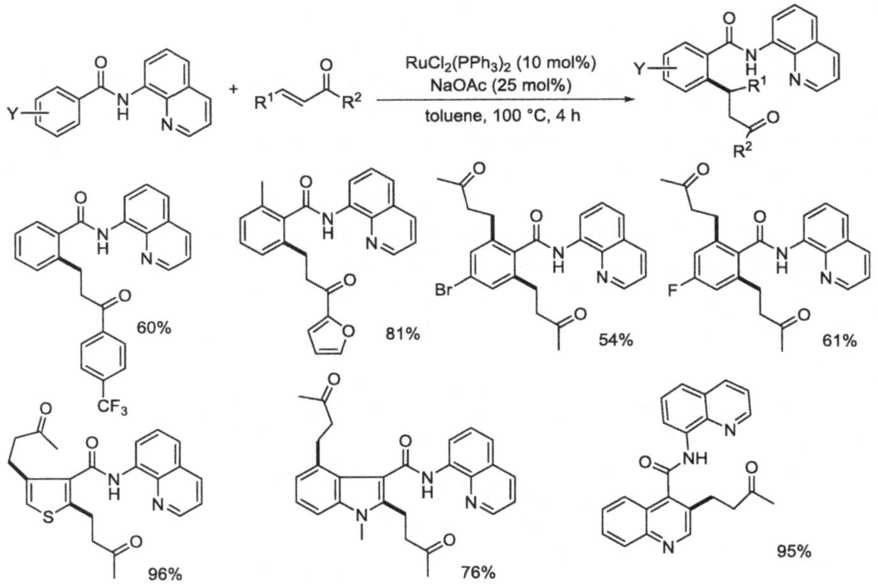

(59)

It is noteworthy that unsaturated ketones containing an electron-withdrawing C=O group facilitate this reaction by contrast to styrene and 1-hexene. The H/D exchange occurred in the cleavage of the *ortho* C–H bond, even in the absence of the alkene, and of the N–H bond indicating that the N–H bond is also involved in the reaction. The mechanism is not clear for this transformation, but the catalytic conditions are closely related to those giving cyclometallate intermediate by C–H bond deprotonation by acetate. Then, the insertion of the ketone double bond into the Ru–C bond could be postulated leading to a rigid strong seven-membered metallacycle, due to strong coordination of the chelating group, from which the β-elimination is not possible. Protonation of the Ru–C bond by the formed AcOH would then release the alkylated product.

4.2 Alkylation of Heteroarenes with Alkenes

4.2.1 Alkylation of Heterocycles with $Ru(O_2CR)_2L_n$ Catalyst

Ackermann has reported that the in situ generated carboxylate–Ru(II) catalyst $Ru(O_2CMes)_2$(*p*-cymene), for the alkylation of arene C–H bonds with unactivated and unstrained alkenes (Eq. 56) [116], could be profitably used for alkylation with vinylsilanes of heterocycles. This alkylation was applied to indole and thiophene derivatives and proceeded in site-selective manner, but it is noteworthy that they were directed by strongly coordinating nitrogen-containing groups such as 2-pyridyl for thiophene (Eq. 60a) and *N*-2-pyrimidyl for indole (Eq. 60b) [116].

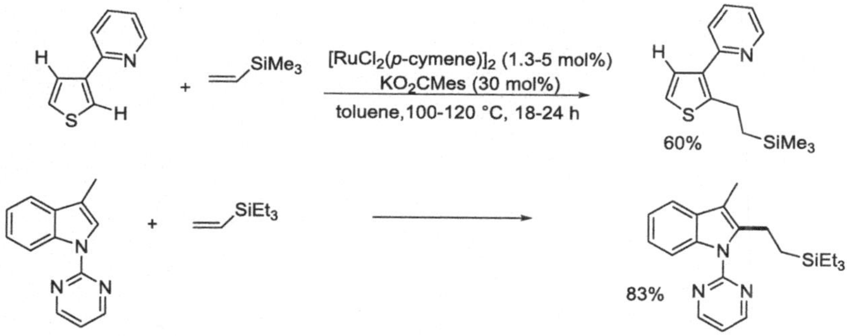

(60a)

(60b)

4.2.2 First Branched Alkylations Observed for Benzofuran

Kommagalla and Ramana have shown a unique way of Ru(II)-catalysed alkylation with acrylates applied to benzofuran directed by a C_2-benzoyl group. They showed that the COPh group does favour the alkylation at carbon C_3, but more importantly they found for the first time the formation of the branched alkylated isomer via hydroarylation of acrylates (Eq. 61) [119]. The catalyst is based on $RuCl_2(PPh_3)_3$ (5 mol%) with 30 mol% of additive $PivCO_2H$ or $MesCO_2H$ with K_2CO_3 (3 equiv.) as a base, but better efficiency was found with AgOAc (30 mol%) as additive. However, with methyl methacrylate, only the linear product was formed; thus, the substituent at alkene C_2 carbon disfavours the alkene C_2-(furan)–C bond formation.

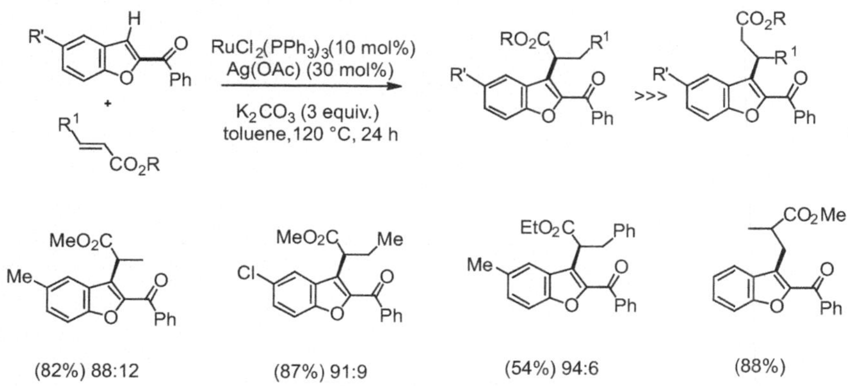

(61)

However, the linear alkylation of the same benzofuran with acrylate can be restored using the $[RuCl_2(p\text{-cymene})]_2$ catalyst (10 mol%) with 30 mol% of PPh_3 with no other additive but with $NaHCO_3$ (5 equiv.) as base and proton source as well (Eq. 62) [119].

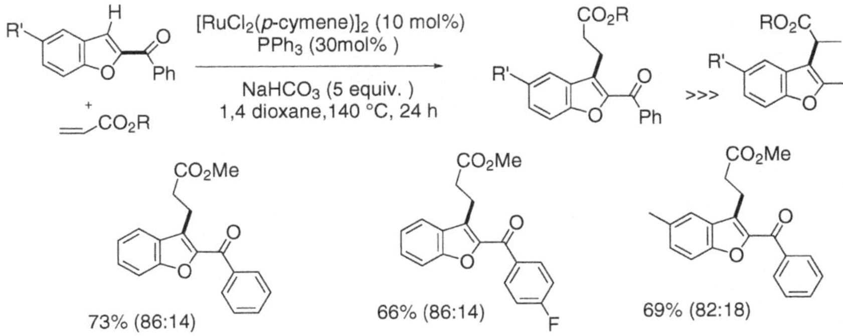

(62)

The rational mechanism (Scheme 9) proposes the initial formation of $Ru(OAc)_2L_3$ catalytic species leading to cyclometallate **I**, accepting the coordination of the olefin that can insert either from conformation **IIa** or **IIb** according to the nature of L_3 being $(PPh_3)_3$ and *p*-cymene, respectively.

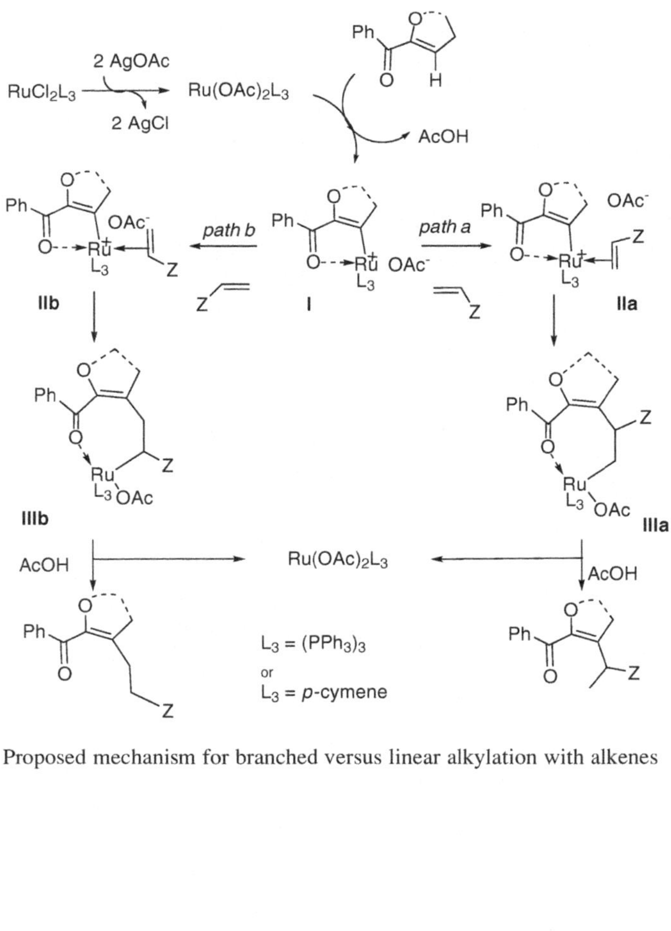

Scheme 9 Proposed mechanism for branched versus linear alkylation with alkenes

It is thus proposed that the steric hindrance of the $(PPh_3)_3$ ligands favours the insertion of alkene with its terminal carbon forming the C–Ru bond in **IIIa** and that its protonation with AcOH leads to the branched alkylated product. Alternatively, with L_3 ligand being *p*-cymene, the electrophilic β-carbon of the acrylate binds preferentially to the carbon of the more polar Ru–C bond in **IIb** than in **IIa** to give **IIIb**, which on protonation by AcOH or $NaHCO_3$ leads to the linear alkylated product (Scheme 9) [119].

5 Conclusion

The direct ruthenium(II)-catalysed oxidative dehydrogenative reaction of simple alkenes with functional arenes, heterocycles and even alkenes and ferrocene derivatives has become in a few years since 2011 a straightforward and inexpensive method to produce multiple functionalised alkenes, and new applications are reported every month [120–123]. Ruthenium(II) catalysts not only offer the advantages of their low-cost metal source, easiness of preparation and stability even in the presence of water which allows their use in water as solvent, but they have been shown efficient actors for innovations in synthesis. Most of the time for alkenylation, the ruthenium(II) catalytic reaction requires $AgSbF_6$, a halide abstractor, plus an oxidant such as $Cu(OAc)_2$ or sometimes Ag(OAc), and these oxidants can be used in catalytic amount but in the presence of air.

The activation of the *ortho* sp^2C–H bond is directed by a variety of strongly coordinating functional groups such as 2-pyridyl or other nitrogen-containing heterocycles. Recently, a variety of weakly coordinating groups, which were not efficient for direct arylation of the same *ortho* C–H bonds, have been shown to direct very efficiently *ortho*-alkenylation. These include ketone, formyl, ester, amide of type CONRR' or NHCOR, amidine, azoxy, sulfonic acid, sulfonamide and sulfoximine, carbamate or carboxylate as phenol protecting groups and even a few examples of free amine, free alcohol and nitrile.

The presence of acetate linked to the oxidant is not innocent as carboxylates are crucial partners of the Ru(II) centre to activate the *ortho* sp^2C–H bond via deprotonation with external carboxylate to produce very easily a five-membered cyclometallate able to insert into its Ru–C bond a variety of functional alkenes from acrylate derivatives to styrenes or non-electrophilic alkenes.

It is shown especially with strongly coordinating directing groups that the formation of the alkenylated product requires the presence of a base, such as carbonate, but sometimes, carboxylate can operate alone. This has been demonstrated in the case of heterocycles. This chapter supports that this observation results from the formation of a rigid stable seven-membered cyclometallate which cannot lead easily to β-elimination, and thus, the alkenylation product arises from β-C–H deprotonation by the base.

The ruthenium(II)-catalysed reaction of similar functional arenes and heteroarenes with alkenes can also lead to *ortho-alkylation* instead of alkenylation, via hydroarylation of the alkene. The first example of formation of the branched alkylation product has been shown with heterocycles. This alkylation is shown to occur with strained alkenes such as methylidene cyclopropanes, but also with strongly coordinating functional groups such as nitrogen-containing heterocycles, and strongly chelating 8-aminoquinoline bidentate groups. It also occurs with amide directing group but in the presence of proton source AcOH, or $NaHCO_3$ with heterocycles. It is demonstrated with heterocycle C–H bond that alkylation product is formed in the absence of base. This chapter supports the point that this observation results from the formation of a rigid stable seven-membered cyclometallate, which cannot lead to β-elimination, and thus, the presence of acid or proton source favours the protonation of Ru–C bond of the metallacycle and releases the C-alkylated product.

The easy ruthenium(II)-catalysed activation of C–H bond offers many challenges to overcome. No doubt that simple or new ruthenium catalysts will be evaluated as an attempt to favour the difficult functionalisation of the Ru–C carbon of the cyclometallate intermediate, and no doubt that future investigations and developments of ruthenium(II)-catalysed activation of C–H bond will deal more with the functionalisation of sp^3C–H bonds.

References

1. Jia C, Kitamura T, Fujiwara Y (2001) Acc Chem Res 34:633
2. Labinger JA, Bercaw JE (2002) Nature 417:507
3. Godula K, Sames D (2006) Science 312:67
4. Bergman RG (2007) Nature 446:391
5. Alberico D, Scott ME, Lautens M (2007) Chem Rev 107:174
6. Seregin IV, Gevorgyan V (2007) Chem Soc Rev 36:1173
7. Mkhalid IAI, Barnard JH, Marder TB, Murphy JM, Hartwig JF (2010) Chem Rev 110:890
8. Song G, Wang F, Li X (2012) Chem Soc Rev 41:3651
9. Campeau LC, Fagnou K (2006) Chem Commun 2006:1253
10. Daugulis O, Zaitsev VG, Shabashov D, Pham QN, Lazareva A (2006) Synlett 2006:3382
11. Beck EM, Gaunt MJ (2010) Top Curr Chem 292:85
12. Engle KM, Mei TS, Wasa M, Yu JQ (2011) Acc Chem Res 45:788
13. Neufeldt SR (2012) Sanford MS Acc Chem Res 45:936
14. Li BJ, Shi ZJ (2012) Chem Soc Rev 41:5588
15. Davies HML, Beckwith REJ (2003) Chem Rev 103:2861
16. Satoh T, Miura M (2010) Chem Eur J 16:11212
17. Wencel-Delord J, Droge T, Liu F, Glorius F (2011) Chem Soc Rev 40:4740
18. Lewis JC, Bergman RG, Ellman JA (2008) Acc Chem Res 41:1013
19. Colby DA, Tsai AS, Bergman RG, Ellman JA (2012) Acc Chem Res 45:814
20. Miura M, Satoh T, Hirano K (2014) Bull Chem Soc Jpn 87:751
21. Murai S, Kakiuchi F, Sekine S, Tanaka Y, Kamatani A, Sonoda M, Chatani N (1993) Nature 366:529
22. Kakiuchi F, Tanaka Y, Sato T, Chatani N, Murai S (1995) Chem Lett 1995:679
23. Kakiuchi F, Murai S (2002) Acc Chem Res 35:826
24. Kakiuchi F, Chatani N (2003) Adv Synth Catal 345:1077
25. Kakiuchi F, Chatani N (2004) In: Bruneau C, Dixneuf PH (eds) Ruthenium catalysts and fine chemistry. Top Organomet Chem 11:45

26. Kakiuchi F, Kochi T (2008) Synthesis 2008:3013
27. Oi S, Fukita S, Hirata N, Watanuki N, Miyano S, Inoue Y (2001) Org Lett 3:2579
28. Oi S, Ogino Y, Fukita S, Inoue Y (2002) Org Lett 4:1783
29. Ritleng V, Sirlin C, Pfeffer M (2002) Chem Rev 102:1731
30. Ackermann L (2011) Chem Rev 111:1315
31. Ackermann L (2010) Chem Commun 46:4866
32. Arockiam PB, Bruneau C, Dixneuf PH (2012) Chem Rev 112:5879
33. Li B, Dixneuf PH (2013) Chem Soc Rev 42:5744
34. Rao Y, Shan G, Yang XL (2014) Sci China Chem 57:930
35. Zheng QZ, Jiao N (2014) Tetrahedron Lett 55:1121
36. Özdemir I, Demir S, Cetinkaya B, Gourlaouen C, Maseras F, Bruneau C, Dixneuf PH (2008) J Am Chem Soc 130:1156
37. Ferrer-Flegeau EF, Bruneau C, Dixneuf PH, Jutand A (2011) J Am Chem Soc 133:10161
38. Fabre I, von Wolff N, Le Duc G, Ferrer Flegeau E, Bruneau C, Dixneuf PH, Jutand A (2013) Chem Eur J 19:7595
39. Fujiwara Y, Moritani I, Matsuda M, Teranishi S (1968) Tetrahedron Lett 1968:633
40. Bras JL, Muzart J (2011) Chem Rev 111:1170
41. Ueyama T, Mochida S, Fukutani T, Hirano K, Satoh T, Miura M (2011) Org Lett 13:706
42. Ackermann L, Pospech J (2011) Org Lett 13:4153
43. Arockiam PB, Fischmeister C, Bruneau C, Dixneuf PH (2011) Green Chem 13:3075
44. Padala K, Jeganmohan M (2011) Org Lett 13:6144
45. Lail M, Arrowood BN, Gunnoe TB (2003) J Am Chem Soc 125:7506
46. Ackermann L (2014) Acc Chem Res 47:281
47. De Sarkar S, Liu W, Kozhushkov SI, Ackermann L (2014) Adv Synth Catal 356:1461
48. Li B, Dixneuf PH (2014) In: Bruneau C, Dixneuf PH (eds) Ruthenium in catalysis. Springer. Top Organomet Chem 48:119
49. Kozhushkov SI, Ackermann L (2013) Chem Sci 4:886
50. Kakiuchi F, Yamauchi M, Chatani N, Murai S (1996) Chem Lett 1996:111
51. Kakiuchi F, Sato T, Yamauchi M, Chatani N, Murai S (1999) Chem Lett 1999:19
52. Ueno S, Chatani N, Kakiuchi F (2007) J Org Chem 72:3600
53. Martinez R, Chevalier R, Darses S, Genet JP (2006) Angew Chem Int Ed 45:8232
54. Martinez R, Simon MO, Chevalier R, Pautigny C, Genet JP, Darses S (2009) J Am Chem Soc 131:7887
55. Ogiwara Y, Tamura M, Kochi T, Matsuura Y, Chatani N, Kakiuchi F (2014) Organometallics 33:402
56. Ogiwara Y, Kochi T, Kakiuchi F (2014) Chem Lett 43:667
57. Farrington EJ, Brown JM, Barnard CFJ, Rowsell E (2002) Angew Chem Int Ed 41:169
58. Farrington EJ, Barnard CFJ, Rowsell E, Brown JM (2005) Adv Synth Catal 347:185
59. Ueno S, Kochi T, Chatani N, Kakiuchi F (2009) Org Lett 11:855
60. Chinnagolla RK, Jeganmohan M (2012) Org Lett 14:5246
61. Heck RF, Nolley JP (1972) J Org Chem 37:2320
62. Weissman H, Song XP, Milstein D (2001) J Am Chem Soc 123:337
63. Yi CS, Lee DW (2009) Organometallics 28:4266
64. Kwon KH, Lee DW, Yi CS (2010) Organometallics 29:5748
65. Li B, Roisnel T, Darcel C, Dixneuf PH (2012) Dalton Trans 41:10934
66. Yamashita M, Hirano K, Satoh T, Miura M (2010) Org Lett 12:592
67. Arockiam PB, Fischmeister C, Bruneau C, Dixneuf PH (2010) Angew Chem Int Ed 49:6629
68. Hashimoto Y, Ueyama T, Fukutani T, Hirano K, Satoh T, Miura M (2011) Chem Lett 40: 1165
69. Arockiam PB, Fischmeister C, Bruneau C, Dixneuf PH (2013) Green Chem 15:67
70. Li B, Devaraj K, Darcel C, Dixneuf PH (2012) Green Chem 14:2706
71. Seki M (2014) RSC Adv 4:29131
72. Hashimoto Y, Ortloff T, Hirano K, Satoh T, Bolm C, Miura M (2012) Chem Lett 41:151
73. Ackermann L (2005) Org Lett 7:3123
74. Li B, Bheeter CB, Darcel C, Dixneuf PH (2011) ACS Catal 1:1221

75. Li B, Devaraj K, Darcel C, Dixneuf PH (2012) Tetrahedron 68:5179
76. Li B, Bheeter CB, Darcel C, Dixneuf PH (2014) Top Catal 57:833
77. Singh KS, Dixneuf PH (2012) Organometallics 31:7320
78. Padala K, Jeganmohan M (2012) Org Lett 14:1134
79. Padala K, Pimparkar S, Madasamy P, Jeganmohan M (2012) Chem Commun 48:7140
80. Graczyk K, Ma W, Ackermann L (2012) Org Lett 14:4110
81. Hu XH, Zhang J, Yang XF, Xu YH, Loh TP (2015) J Am Chem Soc 137:3169
82. Li B, Ma JF, Wang NC, Feng HL, Xu SS, Wang B (2012) Org Lett 14:736
83. Ackermann L, Wang L, Wolfram R, Lygin AV (2012) Org Lett 14:728
84. Qi J, Huang L, Wang Z, Jiang H (2013) Org Biomol Chem 11:8009
85. Manoharan R, Jeganmohan M (2015) Chem Commun 51:2929
86. Zhang J, Loh TP (2012) Chem Commun 48:11232
87. Li J, John M, Ackermann L (2014) Chem Eur J 20:5403
88. Suzuki C, Morimoto K, Hirano K, Satoh T, Miura M (2014) Adv Synth Catal 356:1521
89. Chidipudi SR, Wieczysty MD, Khan I, Lam HW (2013) Org Lett 15:570
90. Li J, Kornhaaβ C, Ackermann L (2012) Chem Commun 48:11343
91. Reddy MC, Jeganmohan M (2013) Eur J Org Chem 2013:1150
92. Li B, Ma J, Liang Y, Wang N, Xu S, Song H, Wang B (2013) Eur J Org Chem 2013:1950
93. Ackermann L, Diers E, Manvar A (2012) Org Lett 14:1154
94. Ma W, Ackermann L (2013) Chem Eur J 19:13925
95. Li H, Xie X, Wang L (2014) Chem Commun 50:4218
96. Ma W, Mei R, Tenti G, Ackermann L (2014) Chem Eur J 20:15248
97. Yadav MR, Rit RK, Shankar M, Sahoo AK (2014) J Org Chem 79:6123
98. Reddy MC, Jeganmohan M (2015) Chem Commun 51:10738
99. Reddy MC, Jeganmohan M (2014) Org Lett 16:4866
100. Mehta VP, Garcia-Lopez JA, Greaney MF (2014) Angew Chem Int Ed 53:1529
101. Lanke V, Prabhu KR (2013) Org Lett 15:2818
102. Li B, Ma J, Xie W, Song H, Xu S, Wang B (2013) J Org Chem 78:9345
103. Zhang LQ, Yang S, Huang X, You J, Song F (2013) Chem Commun 49:8830
104. Lanke V, Prabhu KR (2013) Org Lett 15:6262
105. Kommagalla Y, Mullapudi VB, Francis F, Ramana CV (2015) Catal Sci Technol 5:114
106. Lail M, Bell CM, Conner D, Cundari TR, Gunnoe TB, Petersen JL (2004) Organometallics 23:5007
107. Foley NA, Lail M, Lee JP, Gunnoe TB, Cundari TR, Petersen JL (2007) J Am Chem Soc 129:6765
108. Foley NA, Ke Z, Gunnoe TB, Cundari TR, Petersen JL (2008) Organometallics 27:3007
109. Foley NA, Lee JP, Ke Z, Gunnoe TB, Cundari TR (2009) Acc Chem Res 42:585
110. Andreatta JR, McKeown BA, Gunnoe TB (2011) J Organomet Chem 696:305
111. Oxgaard J, Periana RA, Goddard WA (2004) J Am Chem Soc 126:11658
112. Pittard KA, Lee JP, Cundari TR, Gunnoe TB, Petersen JL (2004) Organometallics 23:5514
113. Kozhushkov SI, Yufit DS, Ackermann L (2008) Org Lett 10:3409
114. Ackermann L, Kozhushkov SI, Yufit DS (2012) Chem Eur J 18:12068
115. Schinkel M, Wallbaum J, Kozhushkov SI, Marek I, Ackermann L (2013) Org Lett 15:4482
116. Schinkel M, Marek I, Ackermann L (2013) Angew Chem Int Ed 52:3977
117. Gonell S, Peris E (2014) ACS Catal 4:2811
118. Rouquet G, Chatani N (2013) Chem Sci 4:2201
119. Kommagalla Y, Srinivas K, Ramana CV (2014) Chem Eur J 20:7884
120. Li XG, Sun M, Liu K, Liu PN (2015) Adv Synth Catal 357:395
121. Tirler C, Ackermann L (2015) Tetrahedron 71:4543
122. Das R, Kapur M (2015) Chem Asian J. doi:10.1002/asia.201500343
123. Sawant D, Singh I, Tulsyan G, Abbagani K, Pardasani RT (2015) Synlett. doi:10.1055/s-0034-1380746

Top Organomet Chem (2016) 55: 189–216
DOI: 10.1007/3418_2015_126

Published online: 14 July 2015

Ruthenium-Catalyzed C–N and C–O Bond-Forming Processes from C–H Bond Functionalization

Suman Dana, M. Ramu Yadav, and Akhila K. Sahoo

Abstract This chapter highlights the robust, cost-effective Ru catalyst-mediated direct functionalization of C–H bonds. The development of Ru-catalyzed novel synthetic methods for the construction of C–N and C–O bonds through the activation of inert sp^2 and sp^3 C–H bonds is enumerated. The effect of a directing group (DG) for the chemo- and regioselective introduction of heteroatoms in the molecule replacing the unactivated C–H bonds is discussed.

Keywords Amidation · C–N and C–O bond · C–H bond functionalization · Directing group · Ruthenium catalysis

Contents

Abbreviations

Ac	Acetyl
Bz	Benzoyl
cat	Catalytic

S. Dana, M.R. Yadav, and A.K. Sahoo (✉)
School of Chemistry, University of Hyderabad, Hyderabad, India
e-mail: akhilchemistry12@gmail.com; akssc@uohyd.ac.in

cod	1,4-Cyclooctadiene
Cp	Cyclopentadienyl
Cy	Cyclohexyl
DCE	1,2-Dichloroethane
DG	Directing group
DMSO	Dimethyl sulfoxide
DPPA	Diphenylphosphoryl azide
equiv	Equivalent
F_{20}-TPP	*meso*-Tetrakis(pentafluorophenyl)porphyrin
Hp	Oxypyridinato
i-Pr	*iso*-Propyl
KIE	Kinetic isotope effect
Me_3tacn	*N*,*N*′,*N*″-Trimethyl-1,4,7-triazacyclononane
Mes	Mesityl
MPS	Methylphenylsulfoximine
NMP	*N*-Methyl-2-pyrrolidinone
Ns	4-Nitrobenzenesulfonyl
OEP	Octaethylporphyrin
t-Bu	*tert*-Butyl
TFA	Trifluoroacetic acid
TFAA	Trifluoroacetic anhydride
THF	Tetrahydrofuran
TM	Transition metal
TMP	*meso*-Tetrakis(mesityl)porphyrin
TPP	Tetraphenylporphyrin
Ts	4-Methylbenzenesulfonyl
TTP	Tetratolylporphyrin

1 Introduction

Construction of amino- and hydroxy-functionalized derivatives is immensely important, as these motifs are widely found in the molecules relevant to biological and pharmaceutical importance and materials [1–7]. Thus, development of convenient straightforward synthetic protocols for the fabrication of C–N and C–O bonds is highly enviable. In general, the conventional methods employed for the C–O and C–N bond formation deliver the products with poor regioselectivity and require harsh reaction conditions (Scheme 1). Furthermore, transition metal (TM)-catalyzed cross-coupling methods readily convert the pre-functionalized arenes, such as aryl halides or pseudohalides and triflates, obtained from arenes through multiple synthetic manipulations, to aniline and phenol derivatives [8–14]. Alternatively, the TM-catalyzed direct functionalization of ubiquitous inert C–H bond offers a step- and atom-efficient way to construct C–N and C–O bonds.

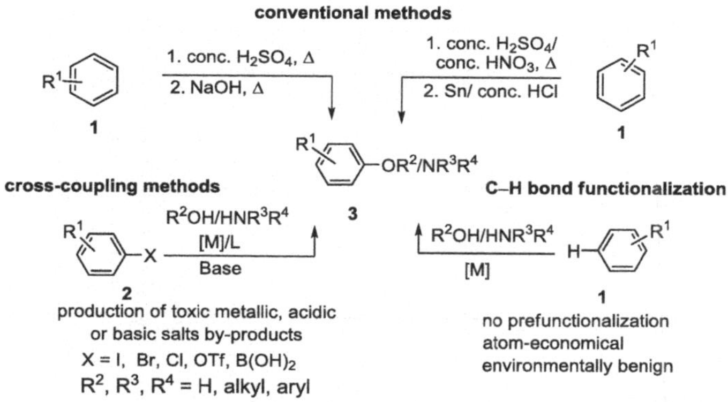

Scheme 1 Strategies for the construction of C(sp^2)–N and C(sp^2)–O bonds on arenes

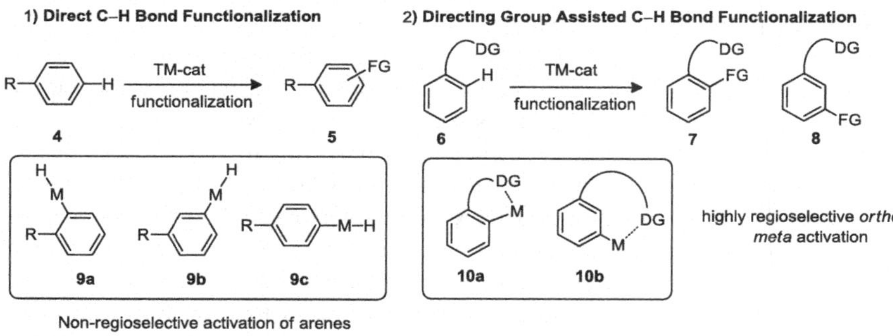

Scheme 2 Non-regioselective and regioselective C–H activation in arenes

The direct functionalization of a C–H bond is possible through the formation of carbometalated intermediates on varieties of C–H bonds in the molecule, producing non-regioselective products. The incorporation of heteroatom-bearing directing groups in the molecule helps to functionalize a particular C–H bond with high site selectivity (Scheme 2).

In 1993, the Murai group discovered Ru-catalyzed carbonyl-assisted regioselective *o*-C–H hydroarylation of arenes with alkenes for the first time (Scheme 3) ([15], [16], and references therein). The reaction begins with the coordination of carbonyl moiety to Ru species followed by the activation of *o*-C–H bond of arenes to form a ruthenacycle **14**. Next, the coordination of alkene to the Ru species through ligand exchange followed by migratory insertion to a C–M bond generates the intermediate **15**. Finally, protodemetalation of **15** culminates the desired product **13** in decent yield.

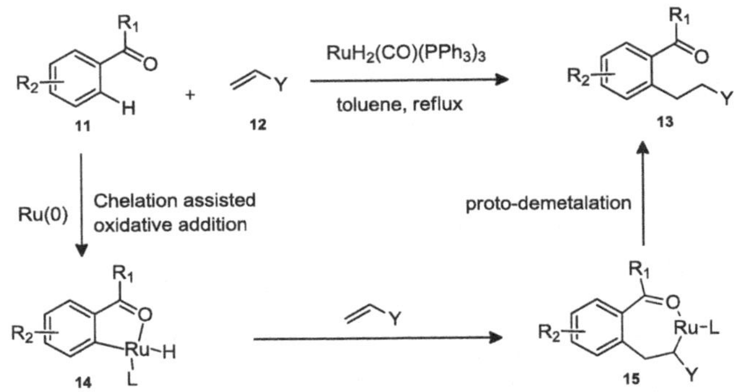

Scheme 3 Ru-catalyzed carbonyl-directed *o*-C–H hydroarylation of olefins

The Ru-catalyzed carbonyl group-assisted hydroarylation of arene C–H bonds opened an arena for the invention of novel methods for TM-catalyzed direct functionalization of inert C–H bonds. Accordingly, various TM catalysts, e.g., Pd, Ru, Rh, Ir, Co, Ni, Fe, and Cu, have subsequently been employed for the activation and functionalization of inert C–H bonds (for selected reviews on the C–H functionalization, see [17–25]). Notably, the use of cost-effective, air- and moisture-insensitive Ru catalysts for inert C–H bond functionalization is always desirable. The common Ru catalysts such as $CpRuCl(PPh_3)_2$, $Cp^*RuCl(PPh_3)_2$, $Cp^*RuCl(COD)$, $RuCl_2(PPh_3)_3$, and $[RuCl_2(p\text{-cymene})]_2$ under the influence of external oxidants (not always necessary) and additives are used for the construction of C–C and C–heteroatom bonds through the replacement of C–H bonds [17–19]. The Ru(0) catalysts usually undergo oxidative addition to C–H bonds through agostic interaction, whereas Ru(II) catalysts are inserted into C–H bond through a base-assisted concerted metalation-deprotonation (CMD) process [18].

The functionalization of C–H bond is well evidenced in the presence of Ru (II) catalysts over Ru(0) species. This chapter enumerates the development of Ru-catalyzed novel synthetic methods for the construction of C–N and C–O bonds.

2 C–N Bond Formation

Transition metal-catalyzed direct C–H amination is an atom-economical method to construct C–N bonds. The importance of amine moiety in a molecule inspires the development of novel methods for the amination of ubiquitous C–H bonds under the influence of Pd, Rh, Ru, Ir, Fe, or Cu catalysts (few reports on C–H amination reactions: [26–42]). Among significant progresses made in the development of novel methods for C–N bond formation, the use of robust and cost-effective Ru catalysts for the C–H amination expands the synthetic utility of the method. The Ru-catalyzed C–N bond-forming synthetic methods are briefly enumerated herewith.

2.1 C(sp^2)–N Bond Formation

The direct activation of unbiased C(sp^2)–H bond leads to non-regioselective products, while the incorporation of a DG can assist to the regioselective construction of C–N bonds. In addition, Ru-catalyzed cross-dehydrogenative couplings and intramolecular C–H activation methods for the construction of C–N bonds are also known.

Nitrenes and electrophilic aminating agents are generally used for the Ru-catalyzed C(sp^2)–N bond formation. Sulfonyl amides and sulfonyl azides are the active precursors of nitrenes; under the influence of oxidants [e.g., $K_2S_2O_8$, *t*-BuOOR^1, PhI(OR)$_2$, etc.], amides readily produce the corresponding nitrenes. The insertion of nitrenes to TM generates metal-nitrenoid species (L_nM=NR) in situ. Whereas the formation of metal-nitrenoid species from azide is possible through the oxidative insertion of metal to azide with the release of N_2 gas, which does not need oxidant. In contrast, oxidative addition of *N*-halo/tosyloxy phthalimide to the metal species occurs readily. Later, reductive elimination of this species leads to the functionalized products.

2.1.1 Direct C–N Bond Formation

The electron-rich heteroarenes directly undergo C–H functionalization under the influence of a TM catalyst. Amidation of furan, pyrrole, and thiophene analogues is satisfactorily performed with the Ru-porphyrin catalyst **18** in the presence of amidating agents such as *N*-(arylsulfonyl)-imino-phenyliodinane (PhI=NR; R=Ts, Ns) or the combination of PhI(OAc)$_2$ and H_2N-R (Scheme 4) [43]. For example, amidation of *N*-arylpyrrole with PhI=NTs in the presence of [Ru^{II}(TTP)(CO)] (**18**) yields C3 and C4 diamidation product **17a**, whereas *N*-arylimidazole affords *N* amidation product **17b**. In contrast, thiophene and benzofuran derivatives produce mono-C2-amidation products **17c** and **17d**, respectively (Scheme 4).

2.1.2 Direct Functionalization of Aldehyde C–H Bonds

Phosphoramidation of aldehyde C–H bond is efficiently conducted with diphenylphosphoryl azide (DPPA; **20**) under the assistance of Ru-porphyrin catalysts, such as [Ru(TTP)CO] (**18**) or [Ru(TTP)Cl$_2$] (**22**) (Scheme 5). The electron-rich aldehydes participated more efficiently under the catalytic condition [44].

Chan and coworkers demonstrated Ru-porphyrin-catalyzed [Ru(TTP)CO] (**18**) direct amidation of aldehyde C–H bonds with tosylamine in the presence of oxidant PhI(OAc)$_2$ (Scheme 6). Amidation of α-[D]-benzaldehyde (**23a**) under the optimized Ru-catalyzed conditions produced *N,N*-[D]-tosylbenzamide (**24a**) (Scheme 6); this observation supports the activation of aldehyde C–H bond [45].

Che group revealed that the amidation of aldehyde C–H bond with PhI=NTs catalyzed by [Ru(TTP)(CO)] (**18**) occurs through the involvement of [Ru$^{(VI)}$(TTP)

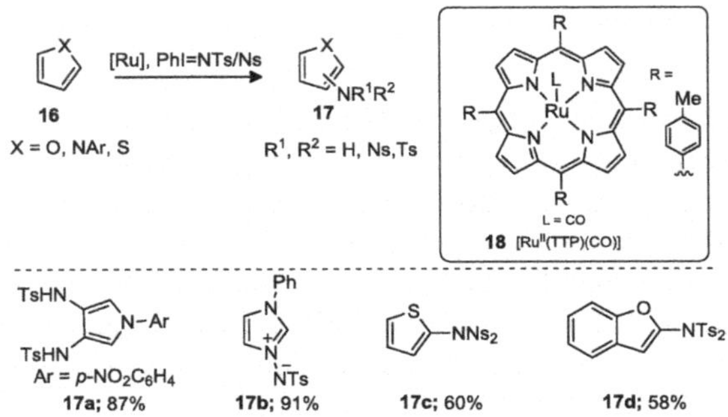

Scheme 4 Direct C(sp^2)−H amidation of heterocycles

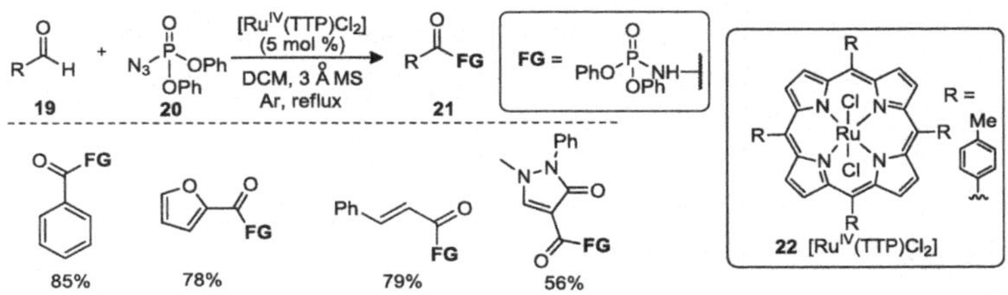

Scheme 5 Phosphoramidation of aldehyde C−H bond

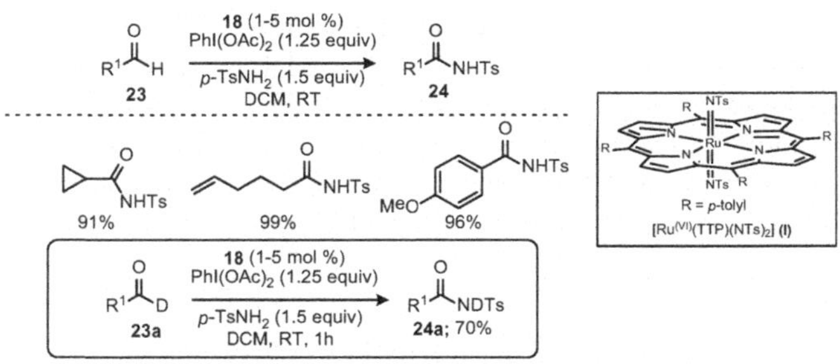

Scheme 6 Direct amidation of aldehyde C−H bond

$(NTs)_2$] (**I**) species (Scheme 6); DFT calculation studies support this fact. For example, amidation of pent-4-enal (having acyl, allylic, and homoallylic C−H bonds) with PhI=NTs in the presence of **18** exclusively provided the acyl C−H amidated product, leaving other C−H bonds unaffected. Aldehyde C−H bonds possess high bond dissociation energy (BDE) than homoallylic or allylic C−H bonds; the minimal steric strain in the corresponding transition state is detrimental to the aldehyde C−H bond amidation [46].

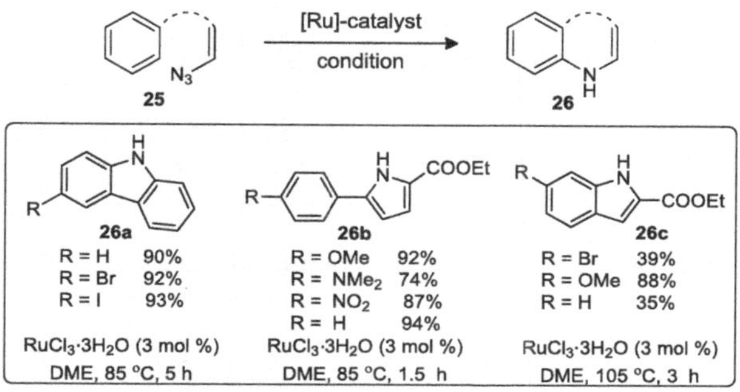

Scheme 7 Synthesis of carbazole, pyrrole, and indole derivatives through intramolecular C(sp^2)–H amination of azides

2.1.3 Intramolecular C–N Bond Formation

$RuCl_3 \cdot 3H_2O$-catalyzed intramolecular C(sp^2)–H amination of organic azides **25** efficiently produces carbazole, pyrrole, and indole derivatives (Scheme 7). For instance, 2-arylphenylazides produce carbazoles (**26a**), whereas 1-azido-4-aryl-1,3-butadienes and 2-azido-3-arylacrylates deliver pyrroles (**26b**) and indoles (**26c**), respectively (Scheme 7) [47].

Pyrazole skeleton is widely found in many molecules of pharmaceutical importance, such as Viagra, Acompila, Celebrex, and so on. Yu Rao and coworkers developed a novel approach for the construction of pyrazole moiety **28** from α,β-unsaturated hydrazones in presence of a Ru catalyst and base in DMSO (Scheme 8) [48]. Kinetic isotope studies ($k_H/k_D = 2.4$) reveal that the C–H bond activation is the rate-determining step.

2.1.4 Directing Group-Assisted C–N Bond Formation

Directing group-assisted TM-catalyzed formation of C–N bonds is well known. Over the years, various DGs have been developed and employed successfully for the Ru-catalyzed amidation of arenes. A general mechanism for the Ru-catalyzed DG-assisted regioselective amination of arene C–H bond is shown in Scheme 9. The reaction initiates with the formation of cyclometalated species **32**, obtained through the coordination of heteroatom in the DG to the metal followed by insertion to the proximal C–H bond. Next, oxidative insertion of electrophilic aminating agents to the metal in **32** delivers intermediates **33**. Finally, reductive elimination of **33** provides the aminated product **30** with the regeneration of catalyst for the next cycle (Scheme 9).

The TM catalyst-aided insertion of azide on C–H bond occurs through stepwise redox-active pathway or concerted redox-neutral pathway. The stepwise redox pathway involves (1) the coordination of azide to TM followed by the formation of metal-nitrenoid species **34** through simultaneous elimination of N_2 with the increase of oxidation state of the TM, (2) migratory insertion of nitrene between M-C bond produces **35**, and finally (3) proto-demetalation to produce amidation product **31** (Scheme 9). Whereas concerted redox-neutral pathway encompasses the

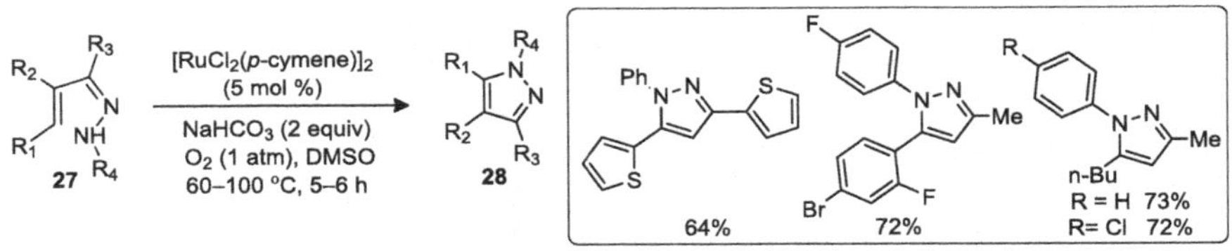

Scheme 8 Direct access to pyrazole derivatives through Ru(II)-catalyzed C(sp^2)–H amidation

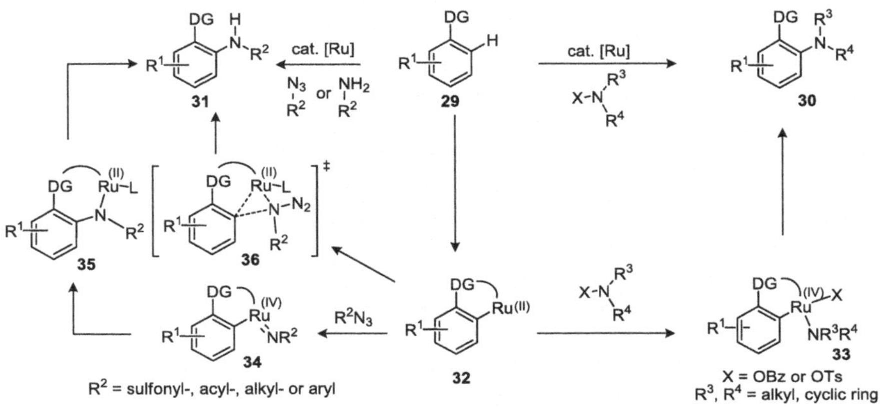

Scheme 9 Ru-catalyzed chelation-assisted formation of C–N bond in arenes

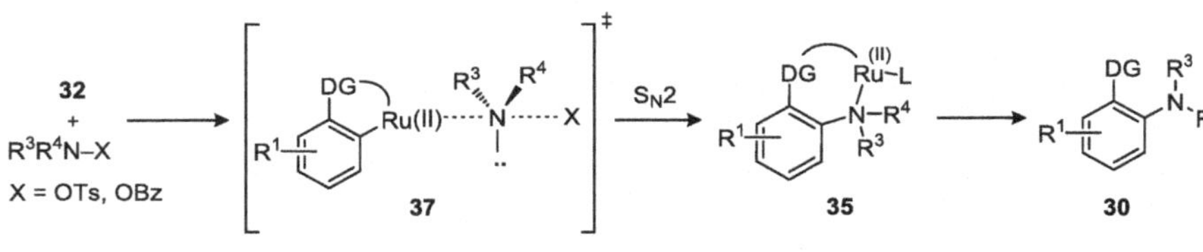

Scheme 10 Probable participation of S_N2 pathway for C–N bond formation

involvement of transition state **36** with the simultaneous coordination of aryl moiety, TM, and the azide *N* atom (Scheme 9). In this case, the oxidation state of the TM remains unchanged [49, 50].

Alternatively, the Ru-catalyzed C–N bond formation under the influence of electrophilic aminating agent may proceed through an S_N2 pathway. The ruthenacycle **37** may displace the leaving group in an S_N2 pathway to form C–N bond (Scheme 10) [51–53].

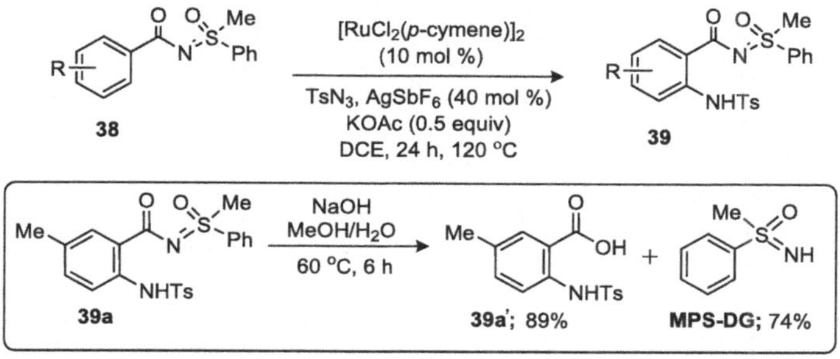

Scheme 11 Sulfoximine-assisted $C(sp^2)$–H amidation of benzoic acid derivatives

Assistance of Carboxylic Acid Derivatives and Ketones

The Ru-catalyzed *o*-C–H amination of benzoic acid derivatives provides anthranilic acids in a straightforward manner. We have demonstrated an exceedingly expedient Ru(II)-catalyzed reusable methyl phenyl sulfoximine (**MPS**)-aided intermolecular *o*-$C(sp^2)$–H amidation of arenes **38** with aryl sulfonyl azides producing **39** (Scheme 11) [54]. The amidation occurs at the less hindered site of *m*-substituted arene carboxylic acid derivatives. Base hydrolysis of the amidation product **39a** provides an anthranilic acid derivative **39a′** with the isolation of **MPS-DG**.

This method is diligently employed for the divergent synthesis of HMR 1766 (**43**). The synthesis of **43** is commenced with the insertion of 5-chlorothiophene-2-sulfonyl amide group at the less hindered *o*-$C(sp^2)$–H bond on **40** for the construction of **41**. Next, the base-mediated hydrolysis of **41** leads to **42.** Finally, amide coupling of the carboxylic acid derivative **42** with the amine produces HMR 1766 (**43**) (Scheme 12) [54].

The *N*-alkyl-benzamide DG-assisted Ru-catalyzed amidation of **44** is successfully accomplished with aryl sulfonyl azides to achieve **45**; the catalytic conditions tolerate bromo, ester, and hydroxyl functional groups, showing a broad scope of the reaction (Scheme 13) [55]. Disappointingly, the aryl azides failed to participate under this catalytic condition.

The Ru-catalyzed amidation of *N*-acyl protected indoline **46** with aryl and heteroaryl sulfonyl azides provides 7-amido-substituted indoline analogues **47** (Scheme 14) [56]. As expected, amidation occurs exclusively at C7. Participation of radical intermediate is ruled out, as the radical quencher 2,6-di-*tert*-butyl-4-methylphenol (BHT) did not inhibit the reaction.

Aryl ketones (**48**) successfully undergo amidation with arenes in the presence of tosyl azide. The combination of catalysts consisting of Ru and Ag salts and Cu $(OAc)_2 \cdot H_2O$ in DCE is suitable for producing the desired product **49** (Scheme 15). Poor performance of electron-deficient arenes over the electron-rich arenes hints the involvement of electrophilic C–H bond activation in this process, although the

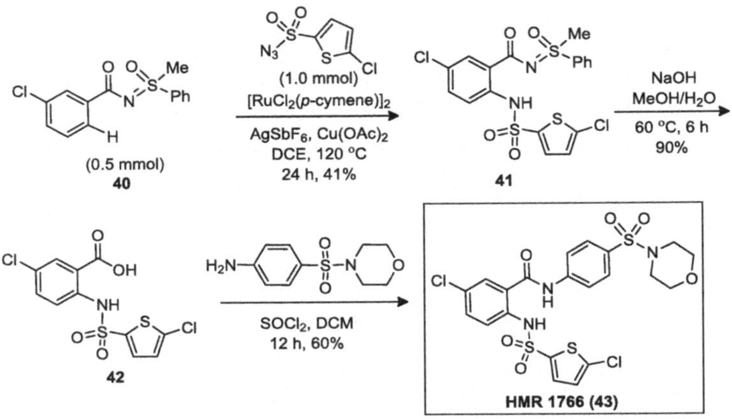

Scheme 12 Synthesis of HMR 1766 through C(sp^2)–H amidation

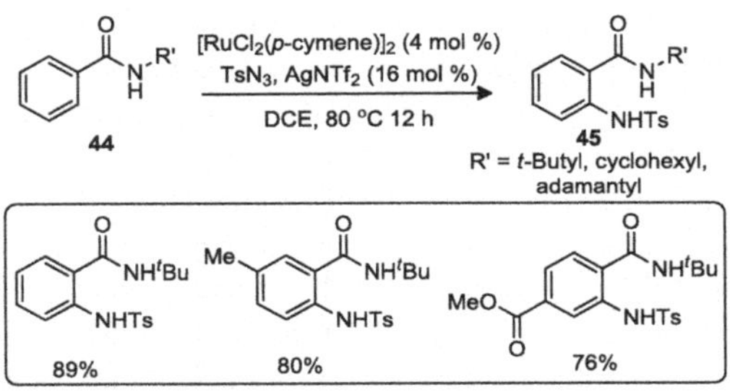

Scheme 13 Amide-assisted amidation of arenes

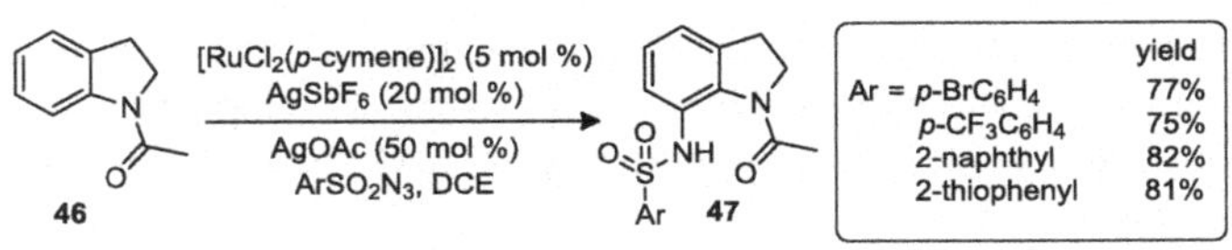

Scheme 14 Ru(II)-catalyzed amidation of indolines

mechanism is yet to be well understood. Accordingly, cationic Ru complex [Ru(*p*-cymene)(OAc)]$^+$, obtained in situ from [RuCl$_2$(*p*-cymene)]$_2$ and AgSbF$_6$ in the presence of Cu(OAc)$_2$ in DCE, is responsible for the activation and amidation of *o*-C–H bond of arenes with sulfonyl azides. The catalytic condition consisting of {[RuCl$_2$(*p*-cymene)]$_2$ and AgNTf$_2$} has also been used for the synthesis of **49** (Scheme 15) [55, 57, 58].

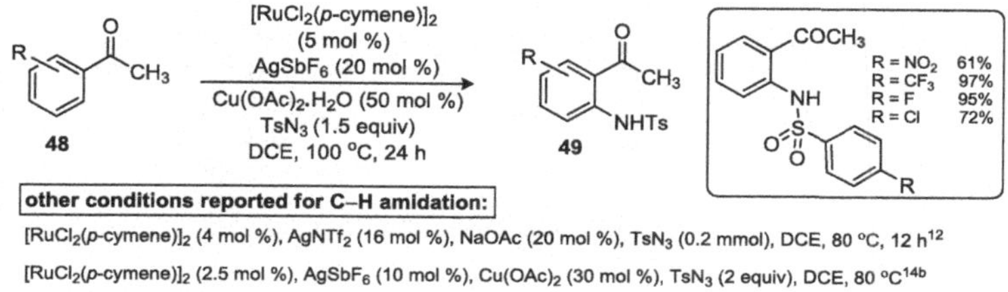

Scheme 15 Amidation of aryl ketones with sulfonyl azides

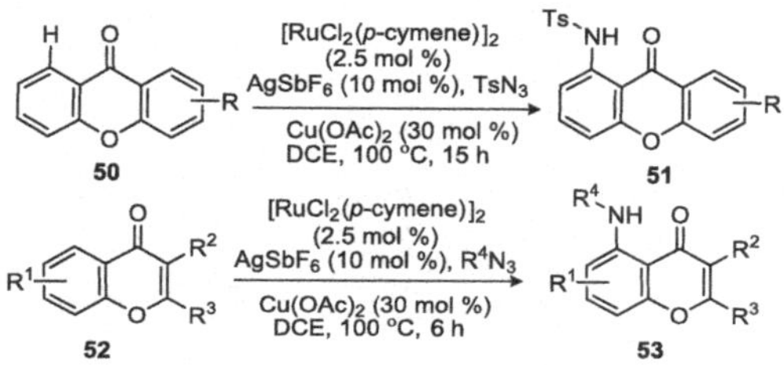

Scheme 16 Amidation of xanthones and chromones with azides

Amidation of xanthones (**50**) and chromones (**52**) successfully leads to 1-aminoxanthones (**51**) and 5-aminochromones (**53**). KIE value of 2.1 supports the possible occurrence of a kinetically relevant C–H bond cleavage step (Scheme 16) [59].

Assistance of Heteroarenes

The Chang group at first investigated 2-pyridyl-directed Ru-catalyzed C(sp^2)–H amidation with aryl sulfonyl azides [55]. This particular work helps to understand the mechanistic insights of Ru-catalyzed amidation reactions. X-ray structural elucidation of the intermediates **55a** and the HRMS data of **55b** helps in sketching the catalytic cycle involved in this reaction (Scheme 17).

Later Liang et al. reported the X-ray structural data of the intermediates **61** and **63**; this information again supports the catalytic cycle involved in the amidation of arenes [50].

The reaction begins with the formation of cationic Ru(II) species **60** from $[RuCl_2(p\text{-cymene})]_2$ and KPF_6. The coordination of pyridyl-*N* atom to Ru-cationic species and subsequent activation of proximal C–H bond lead to ruthenacycle **61**. Next, nitrene insertion to **61** generates Ru-nitrenoid species **62**; migratory insertion of nitrogen to C–Ru bond provides the intermediate **63** (Scheme 18). Finally, protodemetalation of **63** generates the amidation product **58** along with the production of cationic Ru(II) species (Scheme 18).

The competitive experiments suggest that the reaction in the absence of Ru catalyst produces trace amount of aziridination product **67** while the same process

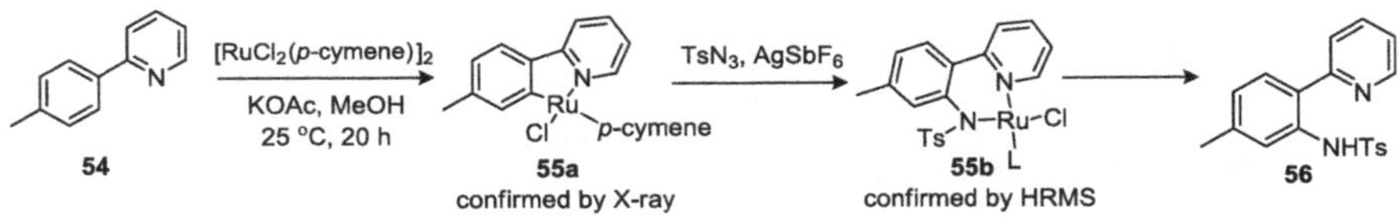

Scheme 17 Chelation assisted C–H amidation with sulfonyl azides

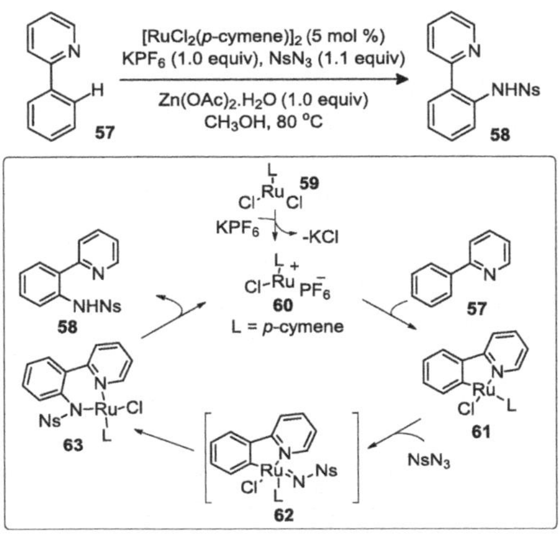

Scheme 18 Mechanistic route for 2-pyridyl DG-assisted C(sp^2)–H amination with aryl sulfonyl azides

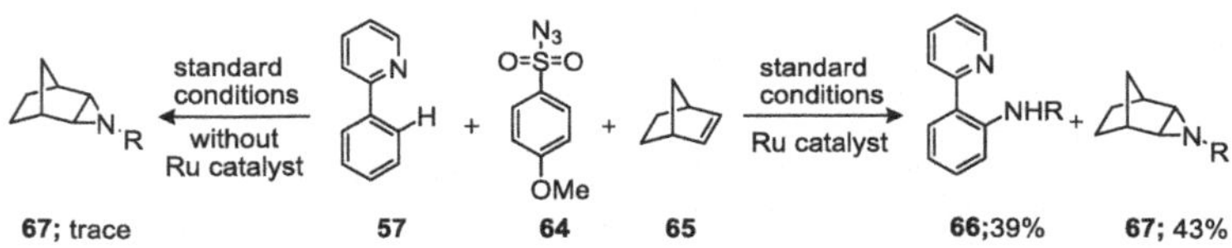

Scheme 19 Competitive experiments for Ru-catalyzed C(sp^2)–H amidation

in the presence of Ru catalyst provides C–H amidation product along with **66** due to the formation of Ru-nitrenoid species (Scheme 19) [50].

Pyrazole and pyrimidine moieties also assist the occurrence of amidation on the less hindered C–H bond of arenes with sulfonyl azides (Scheme 20); the reaction proceeds in the presence of Ru(II) catalyst and $AgSbF_6$. The pyridine-directed functionalization requires acetate base (Scheme 20) [60]. The use of substituted pyrazole and pyrimidine DGs decreases the product yield, while the substituents in sulfonyl azides did not affect the reaction outcome. Kinetic isotope studies reveal that the formation of ruthenacycle is reversible.

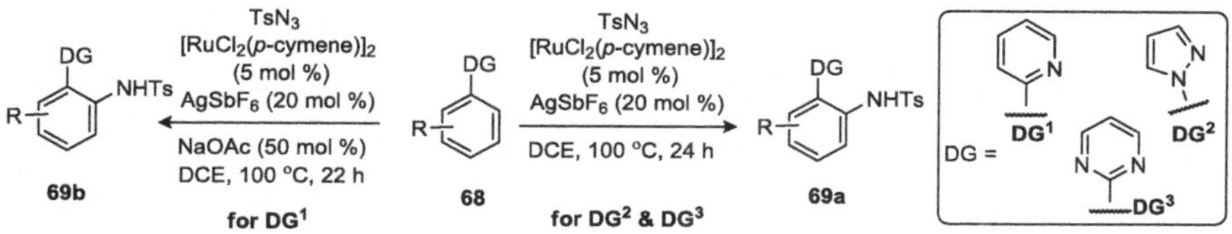

Scheme 20 *N*-heterocycle-assisted amidation of arenes

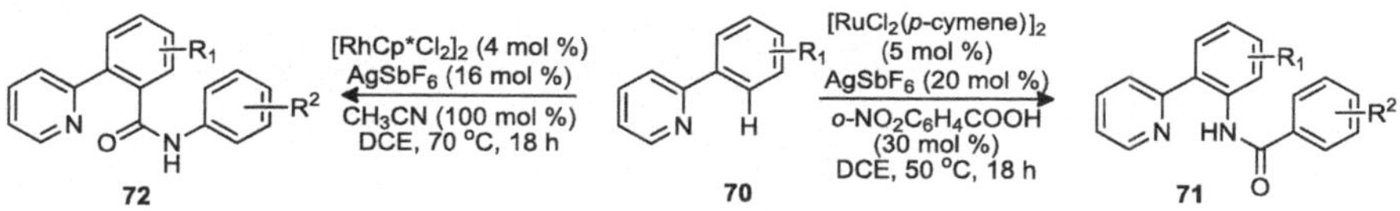

Scheme 21 Selective C–C and C–N bond formation in different conditions

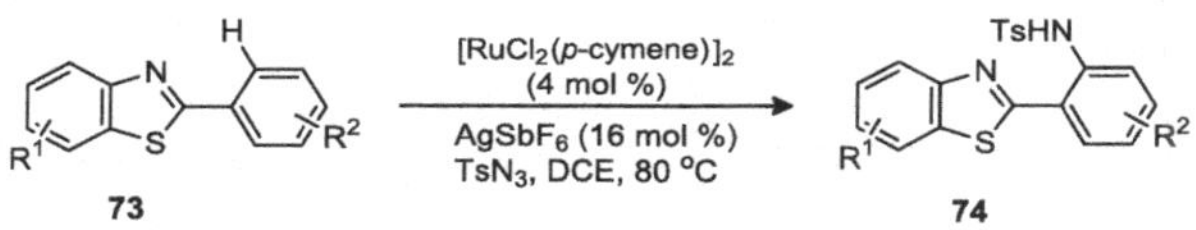

Scheme 22 Amidation of arylbenzo[*d*]-thiadiazoles

Amidation of 2-arylpyridines (**70**) with acyl azides was successfully conducted under Ru catalysis; in contrast, the identical reaction under Rh catalysis in the presence of CH_3CN provides C–C bond (Scheme 21). The low binding affinity of acyl azide to the CH_3CN-bonded Rh catalyst delivers isocyanate in situ, which is responsible for the C–C bond formation. The catalytic conditions used for the kinetically controlled Ru-catalyzed transformation hinders the formation of isocyanate, which in turn leads to the formation of C–N bond. However, the reaction in the presence of isocyanate predominantly delivers C–C bond over C–N (Scheme 21) [61].

Amidation of 2-arylbenzo[d]-thiazoles has successfully been conducted with aryl sulfonyl azides under the combination of catalysts consisting of Ru and Ag salts (Scheme 22) [62].

2.1.5 Use of Electrophilic Reagents for C–N Bond Formation

The *O*-benzoyl hydroxylamine electrophilic aminating agents have successfully been employed for the Ru(II)-catalyzed amination of *o*-C–H arenes in the amide derivative **75** at room temperature (Scheme 23) [63]. First cyclo-ruthenation of **75** generates **77** (observed by ^{1}H NMR). Then oxidative addition of the aminating agent ($\mathbf{A_m}$ = *N*-benzoyloxymorpholine) to **77** produces Ru(IV) species, which in turn functionalizes the proximal C–H bond through reductive elimination engendering the amination product **76**. The possibility of an S_N2 pathway for the product

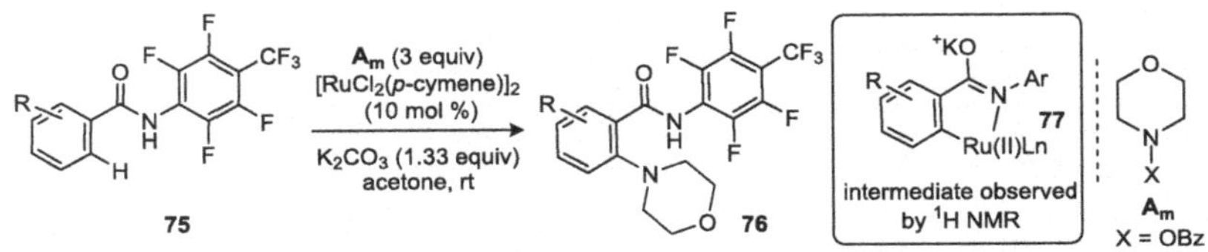

Scheme 23 Ru(II)-catalyzed amination of arenes with *O*-benzoyl hydroxylamine

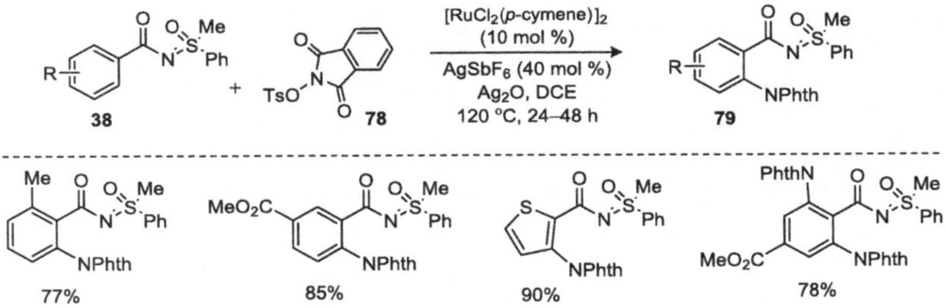

Scheme 24 Sulfoximine-directed imidation of arenes with N-tosyloxyphthalimide

formation also cannot be completely ruled out (Scheme 10) [51]. This process is reliably applied for the C–H amination of pyrazole, thiophene, benzothiophene, furan, benzofuran, and indole-bearing heteroaryl carboxylic acids.

The electrophilic imidating agent N-tosyloxyphthalimide (TsO–N–Phth) (**78**) has successfully been employed for the *o*-C–H imidation of benzoic acid derivatives **38** under the assistance of **MPS-DG** under the catalytic conditions consisting of $[RuCl_2(p\text{-cymene})]_2$, $AgSbF_6$ additive, and Ag_2O oxidant in DCE. The role of Ag_2O oxidant is indistinct. The reaction exhibits a broad substrate scope including arenes and heteroarenes with the tolerance of a wide array of functional groups. This method is applicable to access *o*-C–H di-imidation products. The readily removable **MPS-DG** and easily modifiable phthaloyl moiety make this strategy synthetically viable (Scheme 24) [64].

2.1.6 Cross-Dehydrogenative Coupling

The cross-dehydrogenative coupling (CDC) of C–H and N–H moieties directly allows the formation of C–N bond; the non-requirement of pre-functionalized precursors makes this strategy synthetically important. The cooperative catalytic system involving Ru and Cu species is generally employed for the fabrication of C–N bonds through the CDC of C–H and N–H bonds (Scheme 25) [65]. For example, homo-dimerized carbazole derivatives **81** are rapidly accessed through the oxidative C–N coupling between carbazole derivatives **80** in presence of Ru and Cu catalysts. The absence of either Ru or Cu did not yield the desired product, suggesting the importance of both catalysts in the reaction (Scheme 25).

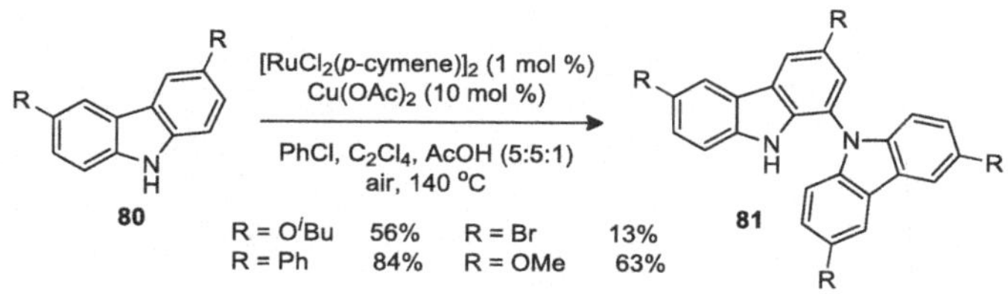

Scheme 25 CDC of C–H and N–H moieties for the construction of C–N bond

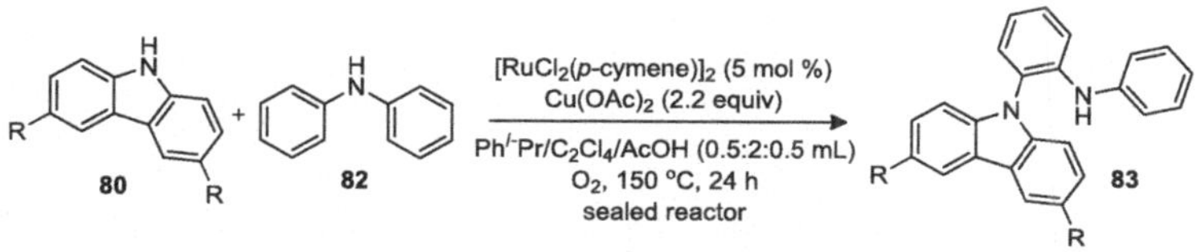

Scheme 26 Ru(II)-catalyzed CDC between carbazoles and diarylamines

The Ru- and Cu-mediated CDC method has successfully been employed for the synthesis of *N*-carbazolate diarylamines **83** [66]. The catalytic conditions did not affect halo moieties and other common functional groups. Comparative studies show that Cu salt and O_2 are not essential in the C–H bond activation, although they play important role in the catalytic cycle (Scheme 26).

2.2 *C(sp^3)–N Bond Formation*

The construction of C(sp^3)–N bond through the activation of C(sp^3)–H bond is always difficult and challenging. The HOMO of low oxidation state TM species or the electron-rich ligated metal species can readily interact with the LUMO of C–H bond, enabling the functionalization of C–H bond. The insertion of Ru(0) to the C(sp^3)–H bond is not facile; in contrast, porphyrin or other strongly donor ligand-bearing Ru(II) catalysts have been employed for the functionalization of C(sp^3)–H bond [67].

Che and coworkers at first reported the amidation of aliphatic (tertiary), benzylic, and allylic C–H bonds with iminoiodinane intermediate in the presence of Ru-porphyrin complex **84** (Scheme 27) [68, 69]. The presence of a double bond in the molecule induces the formation of both aziridine and amidation products (Scheme 27). The mechanistic route involved for allylic/benzylic C–H bond amidation is identical to that of aldehyde C–H bond amidation; this process begins with the formation of [Ru$^{(VI)}$(TTP)(NTs)$_2$] (**I**) in situ (Scheme 6), followed by nitrene insertion to the C–H bond.

The Ru-porphyrin complexes, e.g., [Ru(TPP)CO], [Ru(TMP)CO], and [Ru (OEP)CO], are successfully employed for the intramolecular sulfamidation of C (sp^3)–H bonds (Scheme 28). The [Ru(F_{20}-TPP)CO] (**84**) catalyst in combination

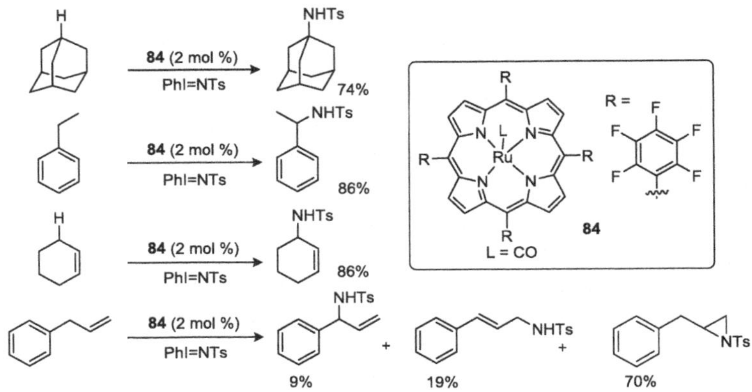

Scheme 27 Tertiary, benzylic, and allylic C(sp^3)–H amidation using Ru-porphyrin catalyst

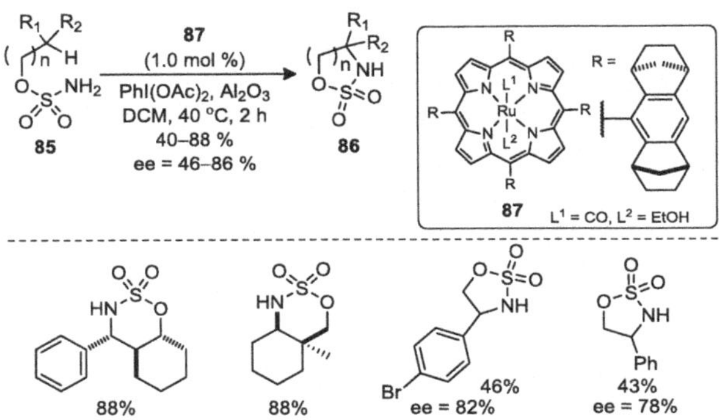

Scheme 28 Ru-porphyrin-catalyzed intramolecular sulfamidation of C(sp^3)–H bond

with $PhI(OAc)_2$ and Al_2O_3 is also used for the intramolecular sulfamidation of benzyl/alkyl C–H bonds in **85** to produce **86** [68, 70]. The electron-donating substituted C–H bond efficiently undergoes sulfamidation, as the electron-rich group in the substrate stabilizes the nitrenoid species. The asymmetric variant of the intramolecular sulfamidation has successfully been realized under the influence of the chiral Ru complex **87** (Scheme 28) [71, 72].

The Ru-pybox complexes [73] and the mixed valence tetrabis(2-oxypyridinato) diruthenium (II, III) chloride [$Ru_2(hp)_4Cl$] species [74] are equally efficient to provide the desired amidation products in good yields. Similarly, in the presence of chiral Ru-porphyrin complex **87**, substrate **88** yields chiral amidation product **89** in moderate yields. The reaction occurs with the insertion of nitrene to the metal complex. Obviously, the order of reactivity of the C–H bonds follows the order 3°>2°>1°>CH_3 (Scheme 29) [75].

Adamantane, allylic, and benzylic C–H bonds are exclusively aminated, when the reaction is conducted in the presence of active Ru-nitrenoid species. The

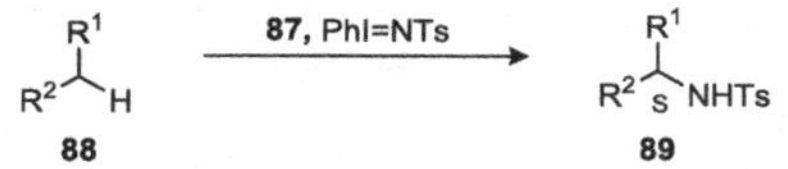

R^1	R^2	yield	*ee* (at 40 °C)
Ph	Me	23%	43%
p-MeOC$_6$H$_4$	Me	45%	33%
2-naphthyl	Me	45%	48%

Scheme 29 C(sp^3)–H amidation using chiral Ru catalyst

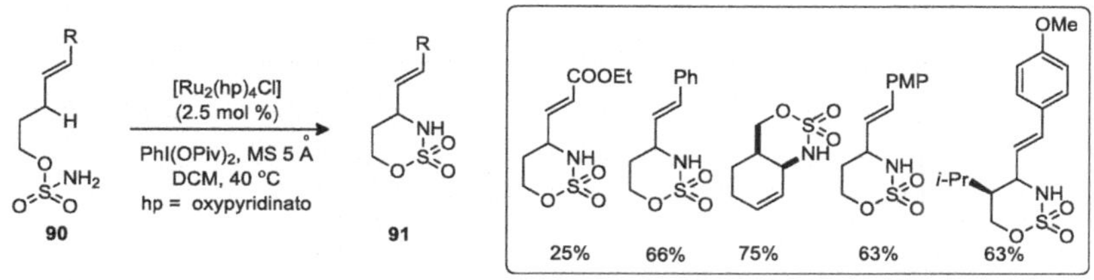

Scheme 30 C(sp^3)–H amidation using Ru-nonporphyrin catalyst

reaction of Ru complexes with the nitrenoid species (PhI=NSO$_2$R), formed in situ from tosyl amine or amide and oxidant PhI(OAc)$_2$, generates the active Ru-nitrenoid species. The use of [Ru$_2$(hp)$_4$Cl] catalyst exclusively forms allylic C–N bond, converting **90** to **91** [74]. The kinetic isotope ($k_H/k_D = 4.9$) studies suggest a stepwise nitrene insertion to the C–H bond; first, metal-nitrenoid species is formed through oxidative insertion of amide species to the TM. Next, interaction of nitrenoid species to the C–H bond yields the desired amidation product [67]. In contrast, Rh-catalyzed nitrene insertion to double bond for the synthesis of aziridine displays KIE = 2.6, indicating the participation of concerted pathway involving three-centered TS (Scheme 30).

Blakey and coworkers demonstrated that chiral Ru-pybox complexes catalyzed enantioselective benzylic and allylic C–H amination in the presence of AgOTf and PhI(O$_2$C^{t}Bu)$_2$ [73]. The use of non-coordinating counteranion-bearing Ag salts (AgBF$_4$, AgSbF$_6$, AgPF$_6$) is effective. The proposed transition states **97** and **98** are probably involved for the formation of **93** and **94**, respectively. Participation of intermediate **97** is more likely, as the presence of two diastereotopic hydrogen atoms proximal to the metal-nitrenoid species favors the bond formation. While the repulsion between phenyl group and the nitrene center disfavors the involvement of intermediate **98**, producing **93** in high enantioselectivity (*ee* = 80–90%) (Scheme 31) [76]. The Ru(II)-bis(imino)pyridyl complexes (**96**) are also successfully applied for benzylic C–H amination (Scheme 31) [77].

3 C–O Bond Formation

Transition metal-catalyzed, DG-assisted oxidation of C–H bonds has emerged as an elegant and powerful tool for the construction of C–O bonds with high chemo- and regioselectivity ([78] and references therein). This method helps in the creation

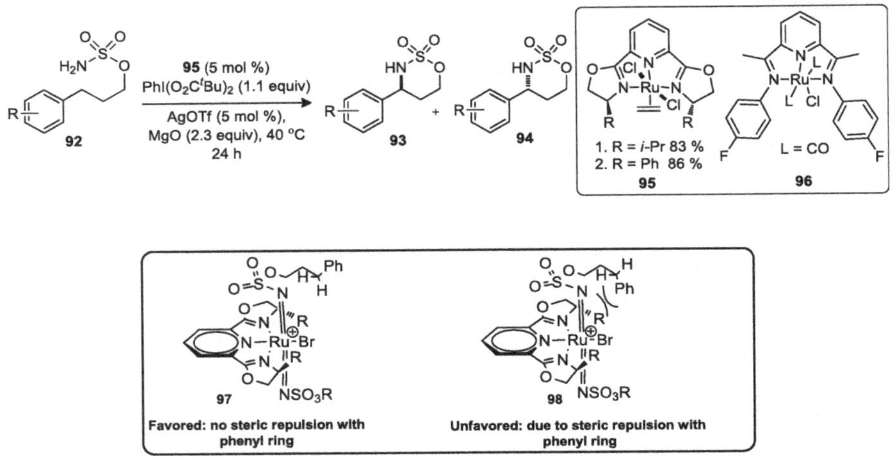

Scheme 31 Benzylic C(sp^3)–N bond formation using chiral Ru-pybox catalyst

of a versatile hydroxyl functional group within a molecule; therefore, it offers to find vast application in synthetic chemistry. As a consequence, a wide range of synthetic strategies [comprising of the use of various DG's in presence of different TMs such as Pd, Rh, Ru, or Cu] have been developed for the fabrication of C–O bond from the ubiquitous C–H bonds. The Ru-catalyzed C–O bond formation methods are briefly discussed herewith.

3.1 *C(sp^2)–O Bond Formation*

The C–H oxidation of arenes directly creates phenol derivatives. The DG-enabled oxidation of arene C–H bond has successfully been performed in the presence of Ru catalysts {e.g., [$RuCl_2$(*p*-cymene)]$_2$, [$Ru(O_2CMes)_2$(*p*-cymene)]$_2$, and [$RuCl_3{\cdot}nH_2O$]} and oxidants [e.g., $PhI(OAc)_2$, $K_2S_2O_8$, $Na_2S_2O_8$, HIO_3, etc.] [19]. A well-accepted mechanism for the Ru-catalyzed DG-enabled oxidation of arene C–H bonds is shown in Fig. 1.

At first, the reaction of [$RuCl_2$(*p*-cymene)]$_2$ with additive $AgSbF_6$ leads to the formation of active catalyst **99** in the presence of a carboxylate source. Next, the coordination of DG heteroatom to complex-**99** [RuOCOR]$^+$ generates complex **100**. The base-assisted deprotonation and metalation then form metallacycle **101**. Insertion of *O*-bearing functional group to the Ru intermediate **101** through ligand displacement creates **102**. Finally, reductive elimination of **102** provides the desired C–O bond-bearing product (**P**) and Ru(0) species. Terminal oxidant helps to regenerate Ru(II) catalyst for the next cycle (Fig. 1). The alternate pathway suggested by Yu Rao implicates the participation of Ru(II)/Ru(IV) catalytic cycle involving C–Ru–OAc intermediate species, which upon reductive elimination

Fig. 1 A general mechanistic cycle for Ru-catalyzed DG-assisted C(sp^2)–O bond formation of arenes

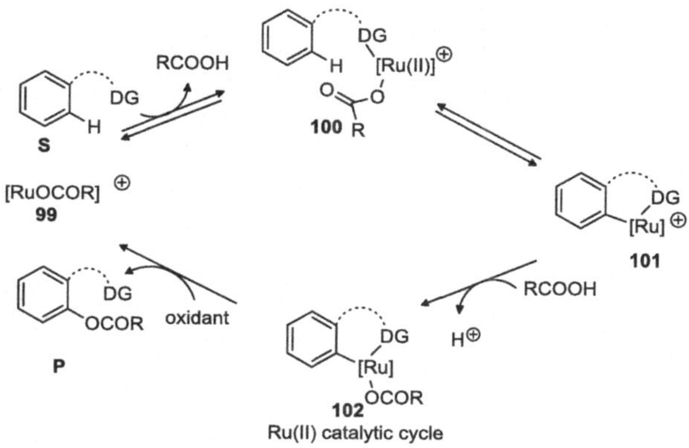

delivers the products [79, 80]. Based on the mechanistic insights, various strategies for the direct construction of phenol derivatives from arenes have been developed.

3.1.1 Assistance of Carbonyl and Carboxylic Acid Derivatives

Yu Rao and coworkers described the Ru-catalyzed *o*-C–H hydroxylation of arenes **103** with the aid of ester DG (Scheme 32) [79]. The optimized condition consisting of $[RuCl_2(p\text{-cymene})]_2$ and $K_2S_2O_8$ in TFA/TFAA enables the hydroxylation of electron-rich and electron-deficient arenes. Other Ru catalysts such as $RuCl_2(PPh_3)_3$, $RuHCl(PPh_3)_3$, and $Ru_3(CO)_{12}$ are ineffective. This protocol has successfully been employed for the efficient synthesis of anti-inflammatory drug Mesalazine (**105**) (Scheme 32).

The isotope labeling studies ($k_H/k_D = 2.3$) suggest that the activation of C–H bond is the rate limiting step. The use of strong oxidant $K_2S_2O_8$ in the reaction hints the participation of intermediate C–Ru(IV)–O, although the possibility of Ru(II)/Ru (0) catalytic system could not be ignored.

Ackermann and coworkers elaborately demonstrated Ru-catalyzed C–H oxidation of benzoic acid derivatives [81]. The *N,N*-dialkylamino, pyrrolidinyl, piperidinyl, and morpholinyl protected benzamides **106** underwent C–H oxidation in the presence of $[Ru(O_2CMes)_2(p\text{-cymene})]_2$ catalyst to provide the corresponding C–O-bearing products **107** (Scheme 33). $[Ru(O_2CMes)_2(p\text{-cymene})]_2$ catalyst is highly efficient; low catalyst loading of 1.0 mol % did not affect the reaction outcome (Scheme 33). Surprisingly, an inexpensive catalyst $[RuCl_3{\cdot}nH_2O]$ in the presence of oxidant $PhI(OAc)_2$ can also be employed to produce the functionalized product.

Weinreb amides are versatile building blocks, useful for the rapid conversion to aldehydes/ketones or alcohols. As a result, these moieties find wide synthetic applications in organic chemistry [82]. The Ackermann group demonstrated Ru-catalyzed *o*-C–H oxygenation of arenes through the assistance of Weinreb amides **108** under mild conditions to provide broad range of *o*-hydroxyaryl amides

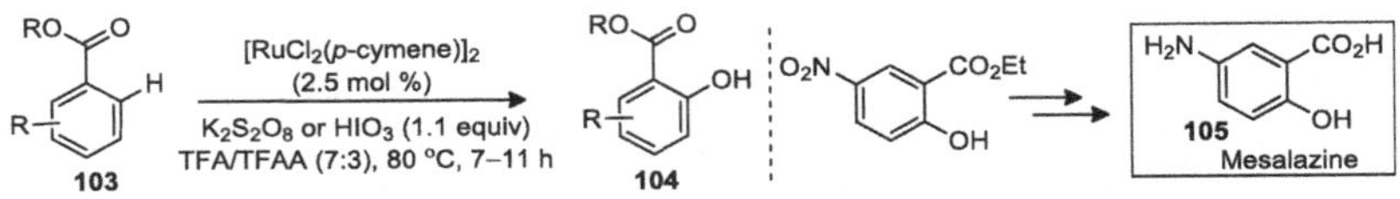

Scheme 32 Ester-directed $C(sp^2)$–O bond formation

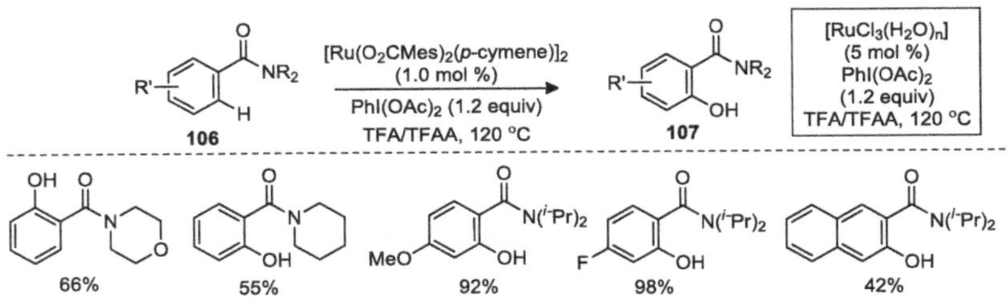

Scheme 33 Amide-directed $C(sp^2)$–O bond formation

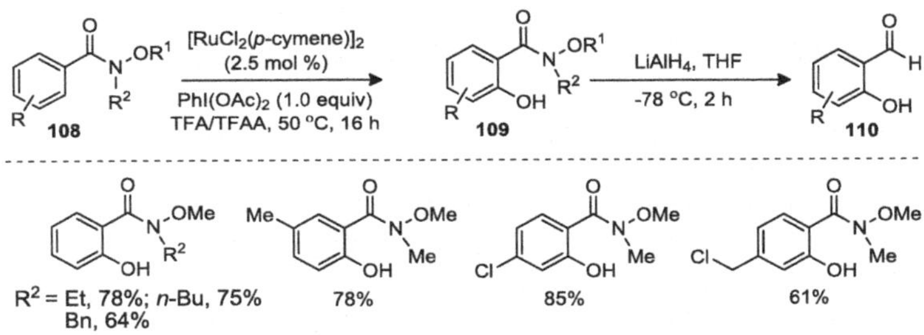

Scheme 34 Weinreb amide-directed C–O bond formation of arenes

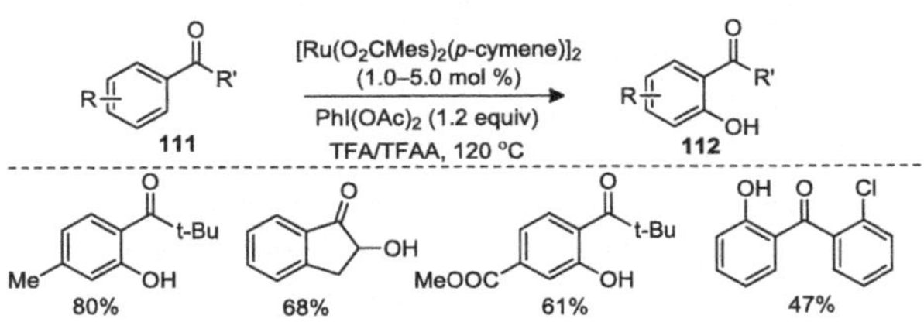

Scheme 35 Ketone-directed $C(sp^2)$–O bond formations

109 (Scheme 34). The $k_H/k_D = 3.0$ suggests the involvement of a rate-determining metalation step. Deuterium-labeling studies conclude that C–H bond activation and cyclometalation step are irreversible. Finally, the reduction of **109** provides *o*-hydroxybenzaldehydes **110** (Scheme 34).

The *o*-C–H hydroxylation of aromatic ketones **111** readily occurred in the presence of $[Ru(O_2CMes)_2(p\text{-cymene})]_2$ or $[RuCl_3{\cdot}nH_2O]$ and oxidant $PhI(OAc)_2$ (Scheme 35) [83]. Interestingly, chemoselective oxidation of $C(sp^3)$–H bond in the

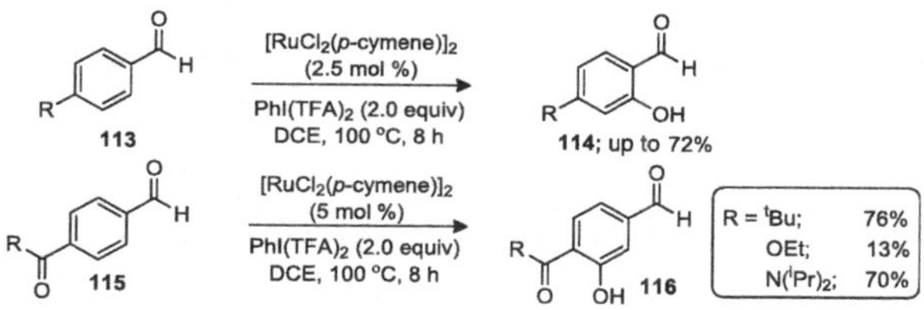

Scheme 36 Aldehyde-directed Ru-catalyzed C–H hydroxylation of arenes

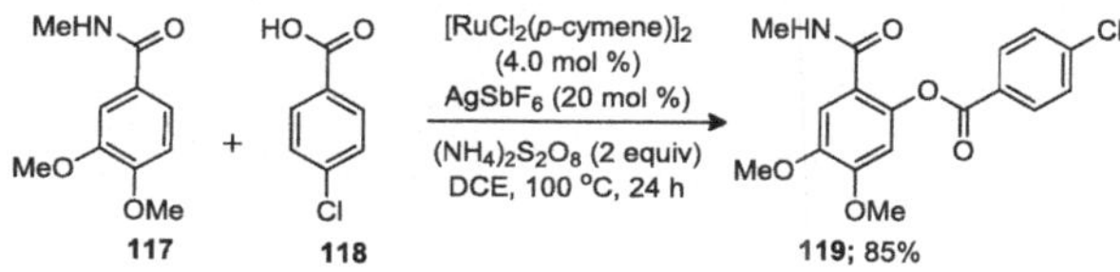

Scheme 37 Ru-catalyzed *o*-C–H benzoxylation of benzamide derivatives

fused aryl-alkyl ketone produces α-hydroxy-ketone. Intramolecular competition experiments show that the DG ability of *t*-butyl ketone is better than ester moiety. As expected, the C–H hydroxylation of electron-rich arenes is facile over electron-deficient arenes (Scheme 35).

A challenging aldehyde-directed Ru-catalyzed C–H oxygenation of arenes has been accomplished by the Ackermann group in the presence of strong oxidant PhI(TFA)$_2$ [84]. Worthy to mention that the easily oxidizable aldehyde group are silent to the oxidizing agent PhI(TFA)$_2$. On the basis of the experimental results, it has been observed that the keto, amide, and ester groups show better directing ability for *o*-C–H oxidation over the aldehyde (Scheme 36).

The reaction between substituted benzamide **117** and benzoic acid **118** in the presence of [RuCl$_2$(*p*-cymene)]$_2$ and oxidant $(NH_4)_2S_2O_8$ provides highly regioselective *o*-C–H benzoxylation product **119** (Scheme 37) [85]. The substituents on DG strongly influence the reaction outcome. For instance, –CONHMe is found to be more effective than –CONHOMe and –CONEt$_2$. Electron-rich and electron-poor benzoic acids are equally reactive and efficiently functionalize the *o*-C–H position of **117** (Scheme 37).

3.1.2 Assistance of *O*- and *N*-Protected Phenol and Aniline Derivatives

The carbamate group unequivocally assists the chemo- and regioselective hydroxylation of *o*-C–H bond of arenes under Ru catalysis to produce **121** (Scheme 38) [86]. The competitive experiments conclude that the carbamate possesses better DG ability over the ester. Finally, cleavage of the carbamate group produces highly substituted catechol derivatives.

In contrast to the carbamate-enabled *o*-C–H hydroxylation of arenes, the electron-rich anisole derivatives **122** surprisingly provide *p*-hydroxyanisole derivatives **123** under the identical catalytic conditions involving [RuCl$_2$(*p*-cymene)]$_2$ and PhI(TFA)$_2$ (Scheme 39) [86]. The inclusion of TEMPO in the reaction

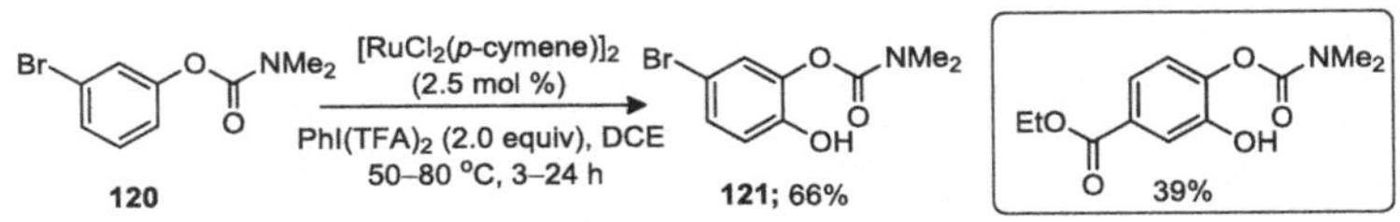

Scheme 38 Carbamate-assisted C–O bond formation of arenes

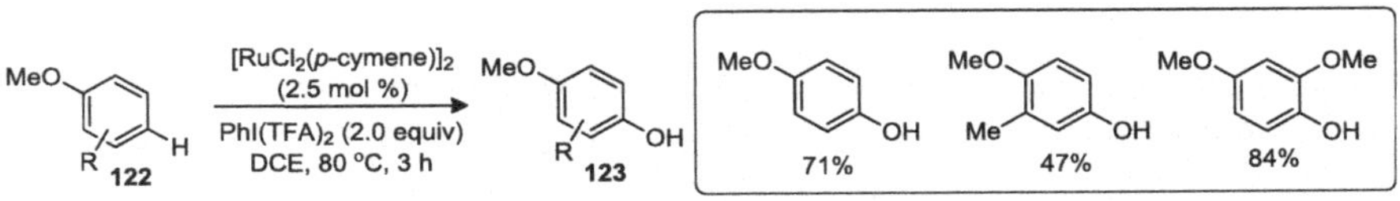

Scheme 39 Ru-catalyzed hydroxylation of anisole

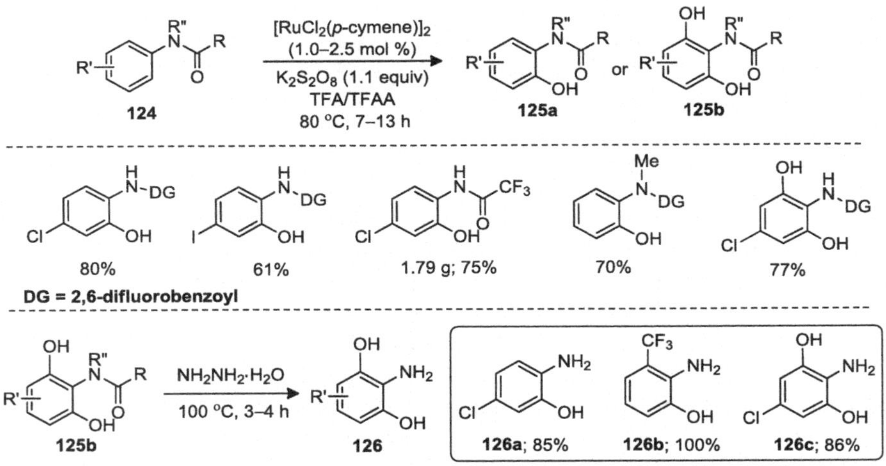

Scheme 40 A modifiable DG-aided oxidation of arene C–H bonds and access to *o*-aminophenols

drastically reduces the product formation, suggesting the involvement of single-electron-transfer oxidation mechanism.

A seminal work describes a modifiable DG-aided Ru(II)-catalyzed selective mono- and di-*o*-C–H hydroxylation of arenes in the presence of $K_2S_2O_8$ as oxidant in TFA/TFAA (Scheme 40) [87]. The reaction shows a broad scope with the tolerance of common functional groups. Notably, the use of a higher amount of oxidant in the reaction leads to the formation of a major amount of di-*o*-C–H hydroxylation products **125b**. Finally, hydrolysis of anilide moiety produces 2-aminophenols **126a** and **126b** and 2-amino-1,3-dihydroxybenzene (**126c**) in good yields (Scheme 40).

This protocol is finally applied to the synthesis of dibenzoxazepine and benzoxazole derivatives from the *o*-C–H hydroxylation compounds involving

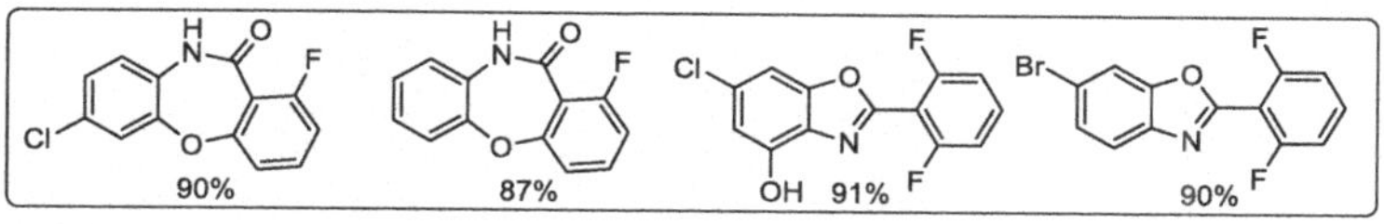

Fig. 2 Library of dibenzoxazepine and benzoxazole heterocycles

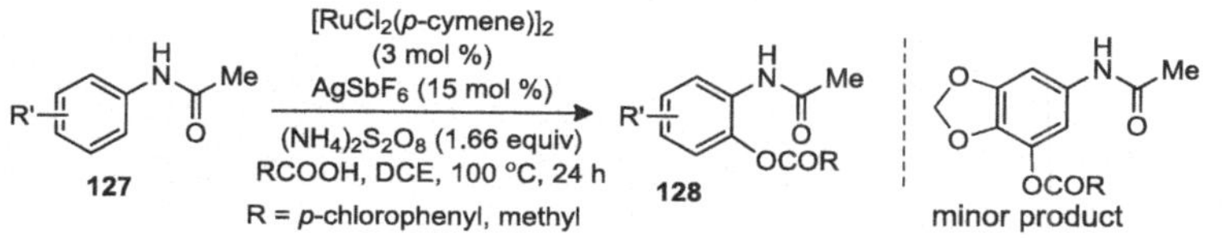

Scheme 41 *o*-Benzoxylation and *o*-acetoxylation of substituted acetanilides

intramolecular nucleophilic aromatic substitution and dehydration reactions, respectively (Fig. 2).

The aid of a similar modifiable DG successfully delivers *o*-benzoxylation and *o*-acetoxylation of aryl-substituted acetanilides **127** under Ru catalysis (Scheme 41). The reaction shows a vast substrate scope with the tolerance of various functionalities [88]. Surprisingly, the occurrence of a minor amount of C–benzoxylation *ortho*- to the ether moiety suggests the directing ability of ether.

3.1.3 Assistance of Heteroarenes

Pyridine or pyrimidine N atoms are readily oxidized to N-oxide in the presence of oxidizing agents. Despite having the inherent challenges for the facile oxidation of N atom in the heteroarenes, the Ackermann group demonstrated pyridine-DG-aided Ru-catalyzed carboxylation of arenes in the presence of halide scavenger $AgSbF_6$ and oxidant $K_2S_2O_8$ to yield **130** (Scheme 42) [89]. Deuterium scrambling experiment suggests the reversible nature of carbometalation. The reaction shows a broad substrate scope; the aliphatic and aromatic carboxylic acids are readily connected to the aryl ring. The removal of a 2-pyridyloxy group leads to catechol derivatives.

3.2 C(sp^3)–O Bond Formation

The oxidation of inert $C(sp^3)$–H bond constructs chemo- and regioselective C–O bonds in aliphatic chain. This method directly produces aliphatic alcohols from alkanes. The Ru-catalyzed DG-aided oxidation of regioselective $C(sp^3)$–H bond remains elusive. In contrast, the RuO_4 catalyst-mediated oxygenation of unactivated tertiary-$C(sp^3)$–H bond is well known.

A notable demonstration from the Du Bois group showcased direct oxidation of 3°-$C(sp^3)$–H bond under the catalytic conditions consisting of $[RuCl_3{\cdot}nH_2O]$ in the

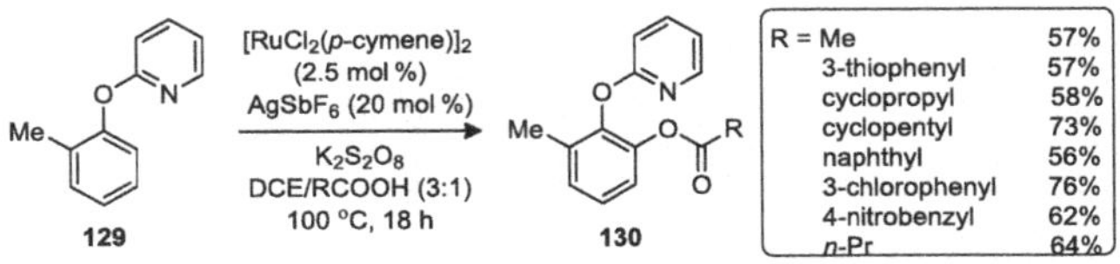

Scheme 42 2-Hydroxypyridine-directed carboxylation of arenes

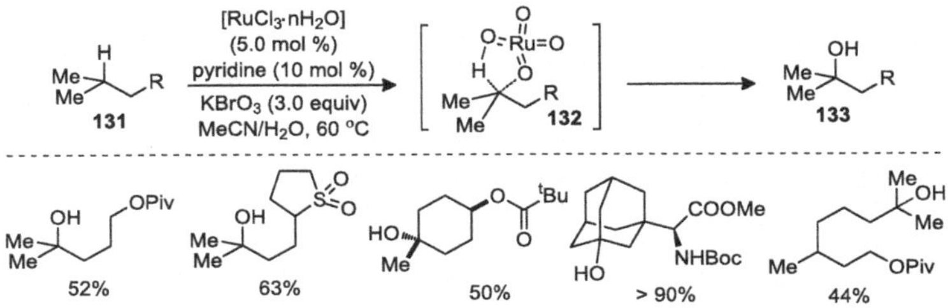

Scheme 43 Ru-catalyzed direct oxidation of 3°-C(sp^3)–H bond

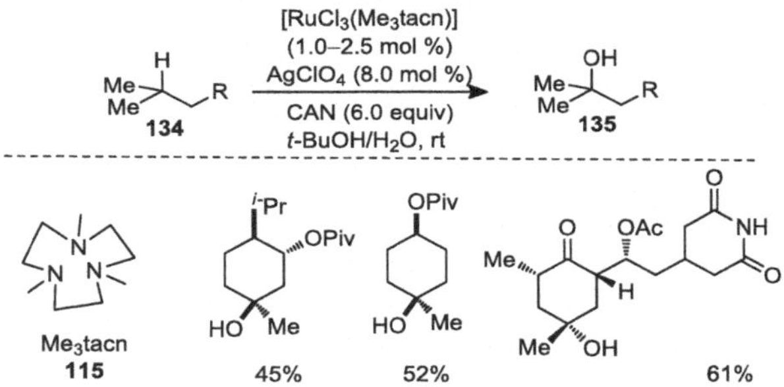

Scheme 44 Ru-catalyzed oxygenation of tertiary and benzylic C–H bonds

presence of oxidizing agent $KBrO_3$ and pyridine (Scheme 43) [90]. The reaction has a broad scope; the presence of various functional groups like sulfones, epoxides, hydroxylated esters, carbamates, and sulfamates in the molecule does not affect the reaction outcome (Scheme 43). The intermediate **132**, obtained through the concerted [3+2]-cycloaddition between RuO_4 and the C–H bond of **131**, is responsible for the formation of C–O bond [91, 92]. Ru-porphyrin complexes have also been used for the oxygenation of 3°-C(sp^3)–H bonds [93].

The oxidation of tertiary and benzylic C–H bonds is possible under mild conditions using low catalyst loading of [(Me_3tacn)Ru(III)Cl_3] and ceric(IV)ammonium nitrate (CAN) as oxidant (Scheme 44). The high k_H/k_D (= 6.7) value suggests that the conversion of **134** to the oxidation product **135** occurs in a stepwise manner involving a sequential abstraction of H atom followed by a fast radical rebound step. The *cis*- and *trans*- dioxoruthenium(VI) species, formed in situ, is able to oxidize the C–H bonds. Interestingly, ligand Me_3tacn inhibits dimerization or multimerization of the catalyst under the reaction conditions [94].

4 Conclusion

This chapter covers the known synthetic methods for the Ru-catalyzed C–N and C–O bond formations. The Ru-catalyzed transformations for the construction of C–N and C–O bond are challenging, as the strong binding ability of electronegative O and N atoms to the active Ru catalyst inhibits the effective reductive elimination. The growing demand of the direct functionalization of inert C–H bond in synthetic chemistry and the inherent challenges involved in the Ru-catalyzed transformations would boost the researchers to develop novel strategies for the Ru-catalyzed C–heteroatom bond formations especially in the unbiased $C(sp^3)$–H bond. The study of mechanistic aspects related to the Ru-catalyzed transformations needs detailed investigations; this would provoke in sketching and executing novel methods. Abundant opportunities remain to unravel novel synthetic paradigm for the efficient fabrication of complex molecular framework involving multiple and sequential Ru-catalyzed functionalization of inert C–H bonds.

References

1. Hili R, Yudin AK (2006) Nat Chem Biol 2:284
2. Ricci A (2007) Amino group chemistry: from synthesis to the life sciences. Wiley-VCH, Weinheim
3. Türmen YE, Aggarwal VK (2014) Science 343:33
4. Grabowski K, Schneider G (2007) Curr Chem Biol 1:115
5. Michael JP (2008) Nat Prod Rep 25:166
6. Doungsoongnuen S, Worachartcheewan A, Pingaew R, Suksrichavalit T, Prachayasittikul S, Ruchirawat S, Prachayasittikul V (2011) EXCLI J 10:155
7. Eschenburg S, Priestman MA, Abdul Latif FA, Delachaume C, Fassy F, Schönbrunn EJ (2005) J Biol Chem 280:14070
8. Evano G, Blanchard N, Toumi M (2008) Chem Rev 108:3054
9. Hartwig JF (2008) Acc Chem Res 41:1534
10. Guram AS, Buchwald SL (1994) J Am Chem Soc 116:5969
11. Wolfe JP, Wagaw S, Marcoux J-F, Buchwald SL (1998) Acc Chem Res 31:805
12. Kakiuchi F, Chatani N (2003) Adv Synth Catal 345:1077
13. Li B, Dixneuf PH (2013) Chem Soc Rev 42:5744
14. Ackermann L (2010) Pure Appl Chem 82:1403
15. Murai S, Kakiuchi F, Sekine S, Tanaka Y, Kamatani A, Sonoda M, Chatani N (1993) Nature 366:529
16. Kakiuchi F, Kochi T, Mizushima E, Murai S (2010) J Am Chem Soc 132:17741
17. Arockiam PB, Bruneau C, Dixneuf PH (2012) Chem Rev 112:5879
18. Kozhushkov SI, Ackermann L (2013) Chem Sci 4:886
19. Thirunavukkarasu VS, Kozhushkov SI, Ackermann L (2014) Chem Commun 50:29
20. Cho SH, Kim JY, Kwak J, Chang S (2011) Chem Soc Rev 40:5068

21. Wencel-Delord J, Dröge T, Liu F, Glorius F (2011) Chem Soc Rev 40:4740
22. Matsuda N, Hirano K, Satoh T, Miura M (2012) Synthesis 12:1792
23. Song W, Kozhushkov SI, Ackermann L (2013) Angew Chem Int Ed 52:6576
24. Zhang XP, Lu H (2011) Chem Soc Rev 40:1899
25. Beccalli EM, Broggini G, Martinelli M, Sottocornola S (2007) Chem Rev 107:5318
26. Neufeldt SR, Sanford MS (2012) Acc Chem Res 45:936
27. Kuhl N, Hopkinson MN, Wencel-Delord J, Glorius F (2012) Angew Chem Int Ed 51:10236
28. Seregin IV, Gevorgyan V (2007) Chem Soc Rev 36:1173
29. Zalatan DN, Du Bois J (2010) Top Curr Chem 292:347
30. Collet F, Dodd RH, Dauban P (2009) Chem Commun 2009:5061
31. Louillat M-L, Patureau FW (2014) Chem Soc Rev 43:901
32. Stokes BJ, Driver TG (2011) Eur J Org Chem 2011:4071
33. Monguchi D, Fujiwara T, Furukawa H, Mori A (2009) Org Lett 11:1607
34. Wang Q, Schreiber SL (2009) Org Lett 11:5178
35. Cho SH, Kim JY, Lee SY, Chang S (2009) Angew Chem Int Ed 48:9127
36. Kim JY, Cho SH, Joseph J, Chang S (2010) Angew Chem Int Ed 49:9899
37. Kawano T, Hirano K, Satoh T, Miura M (2010) J Am Chem Soc 132:6900
38. Zhao H, Wang M, Su W, Hong M (2010) Adv Synth Catal 352:1301
39. Miyasaka M, Hirano K, Satoh T, Kowalczyk R, Bolm C, Miura M (2011) Org Lett 13:359
40. Liu X-Y, Gao P, Shen Y-W, Liang Y-M (2011) Org Lett 13:4196
41. Li Y, Liu J, Xie Y, Zhang R, Jin K, Wang X, Duan C (2012) Org Biomol Chem 10:3715
42. McDonald SL, Hendrick CE, Wang Q (2014) Angew Chem Int Ed 53:4667
43. He L, Hong Chan PW, Tsui WM, Yu WY, Che CM (2004) Org Lett 6:2405
44. Xiao W, Zhou CY, Che CM (2012) Chem Commun 48:5871
45. Chang JWW, Chan PWH (2008) Angew Chem Int Ed 47:1138
46. Guo Z, Guan X, Huang JS, Tsui WM, Lin Z, Che CM (2013) Chem Eur J 19:11320
47. Shou WG, Li J, Guo T, Lin Z, Jia G (2009) Organometallics 28:6847
48. Hu J, Chen S, Sun Y, Yang J, Rao Y (2012) Org Lett 14:5030
49. Park SH, Kwak J, Shin K, Ryu J, Park Y, Chang S (2014) J Am Chem Soc 136:2492
50. Zhang LL, Li LH, Wang YQ, Yang YF, Liu XY, Liang YM (2014) Organometallics 33:1905
51. Berman AM, Johnson JS (2004) J Am Chem Soc 126:5680
52. Campbell MJ, Johnson JS (2007) Org Lett 9:1521
53. Kuhl N, Schröder N, Glorius F (2014) Adv Syn Catal 356:1443
54. Yadav MR, Rit RK, Sahoo AK (2013) Org Lett 15:1638
55. Kim J, Kim J, Chang S (2013) Chem Eur J 19:7328
56. Pan C, Abukader A, Han J, Cheng Y, Zhu C (2014) Chem Eur J 20:3606
57. Bhanuchandra M, Yadav MR, Rit RK, Kuram MR, Sahoo AK (2013) Chem Commun 49:5225
58. Zheng QZ, Liang YF, Qin C, Jiao N (2013) Chem Commun 49:5654
59. Shin Y, Han S, De U, Park J, Sharma S, Mishra NK, Lee EK, Lee Y, Kim HS, Kim IS (2014) J Org Chem 79:9262
60. Thirunavukkarasu VS, Raghuvanshi K, Ackermann L (2013) Org Lett 15:3286
61. Shin K, Ryu J, Chang S (2014) Org Lett 16:2022
62. Zhou X, Luo P, Long L, Ouyang M, Sang X, Ding Q (2014) Tetrahedron 70:6742
63. Shang M, Zeng SH, Sun SZ, Dai HX, Yu JQ (2013) Org Lett 15:5286
64. Ramu Yadav M, Shankar M, Ramesh E, Ghosh K, Sahoo AK (2015) Org Lett 17:1886
65. Louillat ML, Patureau FW (2013) Org Lett 15:164
66. Louillat ML, Biafora A, Legros F, Patureau FW (2014) Angew Chem Int Ed 53:3505
67. Harvey ME, Musaev DG, Du Bois J (2011) J Am Chem Soc 133:17207
68. Yu XQ, Huang JS, Zhou XG, Che CM (2000) Org Lett 2:2233
69. Au SM, Huang JS, Yu WY, Fung WH, Che CM (1999) J Am Chem Soc 121:9120
70. Leung SKY, Tsui WM, Huang JS, Che CM, Liang JL, Zhu N (2005) J Am Chem Soc 127:16629
71. Liang JL, Huang JS, Yu XQ, Zhu N, Che CM (2002) Chem Eur J 8:1563

72. Liang JL, Yuan SX, Huang JS, Yu WY, Che CM (2002) Angew Chem Int Ed 41:3465
73. Musaev DG, Blakey SB (2012) Organometallics 31:4950
74. Harvey ME, Musaev DG, Blakey SB (2011) J Am Chem Soc 133:17207
75. Zhou XG, Yu XQ, Huang JS, Che CM (1999) Chem Commun 1999:2377
76. Milczek E, Boudet N, Blakey S (2008) Angew Chem Int Ed 47:6825
77. Bon JL, Blakey SB (2012) Heterocycles 84:1313
78. Enthaler S, Company A (2011) Chem Soc Rev 40:4912
79. Yang Y, Lin Y, Rao Y (2012) Org Lett 14:2874
80. Rao Y, Yu S, Lin Y, Han X, Shan G (2013) Org Biomol Chem 11:2318
81. Thirunavukkarasu VS, Hubrich J, Ackermann L (2012) Org Lett 14:4210
82. Yang F, Ackermann L (2013) Org Lett 15:718
83. Thirunavukkarasu VS, Ackermann L (2012) Org Lett 14:6206
84. Yang F, Rauch K, Kettelhoit K, Ackermann L (2014) Angew Chem Int Ed 53:11285
85. Padala K, Jeganmohan M (2014) Chem Eur J 20:4092
86. Liu W, Ackermann L (2013) Org Lett 15:3484
87. Yang X, Shan G, Rao Y (2013) Org Lett 15:2334
88. Padala K, Jeganmohan M (2013) Chem Commun 49:9651
89. Raghuvanshi K, Rauch K, Ackermann L (2015) Chem Eur J 21:1790
90. McNeill E, Du Bois J (2010) J Am Chem Soc 132:10202
91. Drees M, Strassner T (2006) J Org Chem 71:1755
92. Bakke JM, Frohaug (1996) J Phys Org Chem 9:310
93. Wang C, Shalyaev KV, Bonchio M, Carofiglio T, Groves JT (2006) Inorg Chem 45:4769
94. McNeill E, Du Bois J (2012) Chem Sci 3:1810

Top Organomet Chem (2016) 55: 217–258
DOI: 10.1007/3418_2015_130

Published online: 2 July 2015

meta- and *para*-Selective C–H Functionalization by C–H Activation

Jie Li, Suman De Sarkar, and Lutz Ackermann

Abstract Transition metal-catalyzed C–H bond functionalization has recently emerged as an indispensable tool for transforming otherwise unreactive C–H bonds. In addition to various strategies for *ortho*-selective functionalization via chelation assistance, organometallic C–H activation has recently enabled novel functionalizations of (hetero)arenes at remote *meta*- or *para*-positions. The *meta*- and *para*-selectivity was governed either by the inherent substrate structure or by the transition metal catalyst. Herein we summarize the rapid recent progress in *meta*- and *para*-selective aromatic C–H functionalization until May 2015.

Keywords C–H activation · Chelation · Remote functionalization · Selectivity · Transition metal catalysis

Contents

J. Li, S. De Sarkar, and L. Ackermann (✉)
Institut für Organische und Biomolekulare Chemie, Georg-August-Universität Göttingen, Tammannstr. 2, 37077 Göttingen, Germany
e-mail: Lutz.Ackermann@chemie.uni-goettingen.de

Abbreviations

Ac	Acetyl
Ar	Aryl
AMLA	Ambiphilic metal-ligand activation
Bn	Benzyl
Bu	Butyl
BQ	1,4-Benzoquinone
cat	Catalytic
cod	1,4-Cyclooctadienyl
Cp^*	Pentamethylcyclopentadienyl
Cy	Cyclohexyl
CMD	Concerted metalation deprotonation
dppb	1,4-Bis(diphenylphosphino)butane
dppe	1,2-Bis(diphenylphosphino)ethane
dtbpy	4,4′-Di-*tert*-butyl-2,2′-dipyridyl
DCE	1,2-Dichloroethane
DG	Directing group
DMF	*N*,*N*-Dimethylformamide
DMPU	1,3-Dimethyl-3,4,5,6-tetrahydro-2(1*H*)-pyrimidinone
equiv.	Equivalent
E	Electrophile
EDG	Electron-donating group
EWG	Electron-withdrawing group
FG	Functional group
Gly	Glycine
Hal	Halogen
Het	Heteroatom
HBCat	Catecholborane
HFIP	Hexafluoroisopropanol
Ind	Indenyl
KIE	Kinetic isotope effect
L	Ligand
m	*meta*
M	Metal
Me	Methyl
Mes	Mesityl
MPAA	Mono-*N*-protected amino acid
MTBE	Methyl *tert*-butyl ether
NMP	*N*-Methylpyrrolidinone
o	*ortho*
p	*para*
Ph	Phenyl
Pin	Pinacol

Piv	Pivaloyl
Pr	Propyl
PEPPSI	[1,3-Bis(2,6-Diisopropylphenyl)imidazol-2-ylidene](3-chloropyridyl) palladium(II) dichloride
T	Template
TBA	Tetrabutylammonium
TBP	Tributyl phosphate
TFA	Trifluoroacetic acid
TM	Transition metal
TMP	2,2,6,6-Tetramethylpiperamidyl

1 Introduction

The site-selective functionalization of aromatic compounds is of major relevance to synthetic organic chemistry with important applications in among others drug discovery and material science as well as pharmaceutical and chemical industries [1–4]. In electrophilic aromatic substitution (S_EAr) reactions, electron-donating groups direct incoming electrophiles preferentially to the *ortho*- and *para*-positions, and difficult-to-separate mixtures of products are often obtained [5, 6]. Arenes containing electron-withdrawing substituents are deactivated for S_EAr-type reactions and undergo *meta*-substitution only under rather harsh reaction conditions (Scheme 1a). Apart from the traditional S_EAr-type transformations, transition metal-catalyzed cross-coupling reactions between organic (pseudo)halides and organometallic reagents represent powerful tools for the assembly of substituted arenes (Scheme 1b) [7–10]. Despite the synthetic utility of this cross-coupling strategy, the requirement of pre-functionalized starting materials imposes a major

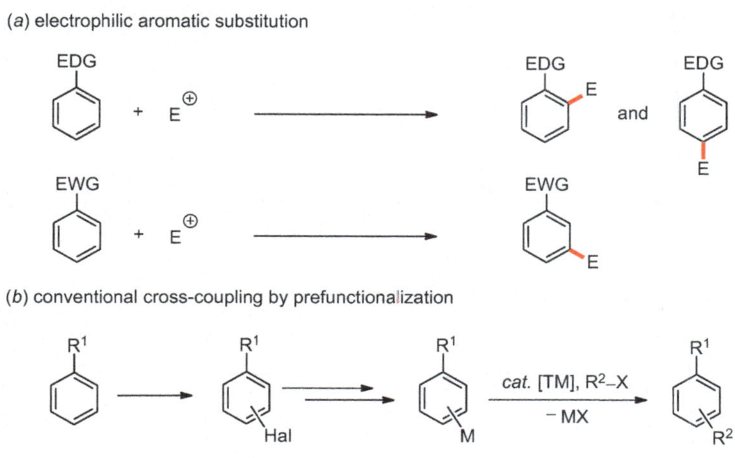

Scheme 1 (**a**) Site selectivity of S_EAr reactions and (**b**) conventional cross-coupling

drawback. As an ecologically benign and economically attractive alternative, aromatic functionalizations via transition metal-catalyzed organometallic C–H bond activation have emerged during the last few decades [11–15].

C–H bonds are ubiquitous in organic molecules. Therefore, controlling the site selectivity of the C–H functionalization process often represents a major obstacle. Pioneering work by Lewis and Smith unraveled the power of chelation assistance for site-selective catalytic functionalizations [16]. Phenol (**1**) was, thus, selectively alkylated at the *ortho*-position by the action of the ruthenium phosphite catalyst **2** (Scheme 2). Mechanistically, the reaction was proposed to proceed via the oxidative addition of the C–H bond to a ruthenium(0) complex by pre-coordination of the in situ formed phosphite ligand.

During the last decades, chelation-assisted C–H metalation has set the stage for catalytic C–H functionalizations with various directing groups (DGs), thereby providing expedient access to *ortho*-functionalized arenes [17–21]. In contrast, developing general protocols for remote *meta*- or *para*-selective C–H functionalizations continues to be extremely demanding (Scheme 3) [22–25].

Herein, we summarize the rapid recent progress in transition metal-catalyzed *meta*- and *para*-selective C–H functionalizations that are enabled by organometal-

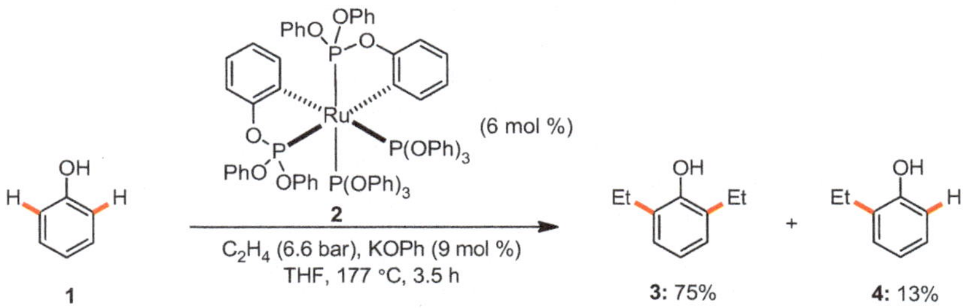

Scheme 2 Ruthenium(0)-catalyzed chelation-assisted C–H alkylation of phenol (**1**)

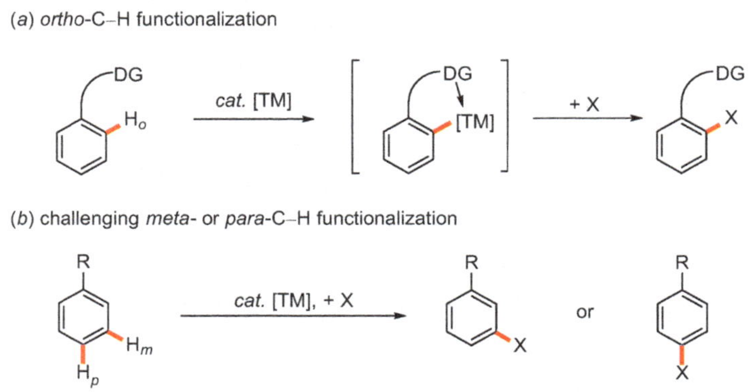

Scheme 3 Selectivity of catalyzed aromatic C–H functionalization

lic C–H activation until May 2015. The catalytic transformations are categorized by the origin of their site selectivity that either arises from (1) the inherent substrate control through steric and/or electronic effects or the use of a directing group or (2) which is imposed by the nature of the transition metal catalyst. In addition, formal *meta*-functionalization using removable directing groups is discussed which can be readily cleaved in a traceless fashion. Only selected examples of norbornene-mediated Catellani-type [26, 27] approaches will be discussed herein.

2 Formal *meta*-C–H Functionalization Using a Removable Directing Group

One approach for achieving the synthesis of *meta*- or *para*-substituted arenes is by employing a removable directing group in disubstituted aromatic substrates [28, 29]. For instance, Brown and co-workers demonstrated that the use of the removable sulfoxide group in substrate **5** gave rise to *meta*-substituted electron-rich arenes **7** within a two-step protocol involving directed *ortho*-metalation (DoM) (Scheme 4) [30]. Since stoichiometric amounts of a strong base were required, a low functional group tolerance and the generation of stoichiometric amounts of undesired metal waste represented the major drawbacks of this approach.

Transition metal-catalyzed C–H functionalizations of carboxylic acids are receiving considerable attention because of the high natural abundance of carboxylic acid derivatives in bioactive compounds and as indispensable building blocks for organic synthesis [31–33]. In addition, the recent development of transition metal-catalyzed decarboxylative coupling reactions [34, 35] illustrated the potential of carboxylic acids as removable directing groups [29, 36] to access *meta*-functionalized products. As illustrated in Scheme 5, transition metal-catalyzed C–H functionalization of *ortho*- or *para*-disubstituted aromatic carboxylic acids delivers *ortho*-functionalized arenes which could subsequently undergo decarboxylation to give the *meta*-substituted products.

Elegant work by Miura, Satoh, and co-workers disclosed a tandem rhodium-catalyzed *ortho*-olefination/decarboxylation sequence of *ortho*-substituted benzoic acids **8**. A series of *meta*-substituted stilbene derivatives **10** were hence accessible by this method (Scheme 6) [37, 38].

The same strategy was nicely applied by Larrosa and co-workers for the palladium-catalyzed direct arylation of benzoic acids **8** with iodoarenes **11** as

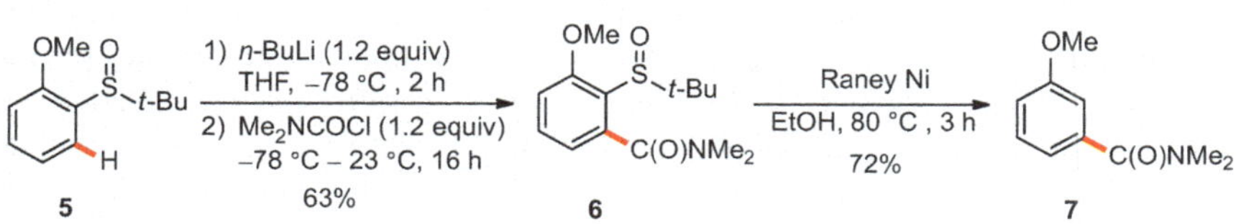

Scheme 4 Formal *meta*-C–H functionalization via DoM with a removable DG

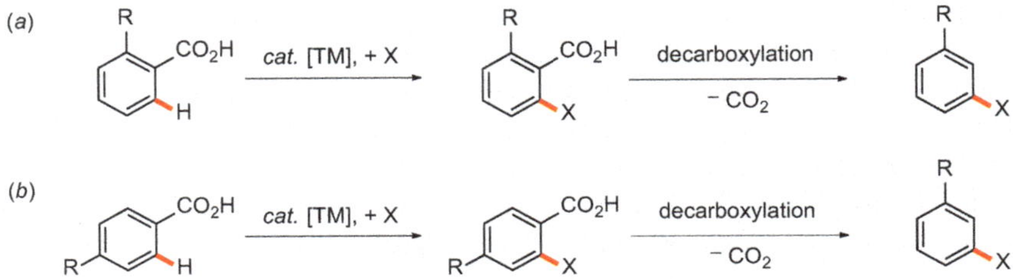

Scheme 5 Transition metal-catalyzed decarboxylative preparation of *meta*-substituted products

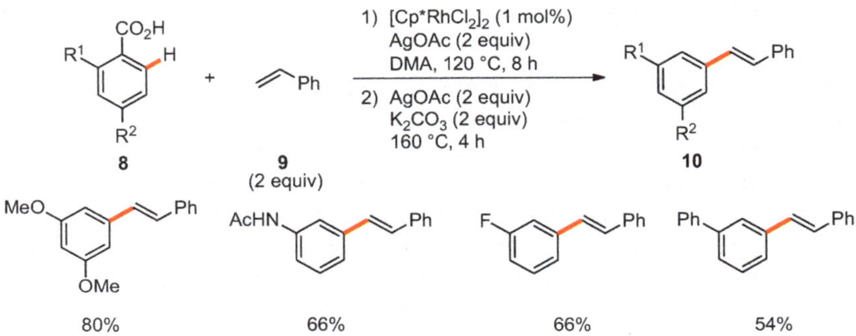

Scheme 6 Rhodium-catalyzed synthesis of *meta*-olefinated products **10**

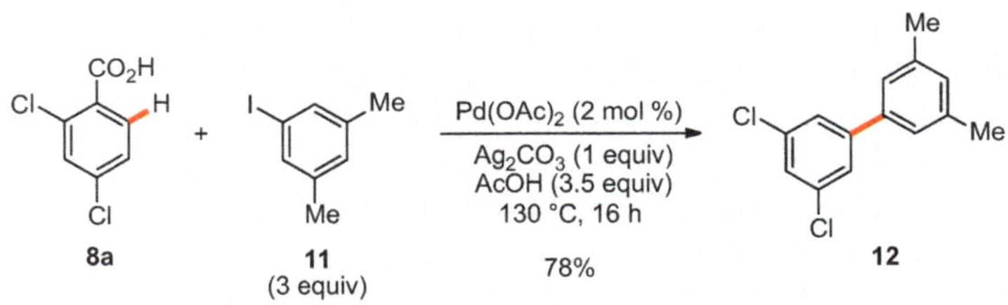

Scheme 7 Palladium(II)-catalyzed preparation of *meta*-arylated arenes **12**

coupling partners [39]. This tandem direct arylation/decarboxylation process provided an efficient alternative route to access *meta*-substituted biaryls **12** (Scheme 7).

In 2014, the same research group modified their methodology by including the installation of the carboxylic acid motif as a traceless directing group onto phenol derivatives **13** via a separate Kolbe–Schmitt reaction [40] and thus enabled the formal palladium-catalyzed *meta*-C–H arylation [41]. This improved method avoided the use of pre-functionalized phenols (Scheme 8). Moreover, the direct arylation of readily available salicylic acids was also described by Larrosa and co-workers [42].

Very recently, You and co-workers developed decarboxylative heteroarylations of aromatic carboxylate acids **8** through rhodium-catalyzed twofold C–H activation. This process provided an alternative synthesis of the *meta*-heteroarylated products **16** from *ortho*-substituted benzoic acids (Scheme 9) [43].

Besides C–C bond formation, this removable C–H functionalization/decarboxylation strategy also allowed for the facile construction of C–Het bonds [44]. Gooßen and co-workers elegantly developed carboxylate-directed *ortho*-alkoxylation of benzoates **17**, which was followed by decarboxylation to furnish the *meta*-substituted aryl ethers **19** (Scheme 10).

More recently, Lee and Chang reported on the iridium-catalyzed mild C–H amidation of benzoic acids **8** with sulfonyl azide **20** [45]. Subsequent palladium-catalyzed decarboxylation of *ortho*-amidated benzoic acid products afforded *meta*-

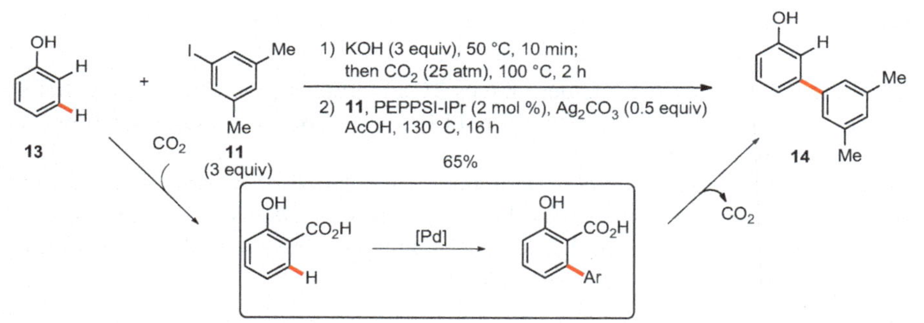

Scheme 8 Palladium-catalyzed sequence for *meta*-arylated arenes **14**

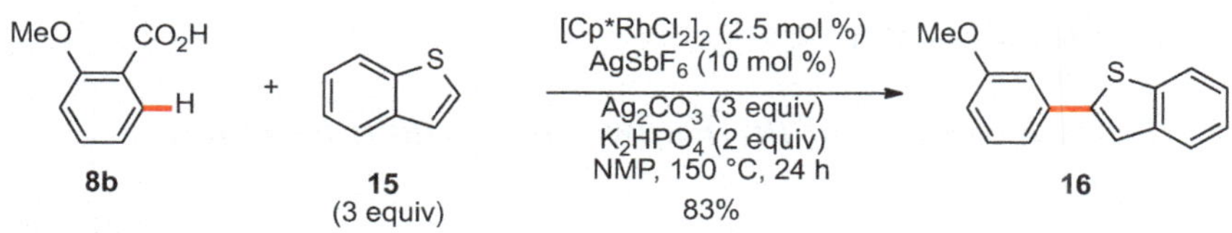

Scheme 9 Rhodium-catalyzed decarboxylative C–H heteroarylation of benzoic acid **8b**

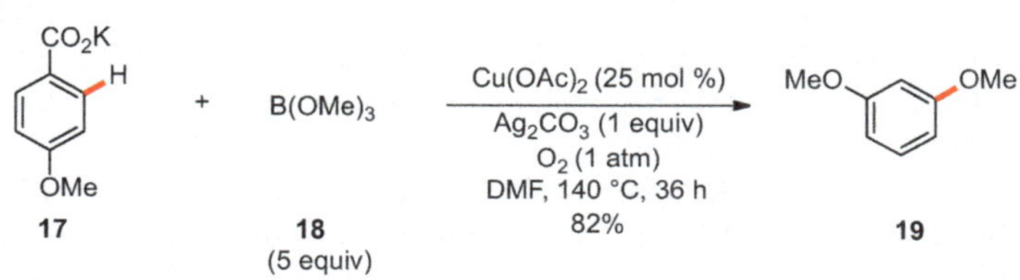

Scheme 10 Copper-catalyzed synthesis of *meta*-alkoxylated product **19**

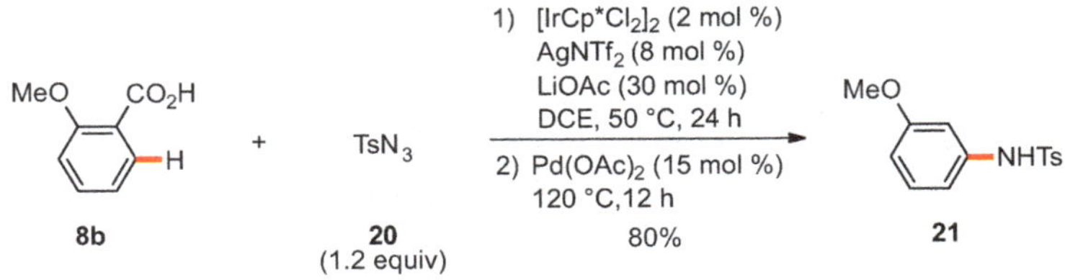

Scheme 11 Iridium-catalyzed synthesis of *meta*-amidated benzoic acid **21**

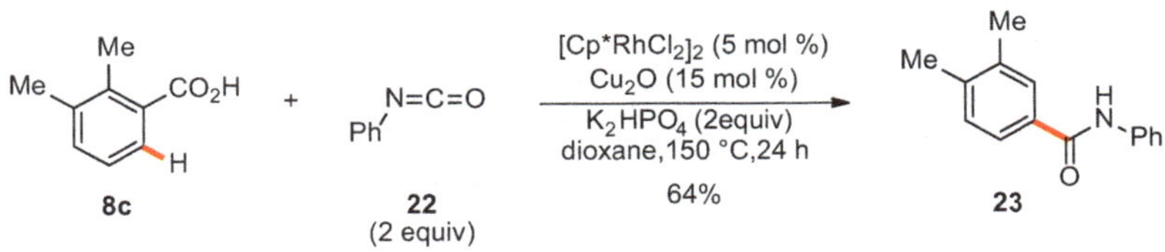

Scheme 12 Rhodium-catalyzed decarboxylative C–H amidation of benzoic acid **8c**

substituted (*N*-sulfonyl)aniline derivatives **21**. In terms of practicability, it is noteworthy that the two transformations could be carried out in a most user-friendly one-pot fashion (Scheme 11).

A rhodium-catalyzed direct C–H aminocarbonylation of benzoic acids **8** with isocyanates followed by decarboxylation was recently devised by Shi, Li, and co-workers (Scheme 12) [46]. Likewise, the authors achieved a related transformation with less expensive ruthenium catalysts [47].

3 Substrate-Controlled *meta*- and *para*-Selective C–H Functionalization

3.1 Early Examples of the Fujiwara–Moritani Reaction

Pioneering work by Fujiwara and Moritani on palladium(II)-catalyzed oxidative twofold C–H functionalizations between arenes and olefins established a step-economical protocol for the synthesis of styrene derivatives [48–50]. Over the last decade, substantial efforts have been devoted to control the site selectivity of the olefination of substituted arenes [51, 52]. The most commonly used strategy involved chelation-assisted *ortho*-olefination, while functionalizations at the remote positions were less explored.

In an early contribution, the inherent site selectivity was studied by Fujiwara and Moritani on the palladium(II)-mediated olefination of monosubstituted arenes **24** (Scheme 13a) [53]. Thus, electron-rich toluene and anisole were converted with considerably high *para*-selectivity (**25a** and **25b**), while the electron-withdrawing

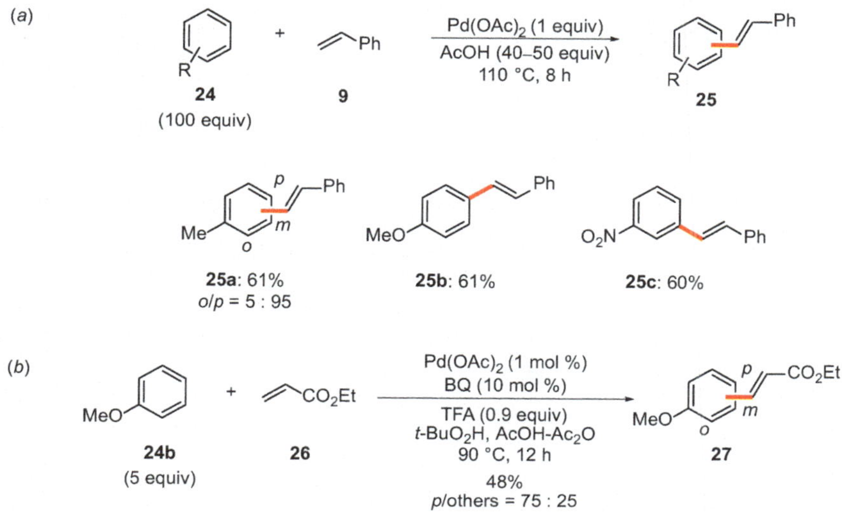

Scheme 13 Early examples of site-selective Fujiwara–Moritani reactions

nitro-substituent furnished the *meta*-product **25c** in high yield. A palladium(II)-catalyzed catalytic version of this transformation was subsequently also achieved with high levels of *para*-selectivity when using anisole (**24b**) as the substrate (Scheme 13b) [54].

3.2 meta-*Selective C–H Functionalization*

The borylation of unactivated C–H bonds is particularly attractive because of the wide synthetic utility of organoboron reagents in transition metal-catalyzed cross-coupling reactions and the facile transformations of C–B bonds into other functional groups [55, 56]. Pioneering studies of Janowicz and Bergman demonstrated that photolysis of complex **28** in a hydrocarbon solvent led to the loss of dihydrogen and subsequent oxidative addition of the solvent to the transition metal center to form complex **30** (Scheme 14a) [57, 58]. Thereafter, Hartwig and co-workers discovered that an iron boryl complex smoothly reacted with benzene to give the phenylboronate ester, again under photochemical reaction conditions [59]. On this basis, the Smith group developed the catalytic borylation of arenes with iridium(III) boryl precatalyst **34** (Scheme 14b) [60]. However, relatively long reaction times were necessary and accompanied by a low turnover number. Only shortly thereafter, Hartwig and co-workers showed that the catalytic efficiency could be significantly improved by using the more reactive rhodium(I) precatalyst **35** (Scheme 14c) [61]. Further investigations revealed that the iridium precatalyst **34** exhibited

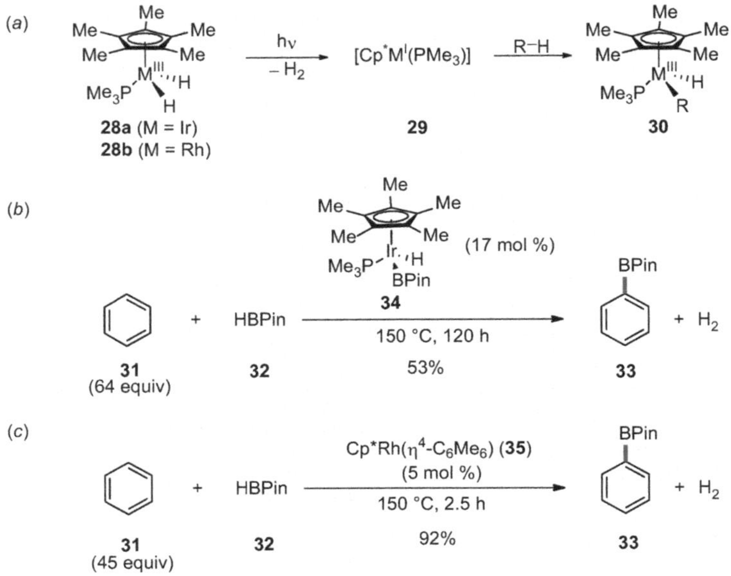

Scheme 14 Stoichiometric C–H activation of hydrocarbons and catalytic C–H borylation of benzene (**31**)

a higher selectivity toward aromatic borylation as compared with the rhodium catalyst **35** [62].

In 2002, Smith reported a remarkably selective and practical method for the borylation of a series of aromatic compounds using the most user-friendly iridium (I) precatalyst **37** (Scheme 15a) [63]. The highly efficient catalytic system exploited the diphosphine dppe as the ligand and allowed for highly *meta*-selective borylations of 1,3-disubstituted benzene derivatives by repulsive steric interactions. However, with monosubstituted arenes **24,** only modest levels of selectivity were observed. The synthetic utility of the protocol was demonstrated by the one-pot C–H borylation followed by palladium-catalyzed arylation using the previously developed Suzuki–Miyaura cross-coupling technology (Scheme 15b).

Furthermore, Ishiyama, Miyaura, and Hartwig showed that more potent C–H borylation catalysts could be generated by employing air-stable, commercially available iridium(I) precursor **41** in combination with the simple 2,2′-bipyridines (bpy) as the ligand [64–66]. Both B_2Pin_2 (**40**) [65] and HBPin (**32**) [66] were successfully used as the borylating reagents. The key role of the 2,2′-bipyridine ligand was identified as to increase the solubility and stability of the iridium catalyst under the reaction conditions. Notably, these advancements allowed the borylation to be achieved at ambient reaction temperature and also dislodged the necessity of using a large excess of the arenes **24** (Scheme 16).

The site selectivity for all the aforementioned catalytic systems was proposed to originate from repulsive steric interaction. Thus, irrespective of the electronic

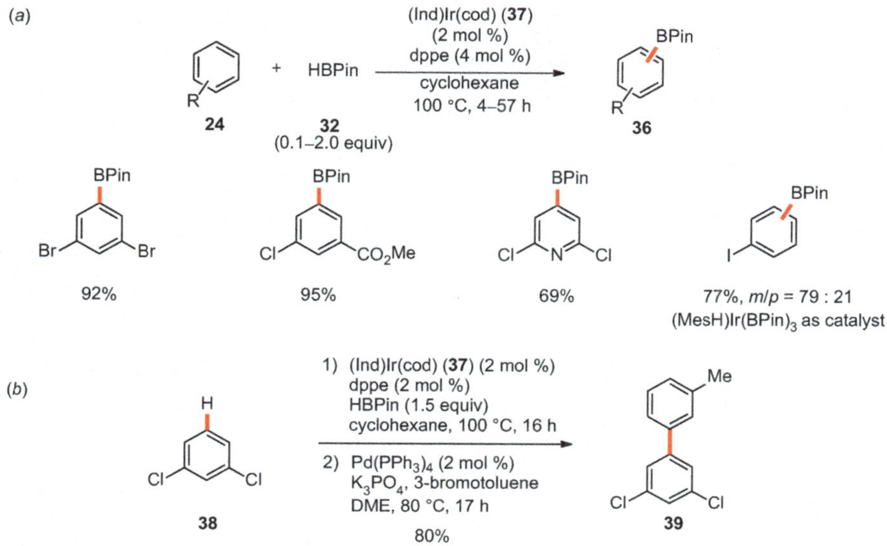

Scheme 15 Iridium-catalyzed *meta*-C–H borylation of arenes **24**

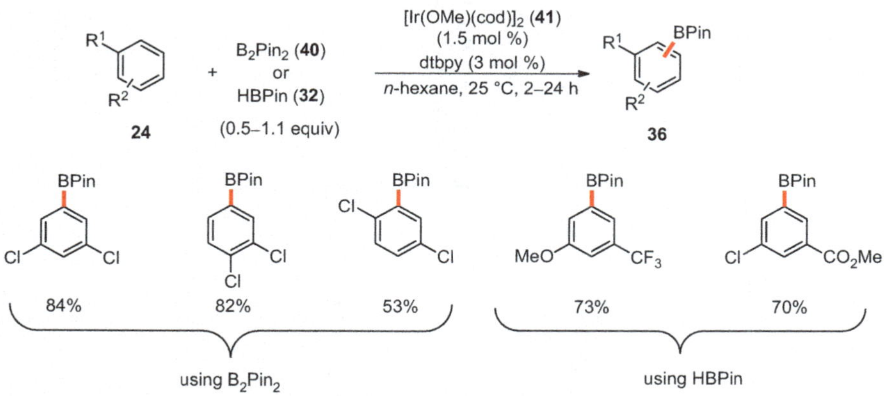

Scheme 16 Iridium-catalyzed *meta*-C–H borylation at ambient reaction temperature

nature, 1,2- and 1,3-disubstituted arenes **24** delivered 1,2,4- or 1,3,5-trisubstituted arenes **36**, respectively, with the borylation occurring at the least hindered site. In agreement with this mechanistic rationale, 1,4-disubstituted arenes were converted with reduced efficacy due to the steric hindrance of the *para*-substituent. Smith and co-workers performed a detailed study on the borylation of 1,4-disubstituted benzonitriles **42**, which unraveled the importance of steric factors (Scheme 17) [67]. Thus, the site-selective *meta*-C–H borylation was observed in case of arenes bearing substituents being larger than the nitrile group.

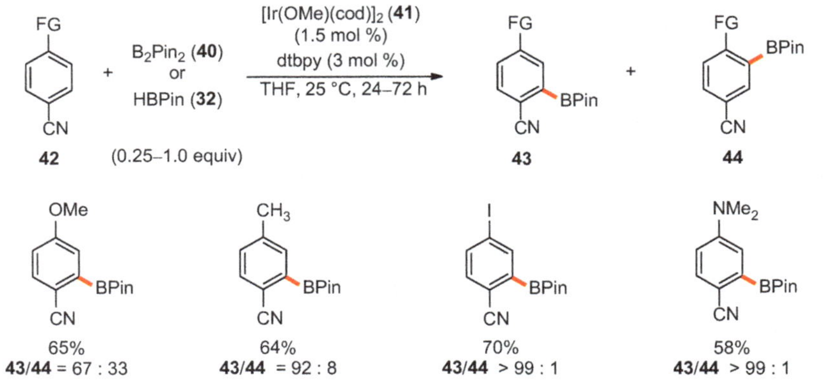

Scheme 17 Sterically controlled C–H borylation

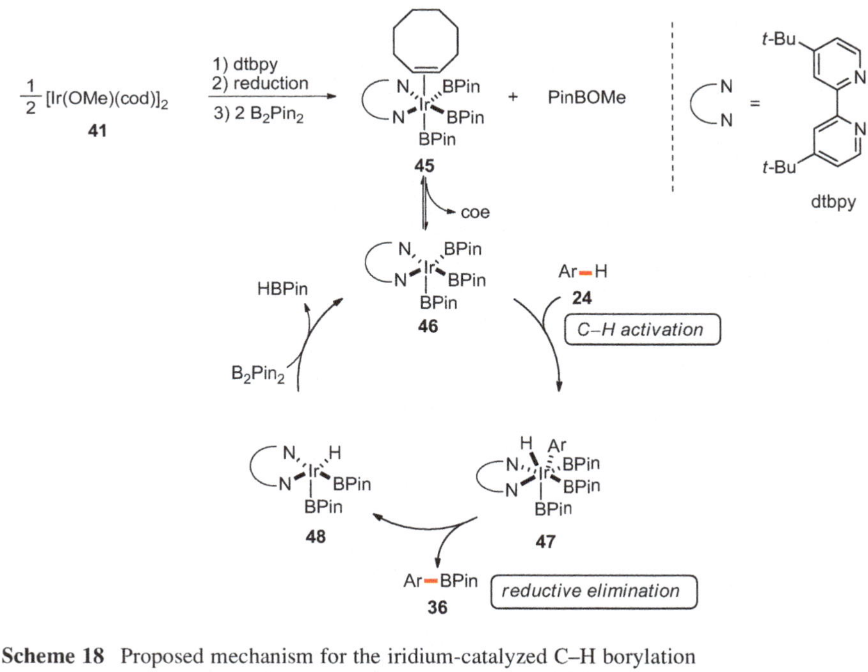

Scheme 18 Proposed mechanism for the iridium-catalyzed C–H borylation

Comprehensive mechanistic studies were performed by Ishiyama, Miyaura, and Hartwig on the catalytic system consisting of $[Ir(OMe)(cod)]_2$ (**41**) and dtbpy, which revealed the catalytically competent iridium(III) trisboryl complex **46** to be formed by the reduction of a cod ligand, along with a subsequent oxidative addition of B_2Pin_2 to the iridium(I) center (Scheme 18) [68]. The C–H activation by oxidative addition produced iridium(IV) complex **47**, which thereafter underwent

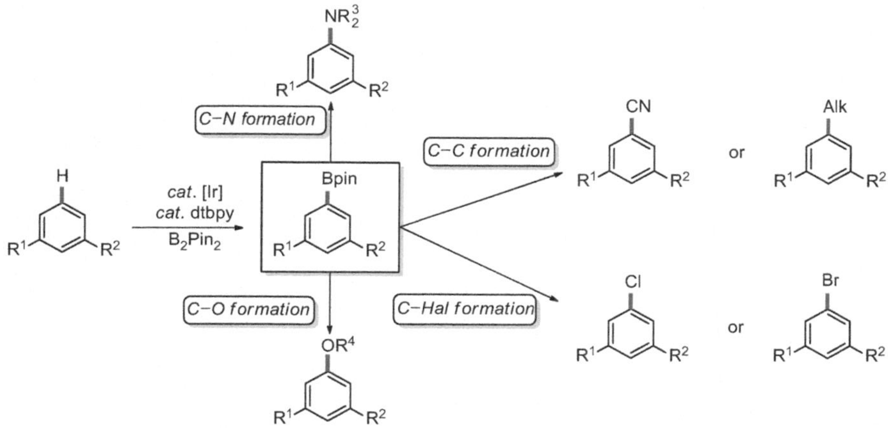

Scheme 19 Synthesis of *meta*-substituted arenes

reductive elimination to deliver the borylated arene **36** and the iridium(III) species **48**. Finally, reaction with B_2Pin_2 regenerated the catalytically active complex **46**.

The unique potential of this approach was illustrated by various applications of the sterically controlled C–H borylation in organic synthesis [55, 69, 70]. Hence, various functional groups were introduced at the *meta*-position of the substituents of arenes via boronates as a temporary functional group (Scheme 19). These transformations were often conducted in a one-pot manner, rendering this strategy operationally simple.

A powerful protocol for the site-selective C–H functionalization by steric interaction through intermolecular silylation reactions was recently disclosed by Chen and Hartwig [71]. The combination of the rhodium(I) catalyst **51** with the 2,2′-biphenyldiphosphine ligand **52** was found to be most effective in the C–H silylation of a series of arenes **24** under rather mild condition (Scheme 20a). A comparison of the steric sensitivity between the rhodium/biphenyldiphosphine and the iridium/dtbpy catalysts in silylation and borylation reactions highlighted the former to be superior for the remote C–H functionalizations (Scheme 20b). For example, *o*-cymene displayed a significant preference for silylation at the position *para* to the larger *i*so-propyl group, whereas the borylation gave almost equimolar amounts of the two isomeric products (**50a** and **36a**). The silylation protocol was also influenced by electronic factors and exhibited a preference for the more electron-rich site. Thus, 4-chloroanisole selectively underwent the silylation at the *ortho*-position of the electron-donating methoxy group, while the borylation of the exact same substrate did not afford a significant selectivity (**50b** and **36b**). A combination of steric and electronic factors was observed with 3-fluorotoluene, where the silylation occurred predominantly at the less hindered and more electron-rich *meta*-position (**50c**).

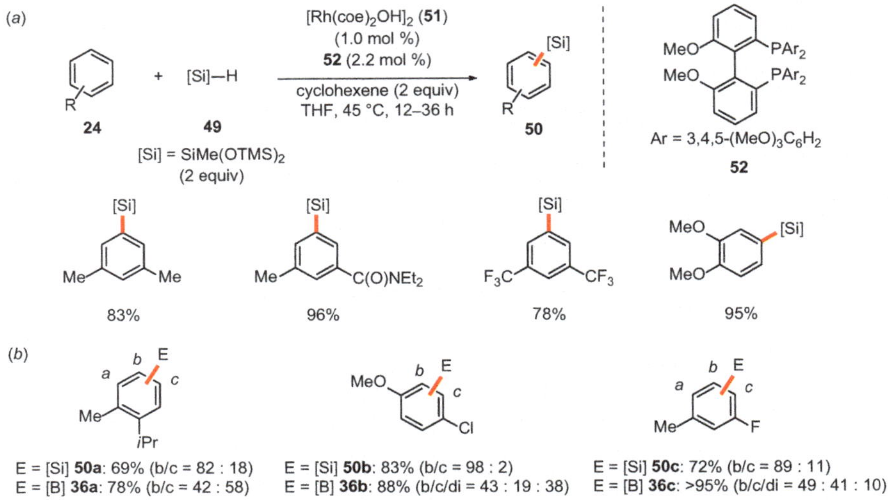

Scheme 20 Sterically controlled C–H silylation of arenes **24**

The site selectivity of the aforementioned silylation process is associated with the combination of the bulky diphosphine ligand **52** and the sterically hindered silylating reagent **49**. Detailed mechanistic studies showed that a distorted squared pyramidal silylrhodium dihydride complex **53** represents the resting state of the catalytic cycle, which is in equilibrium with the active catalyst **54** (Scheme 21) [72]. Hydrogen transfer to cyclohexene via olefin insertion and oxidative addition of the silane delivers the silylrhodium species **55**. C–H activation followed by C–Si bond forming reductive elimination finally regenerates the active catalyst **54**.

Transition metal-catalyzed C–C bond formations via sterically controlled remote C–H activation were thoroughly investigated by different research groups over the last decade [73]. In this context, the synthesis of biaryls by the dehydrogenative hetero-coupling using simple arenes in combination with substrates displaying Lewis-basic directing groups is particularly noteworthy (Scheme 22).

In 2007, Hull and Sanford reported on a palladium-catalyzed oxidative coupling of aromatic C–H bonds [74]. High site selectivity was obtained when 1,3-di- or 1,2,3-trisubstituted arenes **24** were employed as the substrates. The dehydrogenative coupling predominantly took place at the less hindered 5-position. However, more challenging couplings of monosubstituted arenes **24** only resulted in low site selectivities (Scheme 23).

Furthermore, Dong and co-workers disclosed an efficient protocol for the palladium-catalyzed dehydrogenative arylation of phenyl carbamates **60** with simple arenes [75]. 1,2-Disubstituted arenes **61** as the coupling partner led to products **62**, being solely arylated at the less hindered 4-position (Scheme 24).

Based on their previous studies on the rhodium-catalyzed cross-dehydrogenative arylation of benzamides with aryl halides [76], Glorius and co-workers achieved the

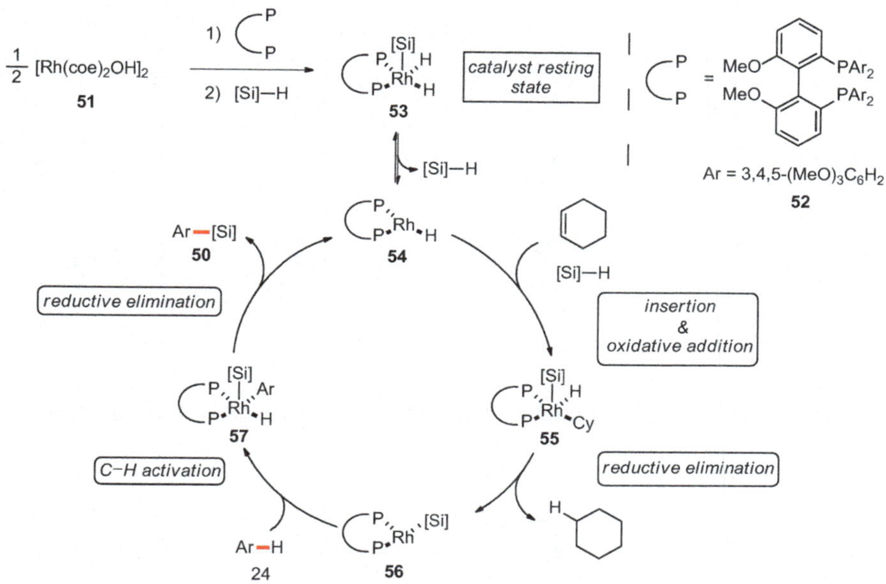

Scheme 21 Proposed mechanism for the rhodium-catalyzed C–H silylation of arenes

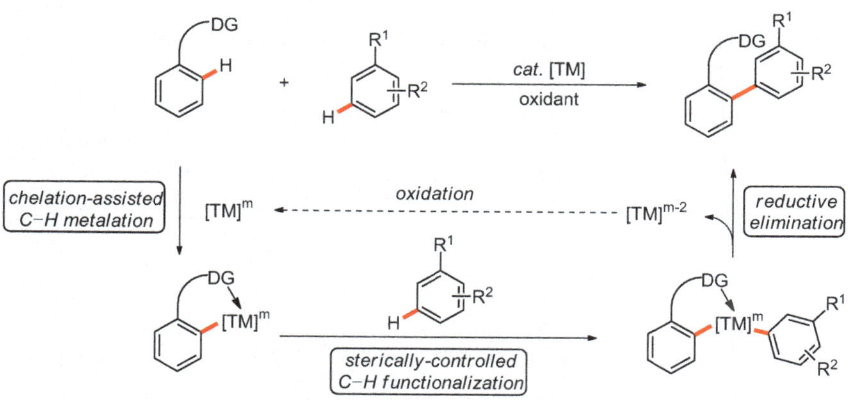

Scheme 22 Biaryl syntheses by sterically controlled C–H activation

coupling between benzamide **63** and arenes **24**, utilizing C_6Br_6 as the sacrificial oxidant [77]. In accordance with the palladium-catalyzed transformations discussed above, the arenes **24** furnished the coupling products at the less hindered position with good site selectivities. In contrast, monosubstituted substrates led to difficult-to-purify mixtures of *meta*- and *para*-arylated products **64** (Scheme 25).

In 2009, the Gaunt group reported on a copper-catalyzed *meta*-selective arylation of anilides **65** using diaryliodonium salts (Scheme 26a) [78]. Notably,

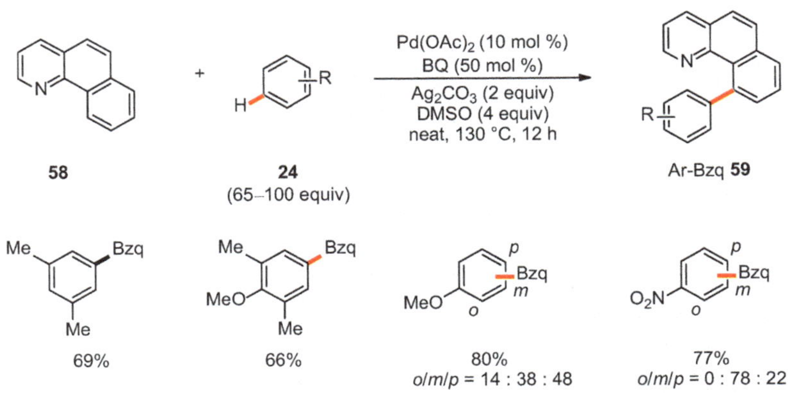

Scheme 23 Palladium(II)-catalyzed dehydrogenative cross-couplings of benzoquinoline **58**

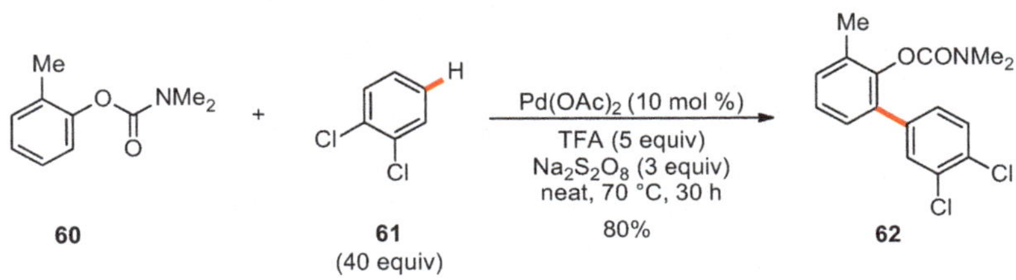

Scheme 24 Palladium(II)-catalyzed dehydrogenative cross-coupling of *o*-tolyl carbamate **60**

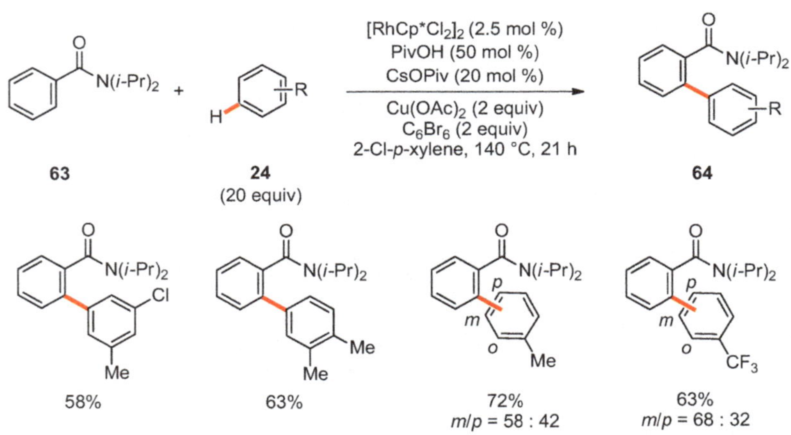

Scheme 25 Rhodium(III)-catalyzed dehydrogenative cross-couplings of benzamides **63**

here, high selectivities for mono-, di-, and trisubstituted anilides **66** were observed. Importantly, *ortho*-arylated product **66e** was isolated in high yield when employing *meta*-methoxy-aniline as the substrate, probably due to the strong *para*-directing effect of the methoxy group within an electrophilic-type activation. Subsequent

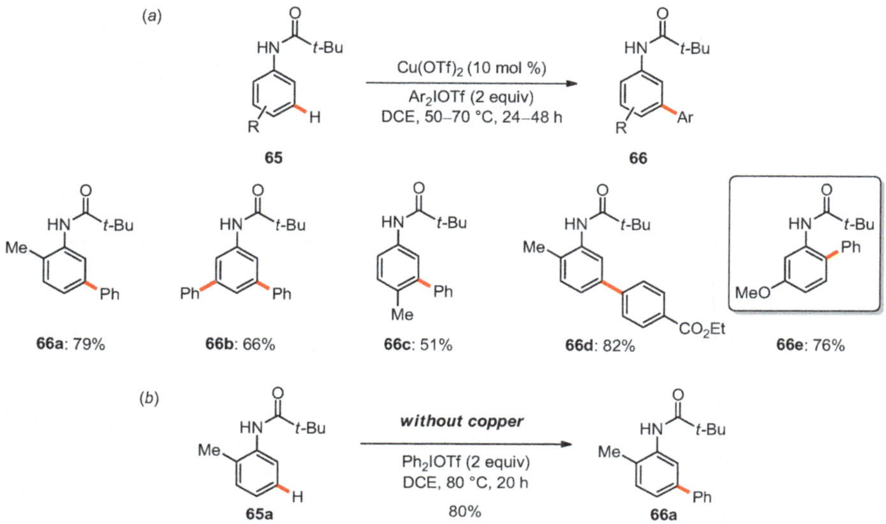

Scheme 26 *meta*-selective C–H arylation of anilides **65**

studies revealed that the *meta*-selective arylation is viable with α-aryl carbonyl compounds, such as amides, esters, and ketones, which, a priori, would be expected to inherently provide the *meta*-substituted products preferentially [79]. Moreover, a recyclable heterogeneous copper catalyst [Cu/AlO(OH)] was employed by Lee and Park in similar transformations of anilides **65** [80]. Consecutive studies showed that the *meta*-selective arylation is feasible in the absence of any copper catalyst at a slightly elevated temperature of 80°C (Scheme 26b) [80]. The reaction most likely occurred through an electrophilic arylation process, while the origin of the *meta*-selectivity still warrants for mechanistic studies [81].

In 2009, Yu and co-workers reported on sterically controlled Fujiwara–Moritani reactions between substituted arenes **24** and activated olefins **67** using $Pd(OAc)_2$ as the catalyst and a bulky 2,6-dialkylpyridine ligand **69** (Scheme 27a) [82]. For 1,3-disubstituted arenes, the olefination exclusively occurred at the least hindered *meta*-position, whereas electron-deficient monosubstituted arenes delivered a product mixture favoring the *meta*-product **68**. The observed site selectivity was attributed to the electronic properties of the substrates in combination with the steric bulk of the ligand. Likewise, Sanford and co-workers revealed the use of acridine **73** as ligand for selective palladium(II)-catalyzed C–H olefinations of electron-rich *o*-xylene **70** (Scheme 27b) [83].

Another recent method for the sterically controlled site-selective C–Het formation by C–H activation was developed by Hartwig and co-workers in amination reactions [84]. The dehydrogenative coupling proceeded using $Pd(OAc)_2$ as the catalyst, $PhI(OAc)_2$ as the oxidant, and the bulky phthalimide **74** as the nitrogen

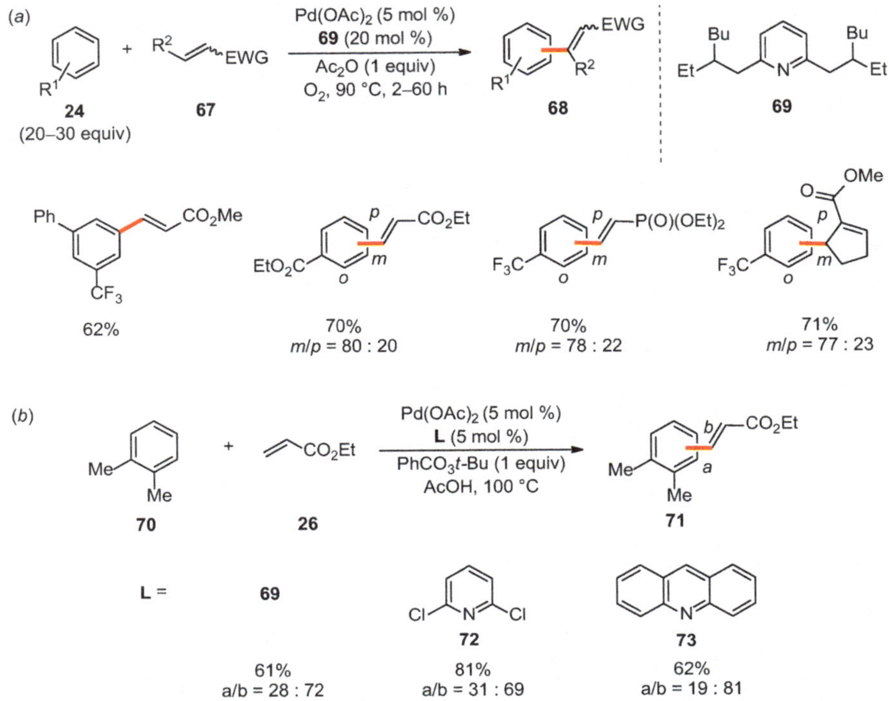

Scheme 27 Palladium(II)-catalyzed *meta*-selective Fujiwara–Moritani reactions

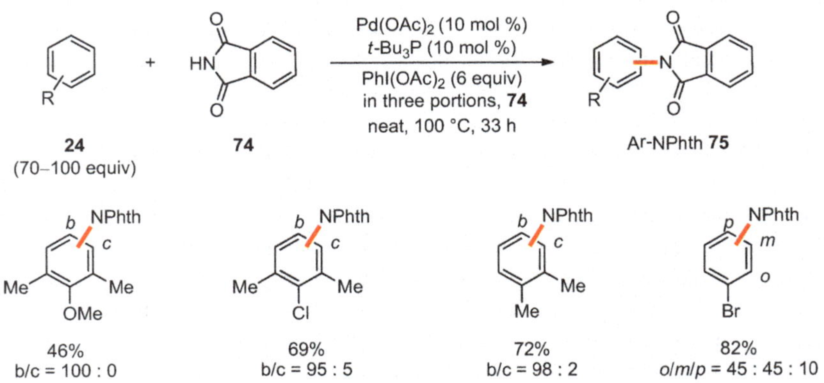

Scheme 28 Sterically controlled C–H amination of arenes **24**

source in an excess of arene **24** as the solvent (Scheme 28). A sterically controlled site selectivity was observed with di- and trisubstituted arenes, whereas monosubstituted substrates **24** delivered a mixture of difficult-to-separate *meta*- and *para*-functionalized products **75**.

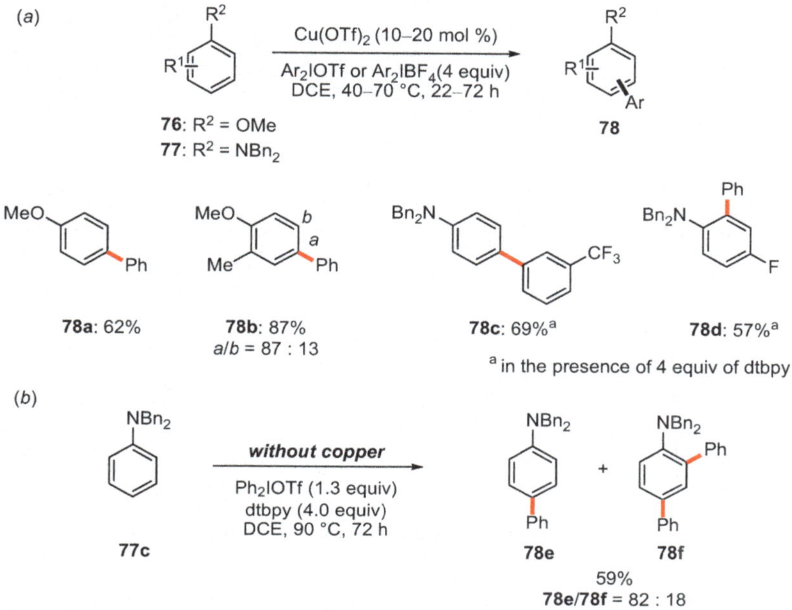

Scheme 29 *para*-selective arylation of electron-rich arenes **76** and **77**

3.3 para-*Selective C–H Functionalization*

Besides steric interactions, electronic factors play a dominant role in determining the site selectivity of transition metal-catalyzed aromatic C–H functionalizations. In a follow-up work of their *meta*-selective arylation, Gaunt and co-workers reported copper-catalyzed *para*-selective arylations of electron-rich arenes **76** and **77** [85]. Strongly electron-donating substituents, such as alkoxy and amino groups, directed the C–H arylation to the *para*-position, an observation that is reminiscent of S_EAr-type transformations (Scheme 29a).

At a slightly elevated temperature of 90 °C, the reaction was found to proceed in the absence of the copper catalyst under otherwise identical reaction conditions but with somewhat reduced efficacy (Scheme 29b). It is noteworthy that the direct *ortho*-arylation took place to yield **78d**, when the *para*-position of aniline **77d** was blocked. This selectivity pattern is consistent with classical electrophilic aromatic substitution. The proposed reaction pathway for the metal-free arylation involves the direct nucleophilic attack of the electron-rich arene at the diaryliodonium salt through an *ipso*-substitution mechanism, with the product formation after rearomatization. A radical pathway was excluded by performing successful coupling reactions in the presence of representative radical scavengers.

A similar strategy exploiting the electronic nature of directing groups was applied in the *para*-arylation of electron-rich phenol derivatives **79** (Scheme 30a) [86]. This palladium-catalyzed method offered high site selectivity at ambient

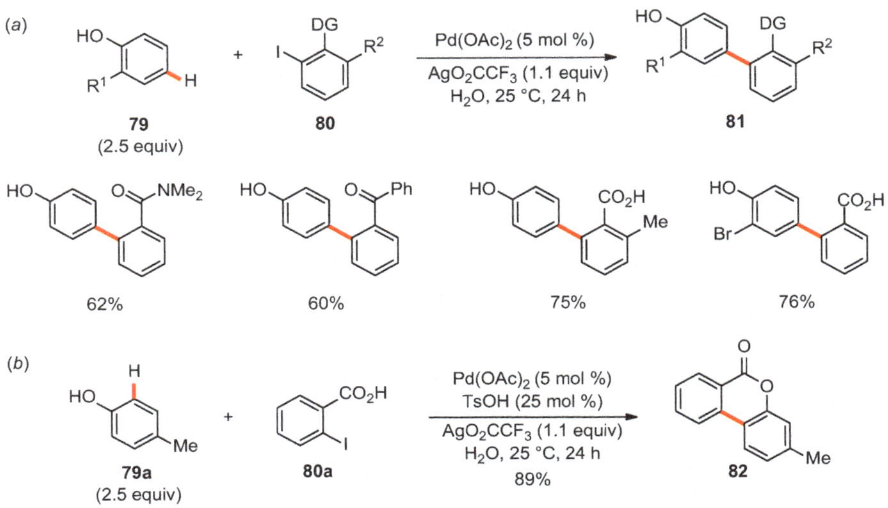

Scheme 30 Palladium(II)-catalyzed *para*- and *ortho*-selective arylation of phenol derivatives **79**

reaction temperature with water as a green solvent [87, 88]. Likewise, *para*-substituted phenol **79a** could be *ortho*-arylated, which after intramolecular condensation led to dibenzopyranone **82** (Scheme 30b). Mechanistic studies featured a KIE of $k_H/k_D = 1.3$. Further, the authors suggested an electrophilic metalation of the phenol derivative followed by an oxidative addition of the aryl iodide **80** to form a high-valent palladium species, which is stabilized by the adjacent coordinating group. Subsequent reductive elimination finally delivered the *para*-substituted product **81** within this S_EAr-type transformation.

Dehydrogenative hetero-couplings by twofold C–H functionalizations between simple arenes **24** and substrates **83** bearing an *ortho*-directing group were realized by Yu and co-workers using palladium(II) catalysis (Scheme 31) [89]. The functionalization of the electron-rich arenes **24** exclusively took place at the *para*-position to the donating substituent. Notably, haloarenes were found to be compatible under the reaction conditions, which should allow for subsequent synthetic elaboration. Strong oxidants, such as NFSI, were used, and a high-valent palladium(IV) species was accordingly suggested to be involved in the catalytic cycle.

Remote C–H functionalizations via radical pathways were investigated by Li and Guo for *para*-selective arene transformations. Thus, the oxidative C–H alkylation of arenes **24** proved viable using $Ru_3(CO)_{12}$ as the catalyst and di-*tert*-butyl peroxide (TBP) as the oxidant (Scheme 32) [90]. Differently substituted arenes **24** with variable electronic properties were alkylated by ruthenium(0) catalysis with good site selectivities. A significant KIE was not observed with chlorobenzene/[D_5]chlorobenzene as the substrates ($k_H/k_D = 1.00$). Thus, the authors proposed that the reaction most likely proceeds through a radical mechanism. The site selectivities could be attributed to the stabilization of the radical intermediates by the *para*-substituents [91].

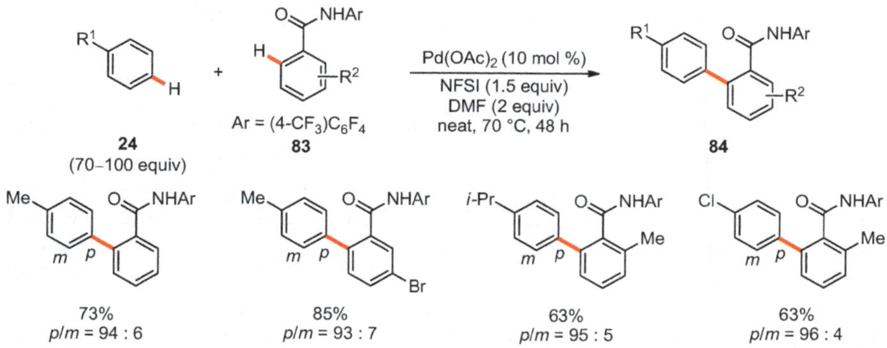

Scheme 31 Palladium(II)-catalyzed *para*-selective cross-dehydrogenative coupling of arenes **24** and **83**

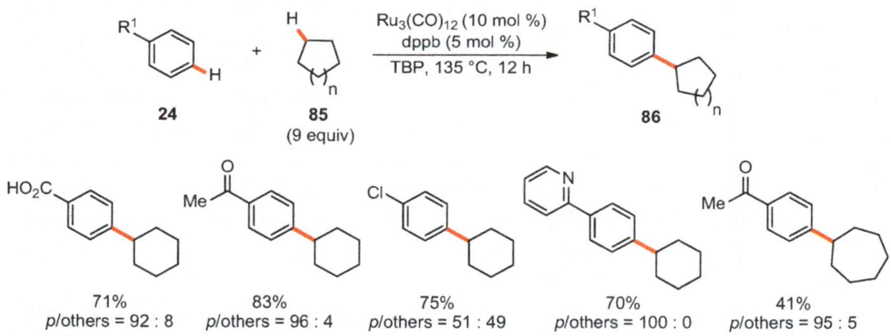

Scheme 32 Ruthenium(0)-catalyzed *para*-selective cross-coupling of arenes **24** and cycloalkanes **85**

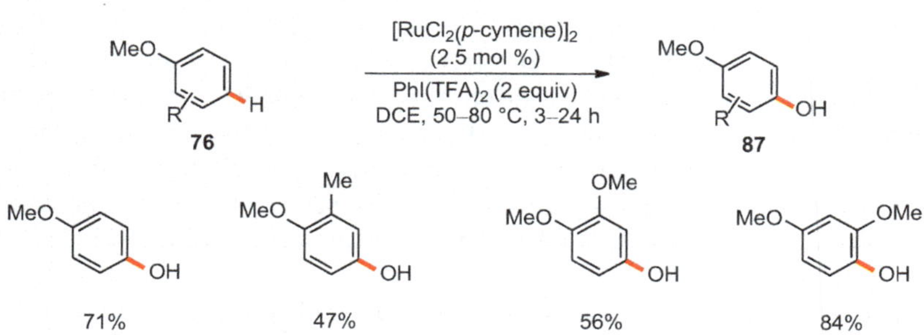

Scheme 33 Ruthenium(II)-catalyzed *para*-C–H oxygenations of anisoles **76**

Liu and Ackermann performed *para*-selective oxygenations [92, 93] of electron-rich anisole derivatives **76** with $PhI(TFA)_2$ as the oxidant in the presence of a ruthenium(II) catalyst (Scheme 33) [94]. Substantial inhibition of the catalytic C–H activation process was observed in the presence of TEMPO as a radical scavenger, thus indicating single-electron transfer (SET) steps likely to be operative.

3.4 meta-*C–H Functionalization Directed by Nitrile-Containing Templates*

Recently, the preparation of *meta*-functionalized arenes has been achieved by utilizing a rationally designed template which facilitates the delivery of the catalyst to the *meta*-C–H bond through the formation of a macropalladacycle. Detailed studies by Yu and co-workers disclosed palladium-catalyzed *meta*-selective C–H Fujiwara–Moritani reaction of arenes **88** by the assistance of nitrile-containing directing groups [95]. The nitrile functionality was presumed to weakly coordinate to the palladium catalyst in an end-on fashion, thus allowing for the remote *meta*-C–H activation via a cyclophane-like pre-transition state.

The first type of auxiliary was attached to the substrate through a removable benzyl ether linkage. Two sterically bulky *tert*-butyl groups were required on the template arene to keep the nitrile group and the *meta*-C–H bond in the same quadrant. Moreover, two *iso*-butyl groups at the α-position of the nitrile group were necessary in the directing group design to facilitate the entropically favored C–H metalation via the Thorpe–Ingold effect [96]. A catalytic system consisting of $Pd(OPiv)_2$ and superstoichiometric amounts of AgOPiv enabled highly *meta*-selective olefinations of electronically and sterically biased substrates **88** [95]. Remarkably, this template-assisted strategy allowed for the facile olefination with disubstituted olefins **67**, which are often problematic in *ortho*-C–H olefination reactions (Scheme 34a). Furthermore, this template could be easily removed via a palladium-mediated hydrogenolysis, thus providing the *meta*-alkenylated toluene derivatives **90** in high yield (Scheme 34b).

The second type of auxiliary contained a benzonitrile moiety which was incorporated into hydrocinnamic acids **92** via a readily cleavable amide linkage [95]. In this case, the mono-*N*-protected amino acid (MPAA) *N*-acetyl glycine (Ac-Gly-OH) was found to considerably accelerate the palladium-catalyzed *meta*-C–H olefination. Again, the U-shaped directing group effectively overrode the electronic effects of electron-donating (**92a**) or electron-withdrawing substituents (**92b**) as well as the steric hindrance in *ortho*-disubstituted substrate **92c**. Moreover, detailed experimental as well as computational studies by Yu and Houk revealed the dual role of the MPAA ligands in this transformation [97]. Hence, besides stabilizing the monomeric palladium complex **94**, the MPAA also served as an internal base for a proton abstraction via a concerted metalation–deprotection (CMD) [11, 98, 99] mechanism (Scheme 35a) [95].

Importantly, the U-shaped template could be readily removed by hydrolysis under basic conditions, providing the *meta*-substituted diacid **95** and the recovered directing group **96** (Scheme 35b) [95].

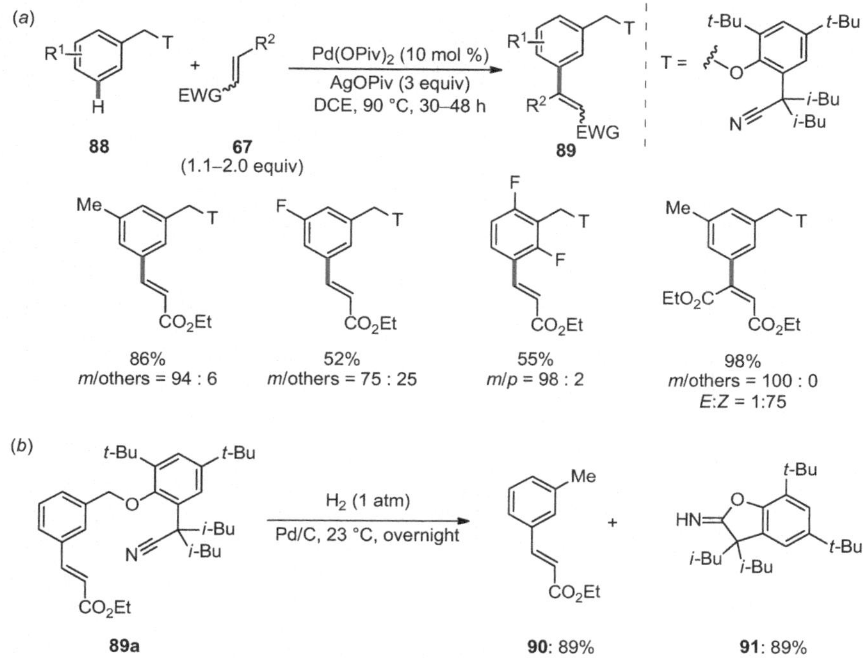

Scheme 34 Template-directed *meta*-C–H olefination of benzylic ethers **88**

An excess of AgOPiv was found to be of crucial importance for the successful *meta*-C–H olefination [100]. Computational studies indicated that the silver(I) salt did not only act as the oxidant but that it was also coordinated by the nitrile, thereby forming the bimetallic intermediate **98**. The (hetero)bimetallic mode of action with participation of $PdAg(OAc)_3$ or $Pd_2(OAc)_4$ species was calculated to require lower activation barriers than either the "monomeric" or the "trimeric" analogous transition state models. This feature was attributed to the significantly reduced ring strain in the macrocyclic transition states (Scheme 36).

In a follow-up contribution, phenol derivatives **99** containing the same U-shaped templates were demonstrated as suitable substrates as well, thus providing an alternative route for the *meta*-functionalization of phenols (Scheme 37) [101].

Furthermore, Deng and Yu discovered that the U-shaped template also enabled the smooth *meta*-C–H olefination of phenylacetic acid derivatives **101** under similar reaction conditions (Scheme 38) [102].

Besides the direct olefination, the synthetic utility of the template-assisted *meta*-functionalization strategy was illustrated by successfully developing palladium(II)-catalyzed direct *meta*-arylation reactions [103]. Indeed, phenylpropanoic acid **92** and phenolic derivatives **99** underwent facile *meta*-selective cross-coupling with aryl boronic esters **103** under palladium(II) catalysis. Notably, these reaction

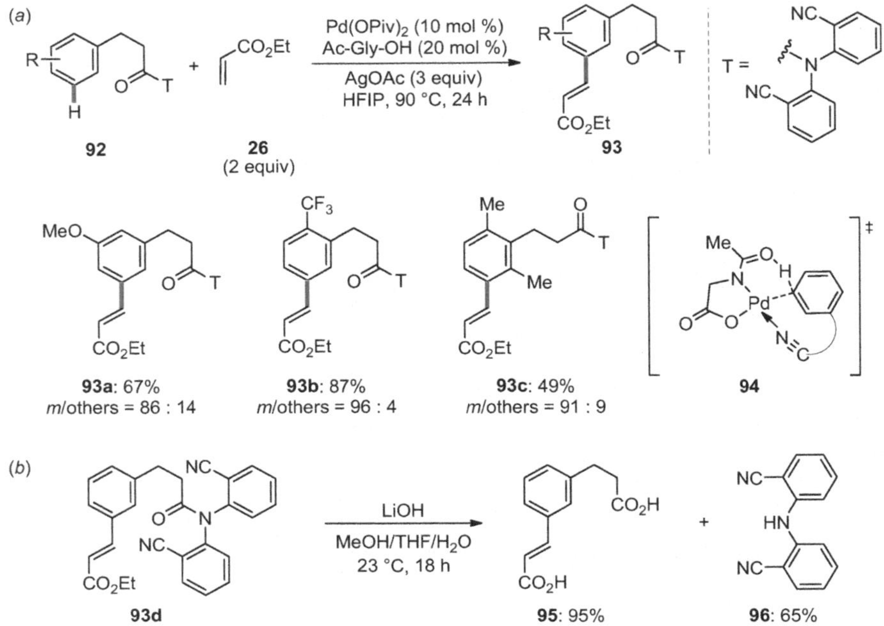

Scheme 35 *meta*-C–H olefination of hydrocinnamic acid derivatives **92**

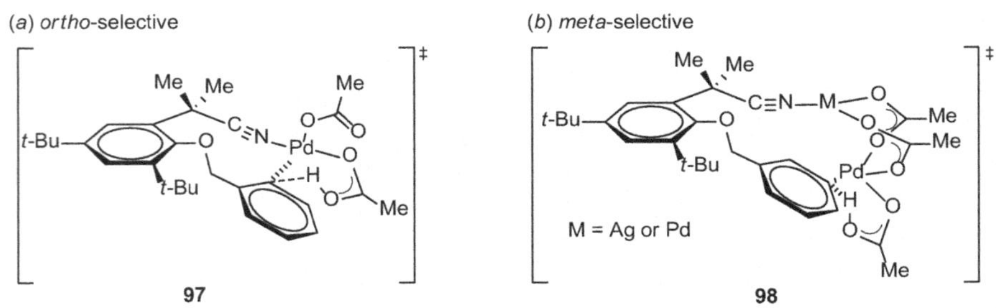

Scheme 36 (**a**) "Monomeric" and (**b**) "dimeric" transition state models **97** and **98**

conditions further enabled the *meta*-methylation when $MeBF_3K$ (**104**) was employed. However, further optimizations are required to extend the scope of alkylborons to those bearing β-hydrogens (Scheme 39).

Subsequently, a template-assisted C–H olefinations of aromatic and benzylic amines **106** proved viable [104]. Here, a fluorine substituent in the auxiliary scaffold was identified to lead to significant changes in the conformation. While the unfluorinated DG directed the activation to the *ortho*-C–H position by the carbonyl group, the fluorinated variant oriented the carbonyl group away from the *ortho*-C–H bond, thus facilitating the approach of the nitrile moiety to the *meta*-C–H bond. The MPAA ligand Ac-Gly-OH was found to play a crucial role in

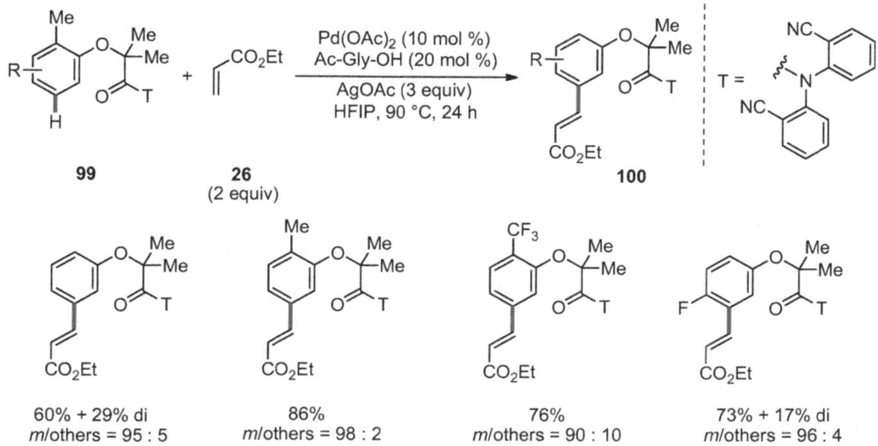

Scheme 37 *meta*-C–H olefination of phenol derivative **99**

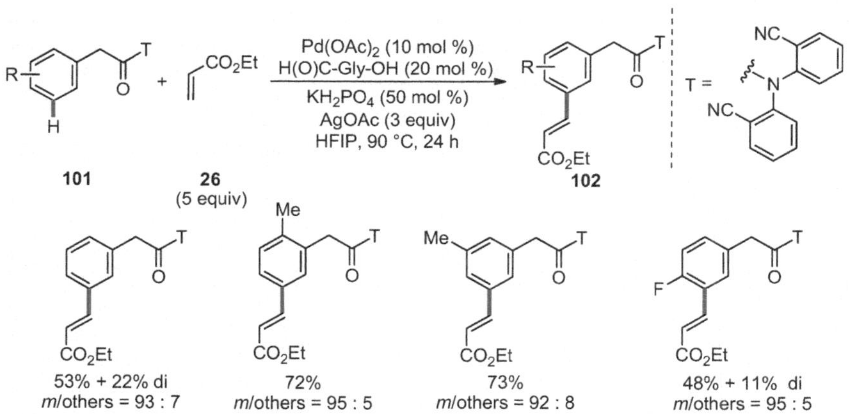

Scheme 38 *meta*-C–H olefination of phenylacetic acid derivative **101**

improving both the reactivity and the site selectivity (Scheme 40a). The template could be easily removed upon hydrolysis, affording the secondary amine **108** and the recovered directing group **109** (Scheme 40b).

Notably, the power of the template-assisted *meta*-C–H activation approach was further highlighted by developing C–Het bond-forming reactions [104]. In this context, a palladium catalyst along with $PhI(OAc)_2$ as the oxidant enabled *meta*-C–H acetoxylation of benzylamine **110** (Scheme 41).

However, the abovementioned auxiliaries were found to be ineffective for the *meta*-C–H activation of indolines, presumably due to the stronger electron-donating effect of the indoline nitrogen. Therefore, an electron-withdrawing sulfonyl-based directing group **112** was established to overcome the intrinsic *ortho*- and *para*-

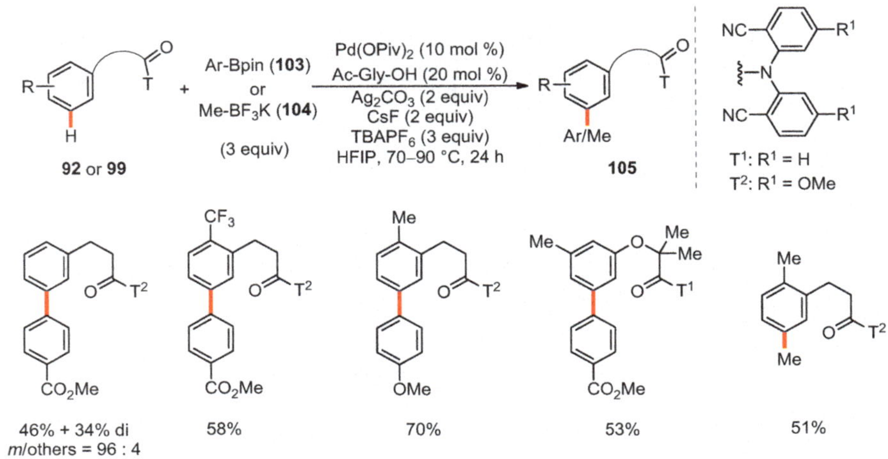

Scheme 39 *meta*-C–H arylation and methylation

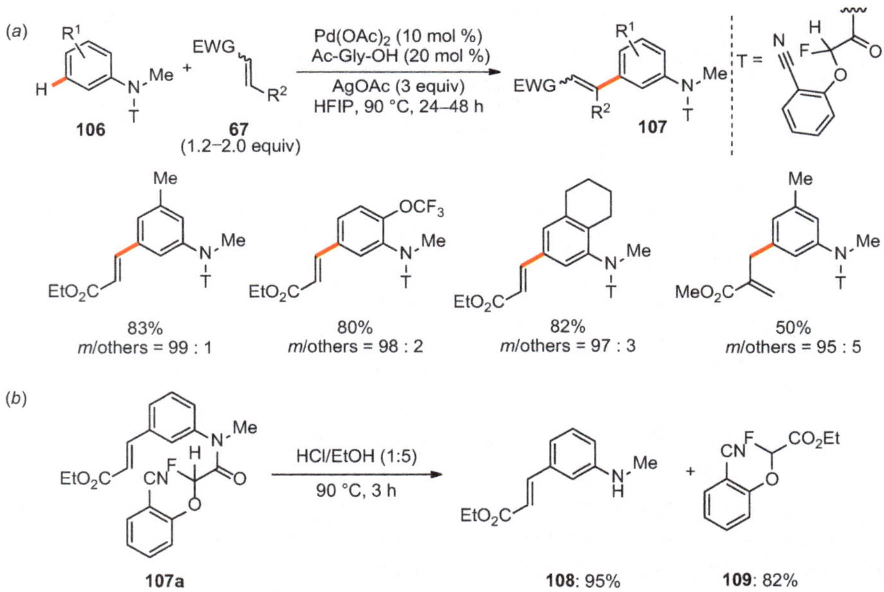

Scheme 40 Template-directed aromatic *meta*-C–H olefination of amines **106**

selectivity [105]. A palladium catalyst in combination with Ac-Gly-OH hence allowed for the facile *meta*-selective olefination of indoline derivatives **112** (Scheme 42).

The versatility of this approach was demonstrated by also performing direct *meta*-selective arylations with various arylboronic esters **103** (Scheme 43)

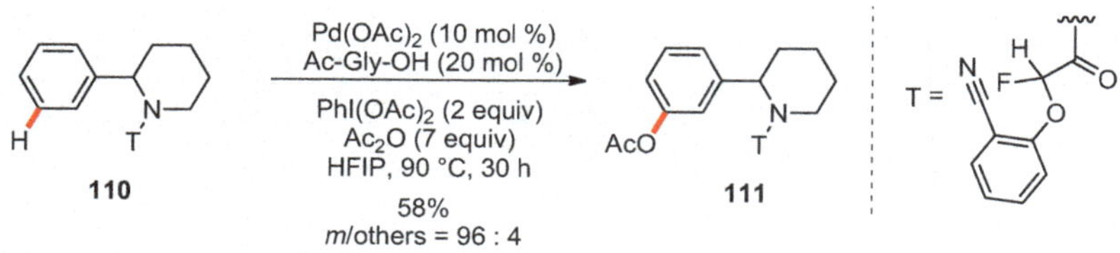

Scheme 41 Aromatic *meta*-C–H acetoxylation of amine **110**

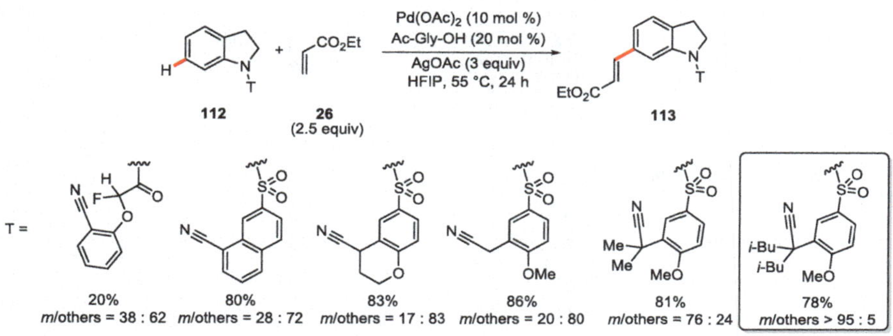

Scheme 42 Influence of auxiliary structure for *meta*-C–H olefination of indoline **112**

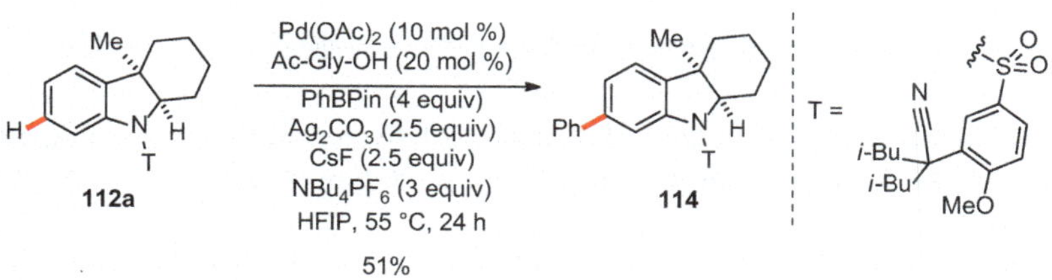

Scheme 43 *meta*-C–H arylation of indoline **112a**

[105]. The NBu_4PF_6 surfactant was added to prevent undesired agglomeration of the palladium(0) species to form palladium black, while the anionic counterion could stabilize the cationic palladium intermediate [103].

Another slightly modified directing group enabled palladium-catalyzed *meta*-C–H acetoxylation of indolines **112** under the previously described reaction conditions (Scheme 44) [105]. However, a substantial amount of the *para*-acetoxylated indoline **115b** was formed due to a competing electrophilic palladation at the *para*-position.

Tan and co-workers reported a synthetically useful silicon-based template **116** which could be easily introduced into the alcohol-based substrates and was removed under standard silyl deprotection conditions [106]. In situ deprotection

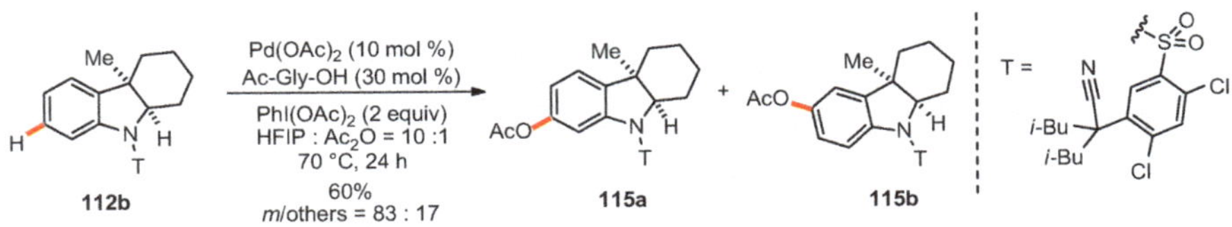

Scheme 44 Directed *meta*-C–H acetoxylation of indoline **112b**

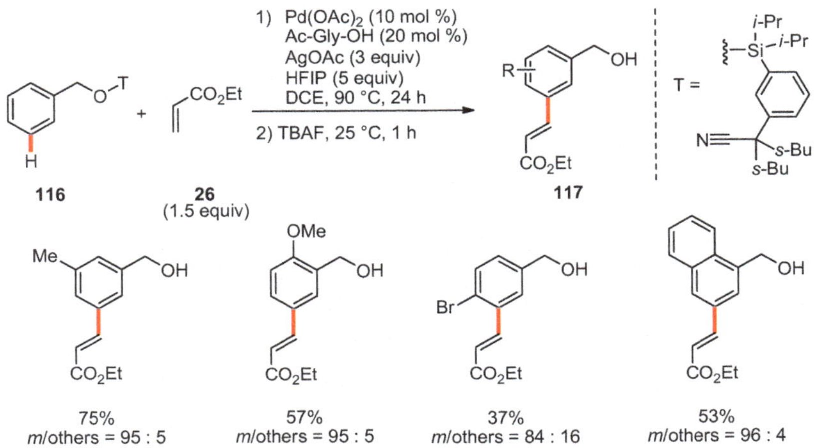

Scheme 45 Template-directed *meta*-C–H olefination of benzyl alcohol derivatives **116**

after palladium-catalyzed *meta*-C–H olefination delivered *meta*-functionalized benzyl alcohol **117** in high yield, thus highlighting the power and practicality of this template (Scheme 45).

With respect to atom and step economy, it is noteworthy that the silicon-based template **119** could be recycled upon deprotection and used for another starting material preparation (Scheme 46) [106].

More recently, Maiti and co-workers disclosed a 2-hydroxybenzonitrile template-directed *meta*-selective C–H olefination of phenylacetic acid derivatives **120** (Scheme 47) [107]. Intriguingly, this considerably simpler DG led to high *meta*-selectivity and effectively suppressed the di-olefination. *trans*-Esterification with the solvent HFIP after the *meta*-olefination directly removed the template under the standard reaction conditions.

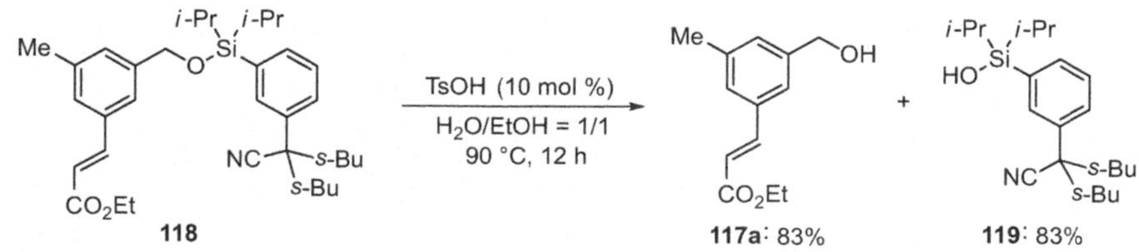

Scheme 46 Removal of the template **119**

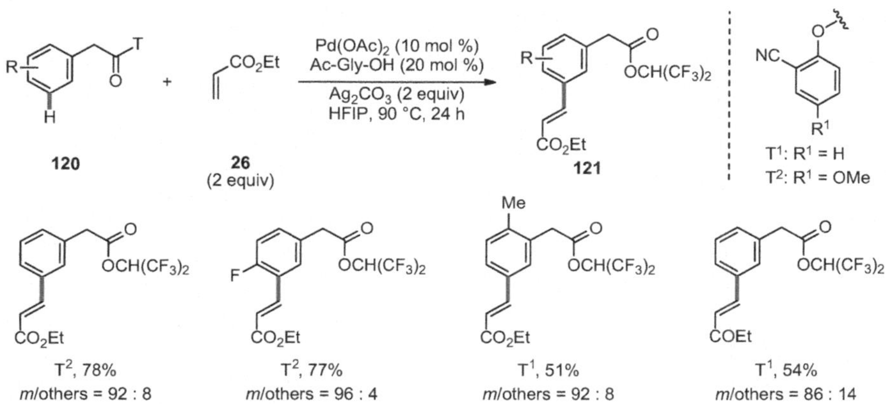

Scheme 47 Template-directed *meta*-C–H olefination of phenylacetic acid derivatives **120**

4 Catalyst-Controlled *meta*-Selective C–H Functionalization

Despite the recent success with nitrile-containing auxiliaries for various palladium-catalyzed *meta*-selective C–H functionalizations, the requirement of preinstalled, engineered directing groups with a high molecular weight constitutes a considerable disadvantage. Therefore, developing new strategies by which the site selectivity of simple aromatic substrates can be readily tuned by catalyst control continues to be challenging.

4.1 Norbornene-Mediated meta-*C–H Functionalization*

A catalytic approach for achieving formal *meta*-selectivity via the Catellani reaction [26, 108] was disclosed by Wilhelm and Lautens [109]. In the presence of alkyl halide **123** and *iso*-propyl boronic acid as the reducing agent, aryl iodides **122** underwent a tandem process to afford *meta*-substituted arenes **124** (Scheme 48). A proposed mechanism for this transformation involves first the oxidative addition of

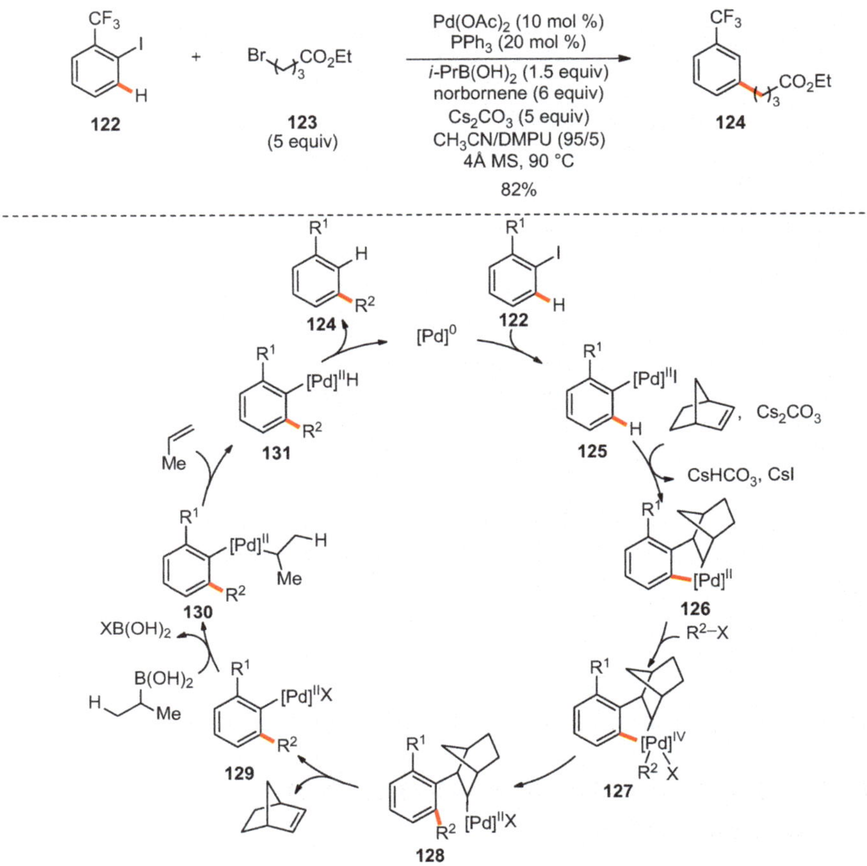

Scheme 48 Synthesis of *meta*-substituted arenes **124** by the Catellani reaction and proposed catalytic cycle

the iodoarene **122** to a palladium(0) species forming complex **125**. This intermediate then undergoes carbopalladation of norbornene and subsequent *ortho*-C–H metalation to form palladacycle **126**. A second oxidative addition of the alkyl halide to palladium(II) is then proposed to generate a putative [110] palladium (IV) species **127**, followed by subsequent reductive elimination and norbornene extrusion to give palladium aryl **129** [109]. Transmetalation of intermediate **129** with alkyl boronic acid and subsequent β-hydride elimination gives rise to the palladium(II) hydride species **131**, which then undergoes reductive elimination to deliver the *meta*-substituted product **124** and regenerates the active palladium (0) catalyst.

Inspired by the success of achieving unique site selectivities through Catellani-type transformations [111–113], a norbornene-mediated palladium-catalyzed *meta*-C–H functionalization with a common *ortho*-directing group **132** was very recently developed (Scheme 49) [114]. Electron-rich pyridine- or quinoline-based ligands **135** were found to be the key for success in improving both the reactivity and site selectivity of the direct *meta*-functionalization. Besides methyl iodide (**133a**), activated ethyl iodoacetate (**133b**) was a competent electrophile. Direct *meta*-benzylation with benzyl bromide **133c** also proved viable. However, the direct *meta*-arylation required the presence of an *ortho*-coordinating group in the arylating reagent **133d** to promote the oxidative addition of the aryl halide **133d**. Direct alkylation with ethyl iodide (**133e**) only delivered an unsatisfactory low yield, thus calling for further developments of the ligands to enable alkylations with β-hydrogen-containing substrates [114, 115].

A proposed working mode for this transformation is shown in Scheme 50. First, a proximity-induced *ortho*-C–H activation takes place and generates the *ortho*-palladacycle **136**. The thus formed palladacycle **136** then undergoes a Catellani-type carbopalladation, and subsequently *meta*-C–H metalation forms the five-membered palladacycle **137** [26]. The formed intermediate **137** further reacts

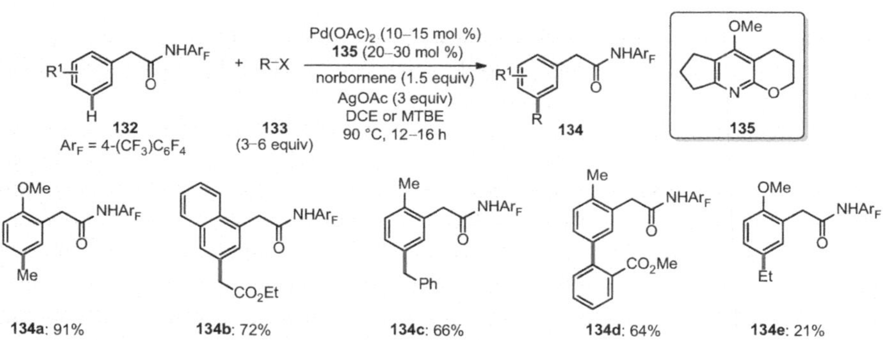

Scheme 49 Norbornene-mediated palladium(II)-catalyzed *meta*-C–H functionalization

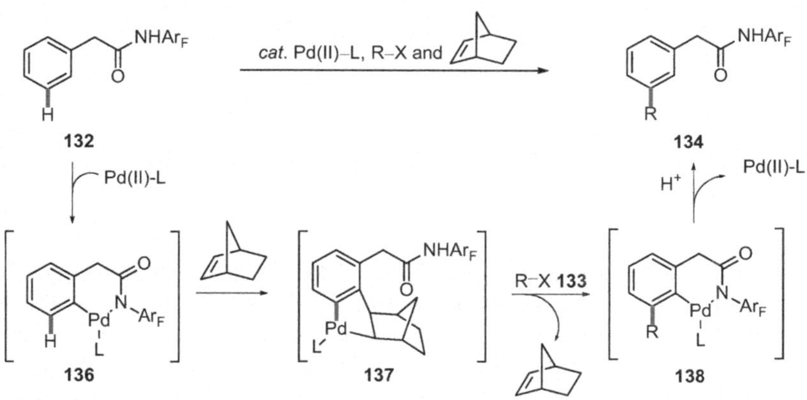

Scheme 50 Proposed pathway for the norbornene-mediated *meta*-C–H activation

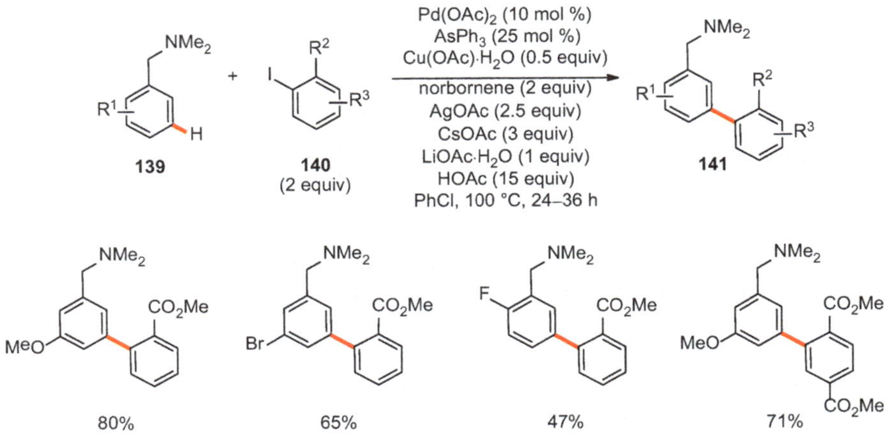

Scheme 51 Norbornene-mediated *meta*-C–H arylation of benzylamines **139**

with the coupling partner **133** to form a new *meta*-C–C bond. Subsequent β-elimination to norbornene and proto-demetalation of the aryl palladium complex **138** regenerates the active palladium catalyst.

Simultaneously, the same approach was exploited by the group of Dong for developing a norbornene-mediated palladium(II)-catalyzed *meta*-C–H arylation with tertiary amine **139** as a simple directing group (Scheme 51) [116]. A catalytic system consisting of $Pd(OAc)_2$ and $AsPh_3$ was employed, along with a combination of acetates as an "acetate cocktail" to improve the reaction rate. The authors proposed that HOAc and LiOAc could aid the dechelation of the DG from palladium after initial C–H metalation, while CsOAc assisted the *ortho*-C–H activation. Moreover, copper(II) acetate could minimize the palladium(0)-mediated reactions by functioning as a palladium(0) scavenger. However, the exact roles of these acetates still warrant further elucidation. Under the optimized reaction conditions, a range of differently substituted benzylamines **139** could be efficiently converted with a range of *ortho*-substituted aryl iodides **140** as the arylation reagents.

4.2 meta-*Selective C–H Functionalization via* ortho-*C–H Metalation*

Early work by Roper and co-workers disclosed the direct nitration of arenes that were σ-bonded to ruthenium(II) without affecting the M–C bond [117]. The Ru–C_{aryl} σ-bond induced a strong *para*-directing effect, introducing the nitro group exclusively at the *para*-position of the arene via a S_EAr-type process (Scheme 52).

The Ru–C_{aryl} σ-bond also exhibited an *ortho*-directing effect when two or more equivalents of copper nitrate were used (Scheme 53). Despite the deactivating

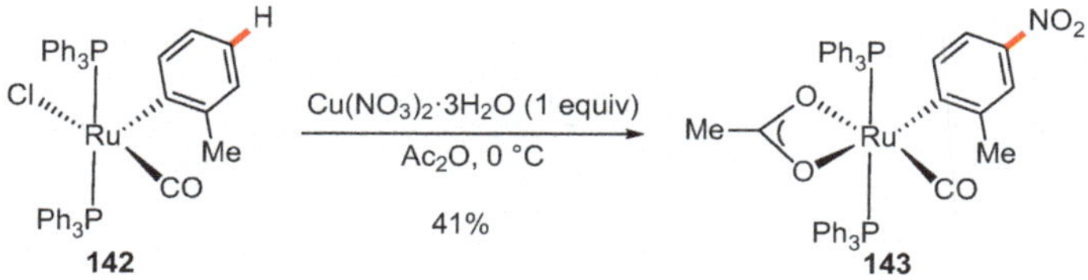

Scheme 52 Metalation-directed *para*-nitration of aryl ruthenium complex **142**

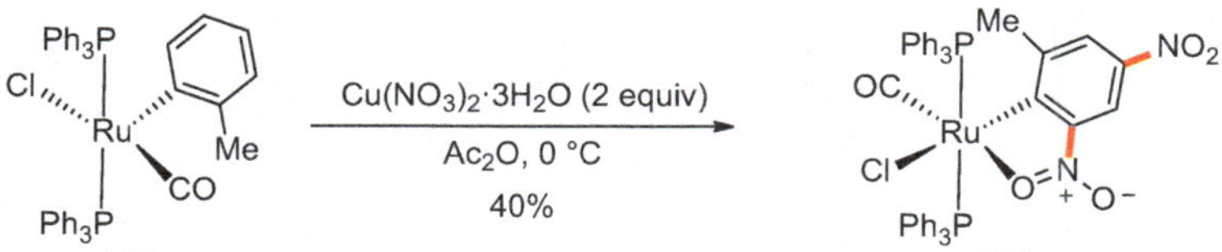

Scheme 53 Metalation-directed *para*- and *ortho*-nitration of ruthenium complex **142**

influence of the first nitro-substituent, further nitration at the *ortho*-position occurred under rather mild reaction conditions. This observation was rationalized in terms of the strong activating effect of the Ru–C_{aryl} σ-bond as well as the formation of the stabilized metallacycle **144**.

In this context, it is noteworthy that various stoichiometric transformations of σ-aryl organometallic complexes have been developed [118]. However, a catalytic version of these transformations remained elusive until very recently.

In conventional cross-coupling chemistry, alkyl electrophiles bearing β-hydrogen atoms had been considered extremely challenging substrates for mainly two reasons. First, the oxidative addition of alkyl C–Hal bonds to a metal center is more difficult than the one of the aryl–Hal analogues because of their more electron-rich nature. Second, the thus formed alkyl metal species are substantially less stable. This instability gives rise to undesired side reactions. Most prominently, β-hydride eliminations can occur, delivering the corresponding β-elimination products [119]. Within our continuous efforts in studying carboxylate-assisted ruthenium(II)-catalyzed C–H functionalization reactions [11, 99], in 2011, Ackermann and co-workers achieved carboxylate-assisted ruthenium-catalyzed *ortho*-C–H alkylation of arenes **145** with unactivated primary alkyl halides **146** (Scheme 54) [120].

Interestingly, *meta*-alkylated by-product **150** was also isolated when performing the direct alkylation with water as the reaction medium (Scheme 55) [121, 122]. Notably, this *meta*-alkylation took place under solvent-free reaction conditions as well.

Frost and co-workers discovered that, in contrast to palladium catalysts [123], ruthenium complexes led to completely different site selectivities in direct C–H sulfonation of 2-arylpyridine derivatives **148** (Scheme 56) [124, 125]. Here, a cyclometalated ruthenium complex containing a Ru–C_{aryl} σ-bond was proposed

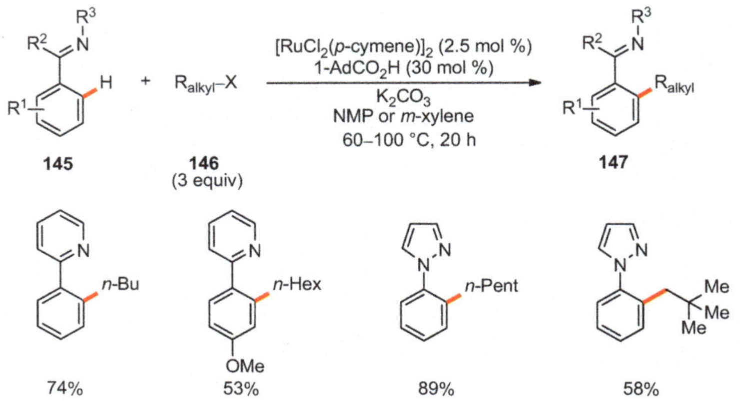

Scheme 54 Ruthenium(II)-catalyzed *ortho*-C–H alkylation with primary alkyl halides **146**

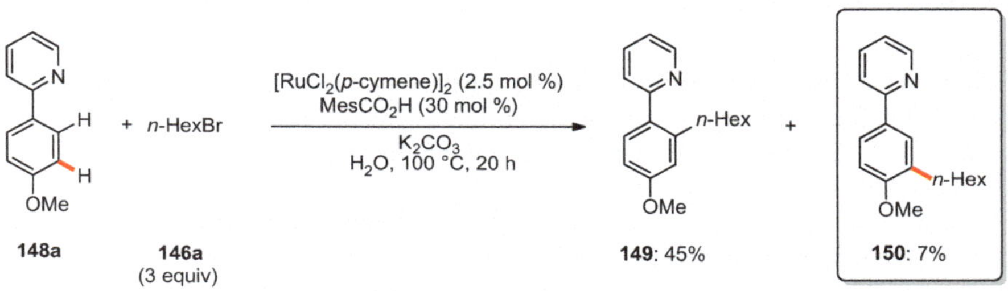

Scheme 55 Observation of ruthenium(II)-catalyzed direct *meta*-alkylation

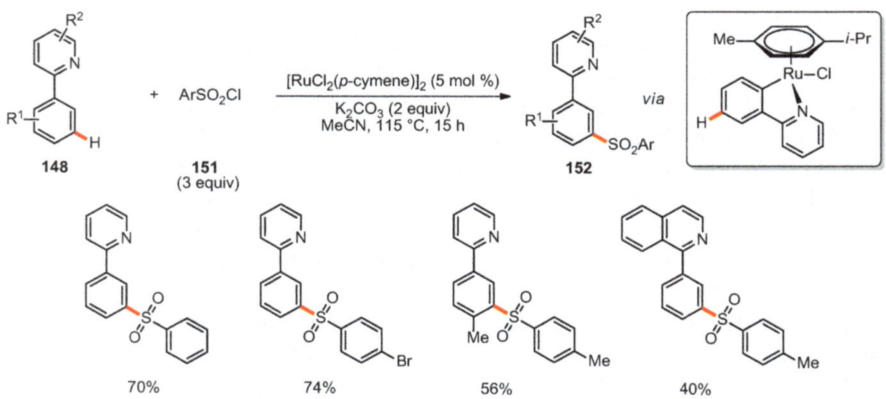

Scheme 56 Ruthenium(II)-catalyzed *meta*-C–H sulfonation of 2-phenylpyridines **148**

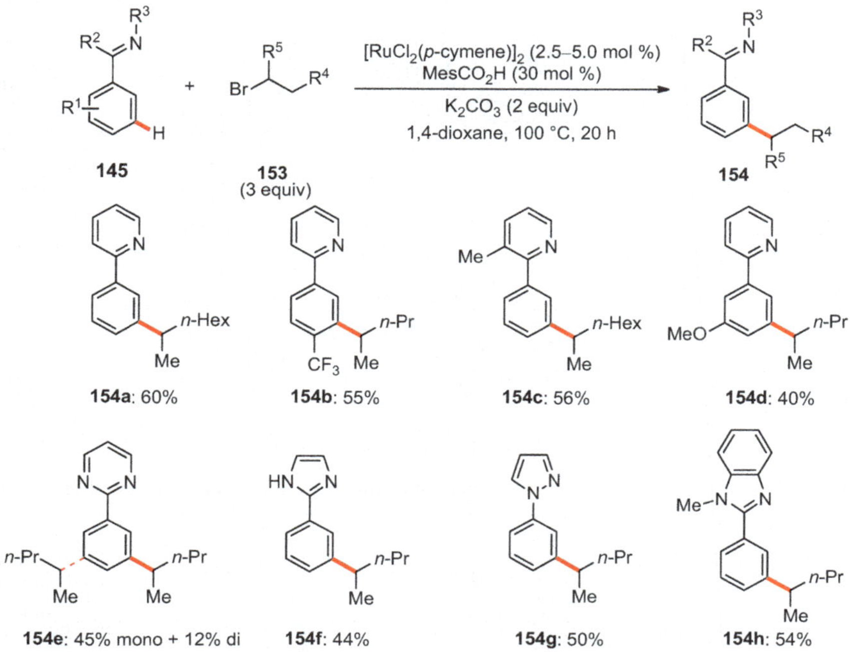

Scheme 57 Ruthenium(II)-catalyzed *meta*-C–H alkylation with secondary alkyl halides **153**

to be initially formed, steering an electrophilic attack to the *para*-position with respect to the ruthenium substituent.

Compared with primary alkyl halides, secondary alkyl halides are more difficult to couple in conventional cross-coupling chemistry due to a more challenging oxidative addition and a facile β-hydride elimination [126, 127]. The Ackermann group disclosed *meta*-selective direct alkylations of 2-phenylpyridines, azole-substituted arenes, and 2-phenylpyrimidines with secondary alkyl halides under the catalysis of $[RuCl_2(p\text{-cymene})]_2$ and 2,4,6-trimethylbenzoic acid ($MesCO_2H$) (Scheme 57) [120, 128]. The direct *meta*-alkylations proceeded smoothly with ample substrate scope and tolerated various valuable functional groups [128].

Intramolecular competition experiments with *ortho*-substituted arenes **148** counterintuitively led to the more hindered products preferentially (Scheme 58) [128].

As shown in Scheme 59, the alkylation of deuterated 2-phenylpyridine ($[D_5]$-**148c**) under the optimized reaction conditions gave rise to a significant D/H exchange in the *ortho*-positions of both the reisolated substrate and the product **154a** [128]. This result provided strong evidence for the C–H bond metalation step to proceed initially in the *ortho*-position of the arene **148** in a reversible fashion.

The independently prepared cyclometalated complex **155b** bearing a carboxylate ligand was found to be catalytically competent [128]. In stark contrast, the corresponding chlororuthenacycle **155c** did not afford the *meta*-alkylated product

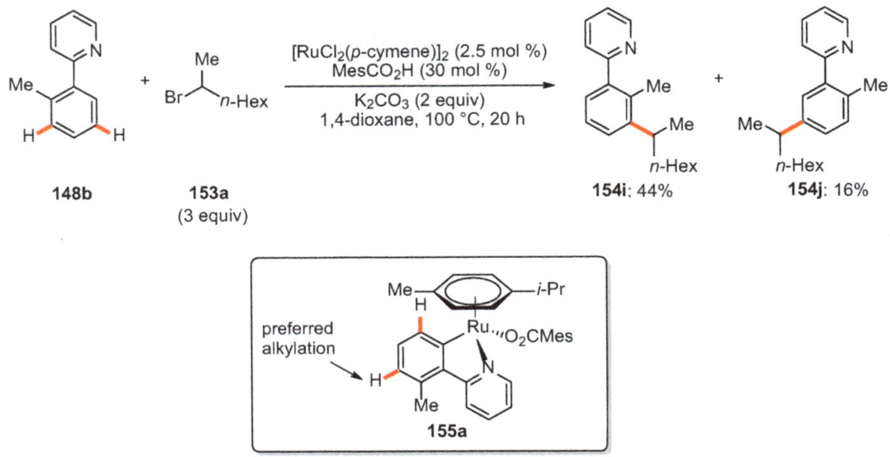

Scheme 58 Intramolecular competition experiment with *ortho*-substituted substrate **148b**

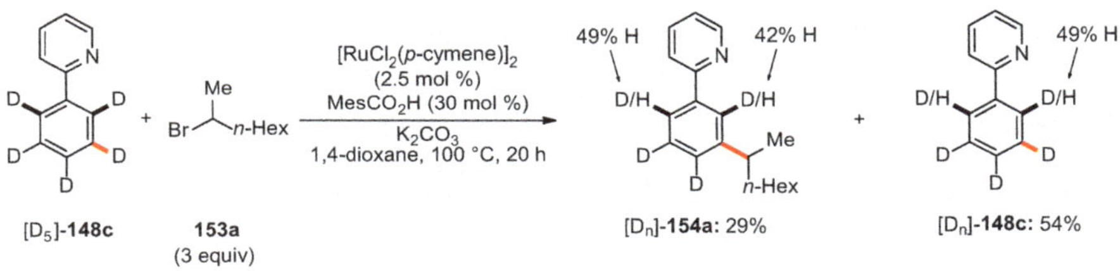

Scheme 59 Studies with isotopically labeled compounds

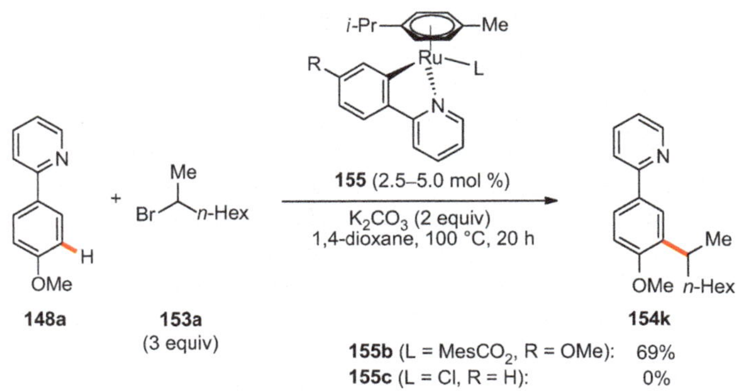

Scheme 60 Catalytic efficiency of well-defined ruthenium complexes **155**

154k. These findings clearly highlighted the importance of carboxylate assistance (Scheme 60) [11].

Based on our mechanistic studies [128], we proposed a catalytic cycle involving the reversible formation of the ruthenacycle **158** via carboxylate-assisted,

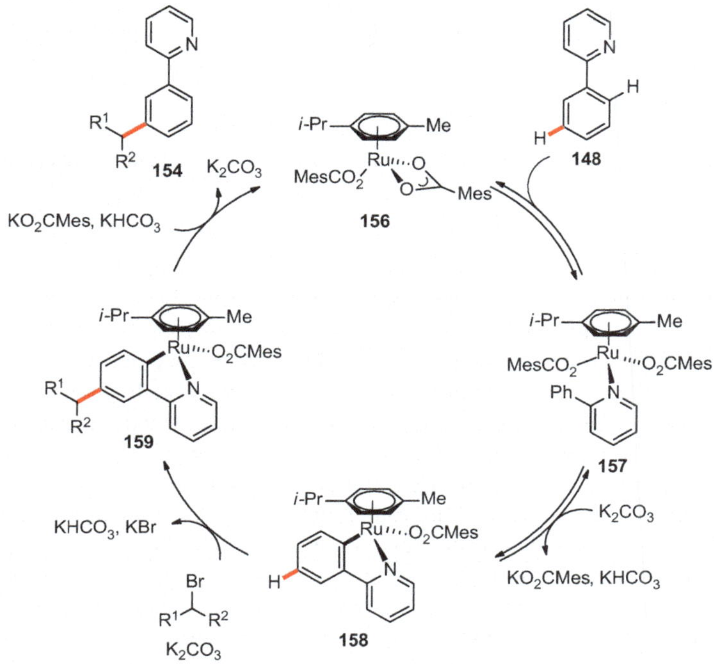

Scheme 61 Plausible catalytic cycle for the ruthenium(II)-catalyzed *meta*-C–H alkylation

proximity-induced C–H activation. Subsequently, the unique directing group effect of the Ru–C(sp^2) σ-bond caused the irreversible alkylation in the *para*-position to the Ru–C(sp^2) σ-bond. Finally, proto-demetalation provided the desired *meta*-substituted product **154** and regenerated the catalytically active complex **156** (Scheme 61).

5 Conclusion

During the past decade, major advances have been realized through chelation-assisted direct arene C–H activation, which set the stage for numerous proximity-induced *ortho*-C–H functionalization methods. However, different strategies have very recently been devised for the transition metal-catalyzed *meta*- and *para*-functionalization of aromatic compounds via remote C–H activation (Scheme 62).

Hence, *meta*-selective borylations and silylations were achieved utilizing the steric interactions between the substrate and the rhodium- or iridium-based catalysts. While this strategy proved powerful for di- or trisubstituted arenes, it remains extremely difficult to site-selectively functionalize monosubstituted substrates. Furthermore, nitrile-containing auxiliaries were successfully exploited in

Scheme 62 Site-selective aromatic C–H functionalization

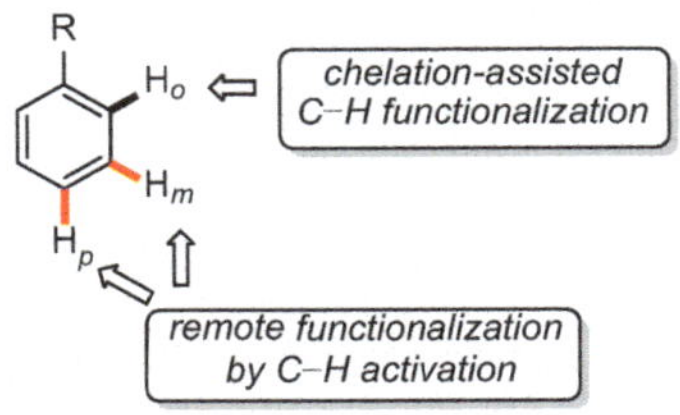

chelation-controlled *meta*-C–H metalation using palladium catalysts. This strategy allowed for *meta*-C–H functionalizations of various arenes but called for a directing group with a molecular weight that usually outcompetes the one of the substrate itself. In addition, most of these templates have to be prepared through lengthy synthetic operations. The Catellani-type *meta*-C–H functionalization was realized with common *ortho*-directing groups. This powerful protocol offered notable possibilities for tuning the site selectivity of known *ortho*-C–H functionalizations to be *meta*-selective. *ortho*-C–H metalation-directed remote functionalization was, on the contrary, accomplished in ruthenium(II)-catalyzed *meta*-alkylation and sulfonation reactions, which bear great potential for future development. The use of removable directing groups also offers means for the synthesis of *meta*-substituted arenes via decarboxylative processes. The functionalization of *para*-C–H bonds largely depends thus far on the electronic nature of the substrate which parallels the electrophilic-type reaction manifold. Thus, only electron-rich arenes were functionalized at the *para*-position employing copper, palladium, or ruthenium catalysts. In consideration of the unique potential of aromatic C–H activation for step-economical organic syntheses, along with the considerable progress in *ortho*-C–H functionalization, exciting future developments in *meta*- and *para*-selective C–H functionalizations are expected in this rapid evolving research area.

References

1. Yamaguchi J, Yamaguchi AD, Itami K (2012) Angew Chem Int Ed 51:8960
2. McMurray L, O'Hara F, Gaunt MJ (2011) Chem Soc Rev 40:1885
3. Brückl T, Baxter RD, Ishihara Y, Baran PS (2011) Acc Chem Res 45:826
4. Ackermann L, Vicente R, Kapdi AR (2009) Angew Chem Int Ed 48:9792
5. Friedel C, Crafts JM (1877) Compt Rend 84:1392
6. Clayden J, Greeves N, Warren SG (2012) Organic chemistry. Oxford University Press, Oxford
7. Ackermann L (2009) Modern Arylation Methods. Wiley, Weinheim
8. Knappke CEI, Jacobi von Wangelin A (2011) Chem Soc Rev 40:4948
9. Martin R, Buchwald SL (2008) Acc Chem Res 41:1461
10. Beller M, Bolm C (2004) Transition Metals for Organic Synthesis. Wiley, Weinheim
11. Ackermann L (2011) Chem Rev 111:1315
12. Wencel-Delord J, Glorius F (2013) Nat Chem 5:369
13. Lyons TW, Sanford MS (2010) Chem Rev 110:1147
14. Colby DA, Bergman RG, Ellman JA (2010) Chem Rev 110:624

15. Daugulis O, Roane J, Tran LD (2015) Acc Chem Res 48:1053
16. Lewis LN, Smith JF (1986) J Am Chem Soc 108:2728
17. Engle KM, Mei T-S, Wasa M, Yu J-Q (2011) Acc Chem Res 45:788
18. Ryabov AD (1990) Chem Rev 90:403
19. De Sarkar S, Liu W, Kozhushkov SI, Ackermann L (2014) Adv Synth Catal 356:1461
20. Colby DA, Tsai AS, Bergman RG, Ellman JA (2012) Acc Chem Res 45:814
21. Ackermann L (2007) Top Organomet Chem 24:35
22. Yang J (2015) Org Biomol Chem 13:1930
23. Julia-Hernandez F, Simonetti M, Larrosa I (2013) Angew Chem Int Ed 52:11458
24. Schranck J, Tlili A, Beller M (2014) Angew Chem Int Ed 53:9426
25. Martínez C, Muñiz K (2013) ChemCatChem 5:3502
26. Martins A, Mariampillai B, Lautens M (2010) Top Curr Chem 292:1
27. Catellani M, Frignani F, Rangoni A (1997) Angew Chem Int Ed 36:119
28. Mulvey RE, Mongin F, Uchiyama M, Kondo Y (2007) Angew Chem Int Ed 46:3802
29. Rousseau G, Breit B (2011) Angew Chem Int Ed 50:2450
30. Flemming JP, Berry MB, Brown JM (2008) Org Biomol Chem 6:1215
31. Satoh T, Miura M (2010) Synthesis 2010:3395
32. Warratz S, Kornhaaß C, Cajaraville A, Niepötter B, Stalke D, Ackermann L (2015) Angew Chem Int Ed 54:5513
33. Shi G, Zhang Y (2014) Adv Synth Catal 356:1419
34. Rodriguez N, Goossen LJ (2011) Chem Soc Rev 40:5030
35. Klotter F, Studer A (2014) Angew Chem Int Ed 53:2473
36. Zhang F, Spring DR (2014) Chem Soc Rev 43:6906
37. Mochida S, Hirano K, Satoh T, Miura M (2010) Org Lett 12:5776
38. Mochida S, Hirano K, Satoh T, Miura M (2011) J Org Chem 76:3024
39. Cornella J, Righi M, Larrosa I (2011) Angew Chem Int Ed 50:9429
40. Lindsey AS, Jeskey H (1957) Chem Rev 57:583
41. Luo J, Preciado S, Larrosa I (2014) J Am Chem Soc 136:4109
42. Luo J, Preciado S, Larrosa I (2015) Chem Commun 51:3127
43. Qin X, Sun D, You Q, Cheng Y, Lan J, You J (2015) Org Lett 17:1762
44. Bhadra S, Dzik WI, Gooßen LJ (2013) Angew Chem Int Ed 52:2959
45. Lee D, Chang S (2015) Chem Eur J 21:5364
46. Shi X-Y, Liu K-Y, Fan J, Dong X-F, Wei J-F, Li C-J (2015) Chem Eur J 21:1900
47. Shi X-Y, Dong X-F, Fan J, Liu K-Y, Wei J-F, Li C-J (2015) Sci China Chem 1
48. Moritani I, Fujiwara Y (1967) Tetrahedron Lett 8:1119
49. Jia C, Piao D, Oyamada J, Lu W, Kitamura T, Fujiwara Y (2000) Science 287:1992
50. Jia C, Kitamura T, Fujiwara Y (2001) Acc Chem Res 34:633
51. Le Bras J, Muzart J (2011) Chem Rev 111:1170
52. Kozhushkov SI, Ackermann L (2013) Chem Sci 4:886
53. Fujiwara Y, Noritani I, Danno S, Asano R, Teranishi S (1969) J Am Chem Soc 91:7166
54. Hong P, Mise T, Yamazaki H (1985) Nippon kagaku zassi 479
55. Hartwig JF (2012) Acc Chem Res 45:864
56. Mkhalid IAI, Barnard JH, Marder TB, Murphy JM, Hartwig JF (2009) Chem Rev 110:890
57. Janowicz AH, Bergman RG (1982) J Am Chem Soc 104:352
58. Arndtsen BA, Bergman RG, Mobley TA, Peterson TH (1995) Acc Chem Res 28:154
59. Waltz KM, He X, Muhoro C, Hartwig JF (1995) J Am Chem Soc 117:11357
60. Iverson CN, Smith MR (1999) J Am Chem Soc 121:7696
61. Chen H, Schlecht S, Semple TC, Hartwig JF (2000) Science 287:1995
62. Cho J-Y, Iverson CN, Smith MR (2000) J Am Chem Soc 122:12868
63. Cho J-Y, Tse MK, Holmes D, Maleczka RE, Smith MR (2002) Science 295:305
64. Ishiyama T, Takagi J, Ishida K, Miyaura N, Anastasi NR, Hartwig JF (2002) J Am Chem Soc 124:390
65. Ishiyama T, Takagi J, Hartwig JF, Miyaura N (2002) Angew Chem Int Ed 41:3056

66. Ishiyama T, Nobuta Y, Hartwig JF, Miyaura N (2003) Chem Commun 2924
67. Chotana GA, Rak MA, Smith MR (2005) J Am Chem Soc 127:10539
68. Boller TM, Murphy JM, Hapke M, Ishiyama T, Miyaura N, Hartwig JF (2005) J Am Chem Soc 127:14263
69. Robbins DW, Hartwig JF (2013) Angew Chem Int Ed 52:933
70. Maleczka RE, Shi F, Holmes D, Smith MR (2003) J Am Chem Soc 125:7792
71. Cheng C, Hartwig JF (2014) Science 343:853
72. Cheng C, Hartwig JF (2014) J Am Chem Soc 136:12064
73. Kuhl N, Hopkinson MN, Wencel-Delord J, Glorius F (2012) Angew Chem Int Ed 51:10236
74. Hull KL, Sanford MS (2007) J Am Chem Soc 129:11904
75. Zhao X, Yeung CS, Dong VM (2010) J Am Chem Soc 132:5837
76. Wencel-Delord J, Nimphius C, Patureau FW, Glorius F (2012) Angew Chem Int Ed 51:2247
77. Wencel-Delord J, Nimphius C, Wang H, Glorius F (2012) Angew Chem Int Ed 51:13001
78. Phipps RJ, Gaunt MJ (2009) Science 323:1593
79. Duong HA, Gilligan RE, Cooke ML, Phipps RJ, Gaunt MJ (2011) Angew Chem Int Ed 50:463
80. Lee EY, Park J (2011) ChemCatChem 3:1127
81. Chen B, Hou XL, Li YX, Wu YD (2011) J Am Chem Soc 133:7668
82. Zhang Y-H, Shi B-F, Yu J-Q (2009) J Am Chem Soc 131:5072
83. Kubota A, Emmert MH, Sanford MS (2012) Org Lett 14:1760
84. Shrestha R, Mukherjee P, Tan Y, Litman ZC, Hartwig JF (2013) J Am Chem Soc 135:8480
85. Ciana C-L, Phipps RJ, Brandt JR, Meyer F-M, Gaunt MJ (2011) Angew Chem Int Ed 50:458
86. Wu Z, Luo F, Chen S, Li Z, Xiang H, Zhou X (2013) Chem Commun 49:7653
87. Li B, Dixneuf PH (2013) Chem Soc Rev 42:5744
88. Ackermann L, Pospech J (2011) Org Lett 13:4153
89. Wang X, Leow D, Yu J-Q (2011) J Am Chem Soc 133:13864
90. Guo X, Li C-J (2011) Org Lett 13:4977
91. Girard SA, Knauber T, Li C-J (2014) Angew Chem Int Ed 53:74
92. Thirunavukkarasu VS, Kozhushkov SI, Ackermann L (2014) Chem Commun 50:29
93. Yang F, Rauch K, Kettelhoit K, Ackermann L (2014) Angew Chem Int Ed 53:11285
94. Liu W, Ackermann L (2013) Org Lett 15:3484
95. Leow D, Li G, Mei TS, Yu J-Q (2012) Nature 486:518
96. Jung ME, Piizzi G (2005) Chem Rev 105:1735
97. Cheng G-J, Yang Y-F, Liu P, Chen P, Sun T-Y, Li G, Zhang X, Houk KN, Yu J-Q, Wu Y-D (2014) J Am Chem Soc 136:894
98. Lapointe D, Fagnou K (2010) Chem Lett 39:1118
99. Ackermann L (2014) Acc Chem Res 47:281
100. Yang YF, Cheng GJ, Liu P, Leow D, Sun TY, Chen P, Zhang X, Yu JQ, Wu YD, Houk KN (2014) J Am Chem Soc 136:344
101. Dai H-X, Li G, Zhang X-G, Stepan AF, Yu J-Q (2013) J Am Chem Soc 135:7567
102. Deng Y, Yu J-Q (2015) Angew Chem Int Ed 54:888
103. Wan L, Dastbaravardeh N, Li G, Yu JQ (2013) J Am Chem Soc 135:18056
104. Tang RY, Li G, Yu J-Q (2014) Nature 507:215
105. Yang G, Lindovska P, Zhu D, Kim J, Wang P, Tang RY, Movassaghi M, Yu JQ (2014) J Am Chem Soc 136:10807
106. Lee S, Lee H, Tan KL (2013) J Am Chem Soc 135:18778
107. Bera M, Modak A, Patra T, Maji A, Maiti D (2014) Org Lett 16:5760
108. Catellani M, Motti E, Della Ca' N (2008) Acc Chem Res 41:1512
109. Wilhelm T, Lautens M (2005) Org Lett 7:4053
110. Cárdenas DJ, Martín-Matute B, Echavarren AM (2006) J Am Chem Soc 128:5033
111. Jiao L, Bach T (2011) J Am Chem Soc 133:12990
112. Jiao L, Herdtweck E, Bach T (2012) J Am Chem Soc 134:14563
113. Dong Z, Dong G (2013) J Am Chem Soc 135:18350

114. Wang X-C, Gong W, Fang L-Z, Zhu R-Y, Li S, Engle KM, Yu J-Q (2015) Nature. doi:10.1038/nature14214
115. Faccini F, Motti E, Catellani M (2004) J Am Chem Soc 126:78
116. Dong Z, Wang J, Dong G (2015) J Am Chem Soc 137:5887
117. Clark GR, Headford CEL, Roper WR, Wright LJ, Yap VPD (1994) Inorg Chim Acta 220:261
118. Gagliardo M, Snelders DJM, Chase PA, Klein Gebbink RJM, van Klink GPM, van Koten G (2007) Angew Chem Int Ed 46:8558
119. Ariafard A, Lin Z (2006) Organometallics 25:4030
120. Ackermann L, Novak P, Vicente R, Hofmann N (2009) Angew Chem Int Ed 48:6045
121. Ackermann L, Hofmann N, Vicente R (2011) Org Lett 13:1875
122. Hofmann N (2013) Carboxylate-assisted ruthenium-catalyzed C – H bond *meta*-alkylations and oxidative annulations *Dissertation*, Georg-August-Universität Göttingen
123. Zhao X, Dimitrijević E, Dong VM (2009) J Am Chem Soc 131:3466
124. Saidi O, Marafie J, Ledger AE, Liu PM, Mahon MF, Kociok-Kohn G, Whittlesey MK, Frost CG (2011) J Am Chem Soc 133:19298
125. Reynolds WR, Liu PM, Kociok-Köhn G, Frost CG (2013) Synlett 24:2687
126. Rudolph A, Lautens M (2009) Angew Chem Int Ed 48:2656
127. Ackermann L (2010) Chem Commun (Camb) 46:4866
128. Hofmann N, Ackermann L (2013) J Am Chem Soc 135:5877

Top Organomet Chem (2016) 55: 259–260
DOI: 10.1007/3418_2015_150

Published online: 8 October 2015

Erratum to: Ruthenium(II)-Catalysed Functionalisation of C–H Bonds with Alkenes: Alkenylation versus Alkylation

Christian Bruneau and Pierre H. Dixneuf

Erratum to: Top Organomet Chem
DOI: 10.1007/3418_2015_134

The positions of substituents at C2 and C3 were originally incorrect. These have been replaced by the correct equations 61, 62 and Scheme 9.

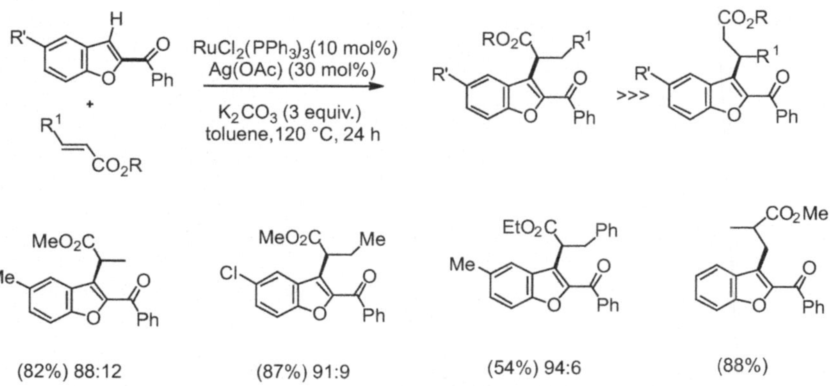

(61)

The online version of the original chapter can be found under
DOI 10.1007/3418_2015_134

C. Bruneau (✉) and P.H. Dixneuf
Institut des Sciences Chimiques de Rennes, UMR 6226 CNRS-Université de Rennes, “Organometallics: Materials and Catalysis” Centre for Catalysis and Green Chemistry, Campus de Beaulieu, 35042 Rennes, France
e-mail: christian.bruneau@univ-rennes1.fr; pierre.dixneuf@univ-rennes1.fr

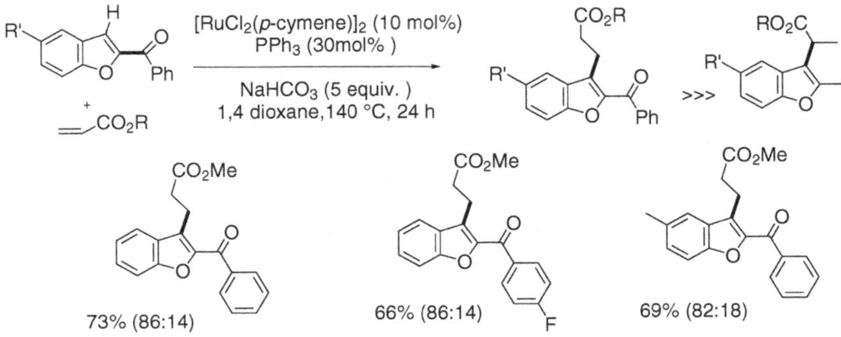

(62)

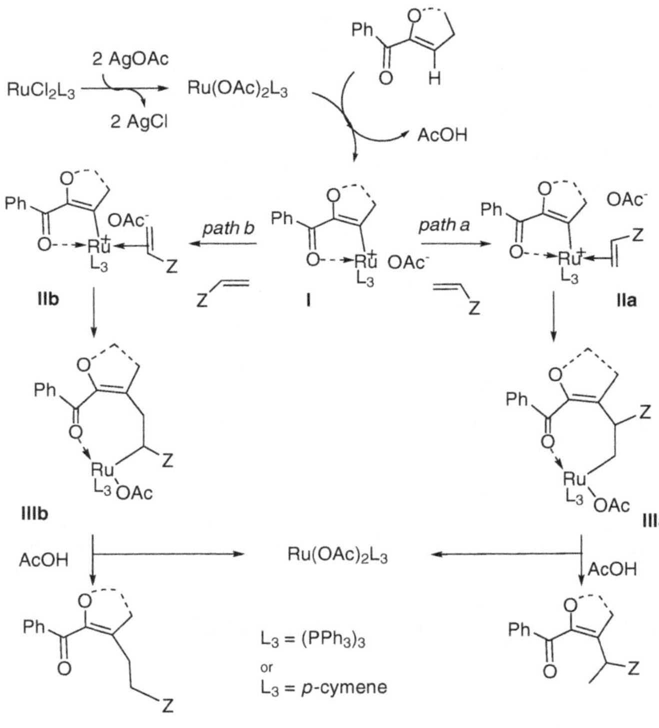

Scheme 9 Proposed mechanism for branched versus linear alkylation with alkenes

Index

The manufacturer's authorised representative in the EU is Springer Nature Customer Service Centre GmbH, Europaplatz 3, 69115 Heidelberg, Germany. If you have any concerns regarding our products, please contact ProductSafety@springernature.com

Printed and bound by CPI Group (UK) Ltd, Croydon, CR0 4YY
25/09/2026
02222859-0014